D1162487

Applied Mathematical Sciences

Volume 1

Applied Mathematical Sciences

1. John: **Partial Differential Equations,** 4th ed.
2. Sirovich: **Techniques of Asymptotic Analysis.**
3. Hale: **Theory of Functional Differential Equations,** 2nd ed.
4. Percus: **Combinatorial Methods.**
5. von Mises/Friedrichs: **Fluid Dynamics.**
6. Freiberger/Grenander: **A Short Course in Computational Probability and Statistics.**
7. Pipkin: **Lectures on Viscoelasticity Theory.**
9. Friedrichs: **Spectral Theory of Operators in Hilbert Space.**
11. Wolovich: **Linear Multivariable Systems.**
12. Berkovitz: **Optimal Control Theory.**
13. Bluman/Cole: **Similarity Methods for Differential Equations.**
14. Yoshizawa: **Stability Theory and the Existence of Periodic Solutions and Almost Periodic Solutions.**
15. Braun: **Differential Equations and Their Applications,** 3rd ed.
16. Lefschetz: **Applications of Algebraic Topology.**
17. Collatz/Wetterling: **Optimization Problems.**
18. Grenander: **Pattern Synthesis: Lectures in Pattern Theory, Vol I.**
20. Driver: **Ordinary and Delay Differential Equations.**
21. Courant/Friedrichs: **Supersonic Flow and Shock Waves.**
22. Rouche/Habets/Laloy: **Stability Theory by Liapunov's Direct Method.**
23. Lamperti: **Stochastic Processes: A Survey of the Mathematical Theory.**
24. Grenander: **Pattern Analysis: Lectures in Pattern Theory, Vol. II.**
25. Davies: **Integral Transforms and Their Applications,** 2nd ed.
26. Kushner/Clark: **Stochastic Approximation Methods for Constrained and Unconstrained Systems.**
27. de Boor: **A Practical Guide to Splines.**
28. Keilson: **Markov Chain Models—Rarity and Exponentiality.**
29. de Veubeke: **A Course in Elasticity.**
30. Sniatycki: **Geometric Quantization and Quantum Mechanics.**
31. Reid: **Sturmian Theory for Ordinary Differential Equations.**
32. Meis/Markowitz: **Numerical Solution of Partial Differential Equations.**
33. Grenander: **Regular Structures: Lectures in Pattern Theory, Vol. III.**
34. Kevorkian/Cole: **Perturbation Methods in Applied Mathematics.**
35. Carr: **Applications of Centre Manifold Theory.**
36. Bengtsson/Ghil/Källén: **Dynamic Meterology: Data Assimilation Methods.**
37. Saperstone: **Semidynamical Systems in Infinite Dimensional Spaces.**
38. Lichtenberg/Lieberman: **Regular and Stochastic Motion.**

(continued after Index)

Fritz John

Partial Differential Equations

Fourth Edition

With 25 Figures

Springer-Verlag
New York Heidelberg Berlin

Fritz John
Courant Institute of
Mathematical Sciences
New York University
New York, NY 10012
USA

Editors

AMS Subject Classifications: 35-02, 35A10, 35EXX, 35L05, 35L10

Library of Congress Cataloging in Publication Data

John, Fritz, 1910-
Partial differential equations.

(Applied mathematical sciences; 1)
Bibliography: p.
Includes index.
1. Differential equations, Partial. I. Title.
II. Series: Applied mathematical sciences (Springer-Verlag New York Inc.); 1.
QA1.A647 vol. 1 1981 [QA374] 510s [515.3′53] 81-16636
ISBN 0-387-90609-6 AACR2

Printed in the United States of America

9 8 7 6 5 4 3 2 (Second printing, 1986)

ISBN 0-387-90609-6 Springer-Verlag New York Heidelberg Berlin
ISBN 3-540-90609-6 Springer-Verlag Berlin Heidelberg New York

Preface to the fourth edition

A considerable amount of new material has been added to this edition. There is an extensive discussion of real analytic functions of several variables in Chapter 3. This permits estimation of the size of the domain of existence in the Cauchy–Kowalevski theorem. A first application of these estimates consists in a rigorous proof of a new version of Holmgren's uniqueness theorem for linear analytic partial differential equations (only sketched in the earlier editions). As another application (following Schauder) we give a second proof for existence of solutions of the initial value problem for symmetric hyperbolic systems in Chapter 5. Chapter 6 now includes a more detailed study of the Hilbert spaces $H_0^\mu(\Omega)$ with applications to the boundary behavior of solutions of the Dirichlet problem in higher dimensions. To Chapter 7 there has been added a proof of Widder's theorem on non-negative solutions of the heat equation. Finally, a new chapter, Chapter 8, contains H. Lewy's construction of a linear differential equation without solutions. There are also more problems, designed, in part, to extend the material discussed in the text.

I am particularly indebted to my colleague Percy A. Deift of the Courant Institute of New York University, to Prof. A. Garder of the Southern Illinois University at Edwardsville, Illinois, and to Dr. George Dassios of the National Technical University of Athens, Greece, for taking the trouble to compile lists of errors in the third edition. I hope that these have all been corrected and not too many new ones added in the present edition.

Beaverbrook,
Wilmington, New York

FRITZ JOHN

Contents

Chapter 6

Higher-Order Elliptic Equations with Constant Coefficients 185

Chapter 7

Parabolic Equations 206

The Single First-Order Equation* 1

1. Introduction

A partial differential equation (henceforth abbreviated as P.D.E.) for a function $u(x,y,\ldots)$ is a relation of the form

$$F(x,y,\ldots,u,u_x,u_y,\ldots,u_{xx},u_{xy},\ldots)=0, \tag{1.1}$$

where F is a given function of the independent variables $x,y,\ldots$, and of the "unknown" function u and of a *finite* number of its partial derivatives. We call u a *solution* of (1.1) if after substitution of $u(x,y,\ldots)$ and its partial derivatives (1.1) is satisfied identically in $x,y,\ldots$ in some region Ω in the space of these independent variables. Unless the contrary is stated we require that $x,y,\ldots$ are real and that u and the derivatives of u occurring in (1.1) are continuous functions of $x,y,\ldots$ in the real domain Ω.† Several P.D.E.s involving one or more unknown functions and their derivatives constitute a *system*.

The *order* of a P.D.E. or of a system is the order of the highest derivative that occurs. A P.D.E. is said to be *linear* if it is linear in the unknown functions and their derivatives, with coefficients depending on the independent variables $x,y,\ldots$. The P.D.E. of order m is called *quasi-linear* if it is linear in the derivatives of order m with coefficients that depend on $x,y,\ldots$ and the derivatives of order $<m$.

*([6], [13], [27], [31])

†For simplicity we shall often dispense with an explicit description of the domain Ω. Statements made then apply "locally," in a suitably restricted neighborhood of a point of $xy\ldots$-space.

2. Examples

Partial differential equations occur throughout mathematics. In this section we give some examples. In many instances one of the independent variables is the time, usually denoted by t, while the others, denoted by $x_1, x_2, \ldots, x_n$ (or by x,y,z when $n \leqslant 3$) give position in an n-dimensional space. The space differentiations often occur in the particular combination

$$\Delta = \frac{\partial^2}{\partial x_1^2} + \cdots + \frac{\partial^2}{\partial x_n^2} \tag{2.1}$$

known as the *Laplace operator*. This operator has the special property of being invariant under rigid motions or equivalently of not being affected by transitions to other cartesian coordinate systems. It occurs naturally in expressing physical laws that do not depend on a special position.

(i) The *Laplace equation* in n dimensions for a function $u(x_1, \ldots, x_n)$ is the linear second-order equation

$$\Delta u = u_{x_1x_1} + u_{x_2x_2} + \cdots + u_{x_nx_n} = 0. \tag{2.2}$$

This is probably the most important individual P.D.E. with the widest range of applications. Solutions u are called *potential* functions or *harmonic* functions. For $n=2$, $x_1 = x$, $x_2 = y$, we can associate with a harmonic function $u(x,y)$ a "conjugate" harmonic function $v(x,y)$ such that the first-order system of *Cauchy–Riemann* equations

$$u_x = v_y, \qquad u_y = -v_x \tag{2.3}$$

is satisfied. A real solution (u,v) of (2.3) gives rise to the *analytic* function

$$f(z) = f(x+iy) = u(x,y) + iv(x,y) \tag{2.4}$$

of the complex argument $z = x + iy$. We can also interpret $(u(x,y), -v(x,y))$ as the velocity field of an irrotational, incompressible flow. For $n=3$ equation (2.2) is satisfied by the velocity potential of an irrotational incompressible flow, by gravitational and electrostatic fields (outside the attracting masses or charges), and by temperatures in thermal equilibrium.

(ii) The *wave equation* in n dimensions for $u = u(x_1, \ldots, x_n, t)$ is

$$u_{tt} = c^2 \Delta u \tag{2.5}$$

($c = \text{const.} > 0$). It represents vibrations of strings or propagation of sound waves in tubes for $n=1$, waves on the surface of shallow water for $n=2$, acoustic or light waves for $n=3$.

(iii) *Maxwell's equations* in vacuum for the electric vector $E = (E_1, E_2, E_3)$ and magnetic vector $H = (H_1, H_2, H_3)$ form a linear system of essentially 6 first-order equations

$$\varepsilon E_t = \operatorname{curl} H, \qquad \mu H_t = -\operatorname{curl} E \tag{2.6a}$$

$$\operatorname{div} E = \operatorname{div} H = 0 \tag{2.6b}$$

with constants ε, μ. (If relations (2.6b) hold for $t=0$, they hold for all t as a consequence of relations (2.6a)). Here each component E_j, H_k satisfies the wave equation (2.5) with $c^2=1/\varepsilon\mu$.

(iv) *Elastic waves* are described classically by the linear system

$$\rho\frac{\partial^2 u_i}{\partial t^2}=\mu\,\Delta u_i+(\lambda+\mu)\frac{\partial}{\partial x_i}(\operatorname{div} u) \tag{2.7}$$

$(i=1,2,3)$, where the $u_i(x_1,x_2,x_3,t)$ are the components of the displacement vector u, and ρ is the density and λ, μ the Lamé constants of the elastic material. Each u_i satisfies the fourth-order equation

$$\left(\frac{\partial^2}{\partial t^2}-\frac{\lambda+2\mu}{\rho}\Delta\right)\left(\frac{\partial^2}{\partial t^2}-\frac{\mu}{\rho}\Delta\right)u_i=0, \tag{2.8}$$

formed from two different wave operators. For *elastic* equilibrium $(u_t=0)$ we obtain the *biharmonic* equation

$$\Delta^2 u=0. \tag{2.9}$$

(v) The equation of *heat conduction* ("heat equation")

$$u_t=k\,\Delta u \tag{2.10}$$

$(k=\text{const.}>0)$ is satisfied by the temperature of a body conducting heat, when the density and specific heat are constant.

(vi) *Schrödinger's wave equation* $(n=3)$ for a single particle of mass m moving in a field of potential energy $V(x,y,z)$ is

$$i\hbar\psi_t=-\frac{\hbar^2}{2m}\Delta\psi+V\psi, \tag{2.11}$$

where $h=2\pi\hbar$ is Planck's constant.

The equations in the preceding examples were all linear. Nonlinear equations occur just as frequently, but are inherently more difficult, hence in practice they are often approximated by linear ones. Some examples of nonlinear equations follow.

(vii) A *minimal surface* $z=u(x,y)$ (i.e., a surface having least area for a given contour) satisfies the second-order quasi-linear equation

$$\left(1+u_y^2\right)u_{xx}-2u_xu_yu_{xy}+\left(1+u_x^2\right)u_{yy}=0. \tag{2.12}$$

(viii) The *velocity potential* $\phi(x,y)$ (for velocity components ϕ_x, ϕ_y) of a two-dimensional steady, adiabatic, irrotational, isentropic flow of density ρ satisfies

$$\left(1-c^{-2}\phi_x^2\right)\phi_{xx}-2c^{-2}\phi_x\phi_y\phi_{xy}+\left(1-c^{-2}\phi_y^2\right)\phi_{yy}=0, \tag{2.13}$$

where c is a known function of the speed $q=\sqrt{\phi_x^2+\phi_y^2}$. For example

$$c^2=1-\frac{\gamma-1}{2}q^2 \tag{2.14}$$

for a polytropic gas with equation of state

$$p=A\rho^{\gamma}. \tag{2.15}$$

(ix) The *Navier–Stokes* equations for the viscous flow of an incompressible liquid connect the velocity components u_k and the pressure p:

$$\frac{\partial u_i}{\partial t}+\sum_k \frac{\partial u_i}{\partial x_k}u_k=-\frac{1}{\rho}\frac{\partial p}{\partial x_i}+\gamma\,\Delta u_i, \tag{2.16a}$$

$$\sum_k \frac{\partial u_k}{\partial x_k}=0, \tag{2.16b}$$

where ρ is the constant density and γ the kinematic viscosity.

(x) An example of a third-order nonlinear equation for a function $u(x,t)$ is furnished by the *Korteweg–de Vries* equation

$$u_t+cuu_x+u_{xxx}=0 \tag{2.17}$$

first encountered in the study of water waves.

In general we shall try to describe the *manifold of solutions* of a P.D.E. The results differ widely for different classes of equations. Meaningful "well-posed" problems associated with a P.D.E. often are suggested by particular physical interpretations and applications.

3. Analytic Solution and Approximation Methods in a Simple Example*

We illustrate some of the notions that will play an important role in what follows by considering one of the simplest of all equations

$$u_t+cu_x=0 \tag{3.1}$$

for a function $u=u(x,t)$, where $c=\text{const.}>0$. Along a line of the family

$$x-ct=\text{const.}=\xi \tag{3.2}$$

("characteristic line" in the xt-plane) we have for a solution u of (3.1)

$$\frac{du}{dt}=\frac{d}{dt}u(ct+\xi,t)=cu_x+u_t=0.$$

Hence u is constant along such a line, and depends only on the parameter ξ which distinguishes different lines. The general solution of (3.1) then has the form

$$u(x,t)=f(\xi)=f(x-ct). \tag{3.3}$$

Formula (3.3) represents the general solution u uniquely in terms of its *initial values*

$$u(x,0)=f(x). \tag{3.4}$$

Conversely every u of the form (3.3) is a solution of (3.1) with initial values f provided f is of class $C^1(\mathbb{R})$. We notice that the value of u at any point

*([18], [20], [25], [29])

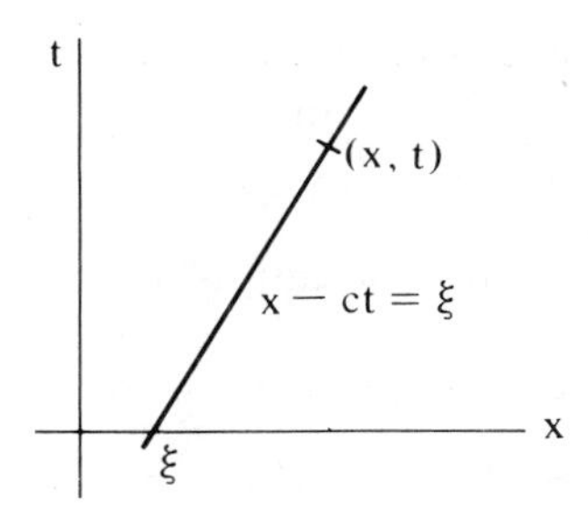

Figure 1.1

(x,t) depends only on the initial value f at the single argument $\xi = x - ct$, the abscissa of the point of intersection of the characteristic line through (x,t) with the initial line, the x-axis. The *domain of dependence* of $u(x,t)$ on the initial values is represented by the single point ξ. The *influence* of the initial values at a particular point ξ on the solution $u(x,t)$ is felt just in the points of the characteristic line (3.2). (Fig. 1.1)

If for each fixed t the function u is represented by its graph in the xu-plane, we find that the graph at the time $t = T$ is obtained by translating the graph at the time $t=0$ parallel to the x-axis by the amount cT:

$$u(x,0) = u(x+cT,T) = f(x).$$

The graph of the solution represents a *wave* propagating to the right with velocity c without changing shape. (Fig. 1.2)

We use this example with its explicit solution to bring out some of the notions connected with the numerical solution of a P.D.E by the *method of finite differences*. One covers the xt-plane by a rectangular grid with mesh size h in the x-direction and k in the t-direction. In other words one considers only points (x,t) for which x is a multiple of h and t a multiple of k. It would seem natural for purposes of numerical approximation to replace the P.D.E. (3.1) by the difference equation

$$\frac{v(x,t+k) - v(x,t)}{k} + c\frac{v(x+h,t) - v(x,t)}{h} = 0. \tag{3.5}$$

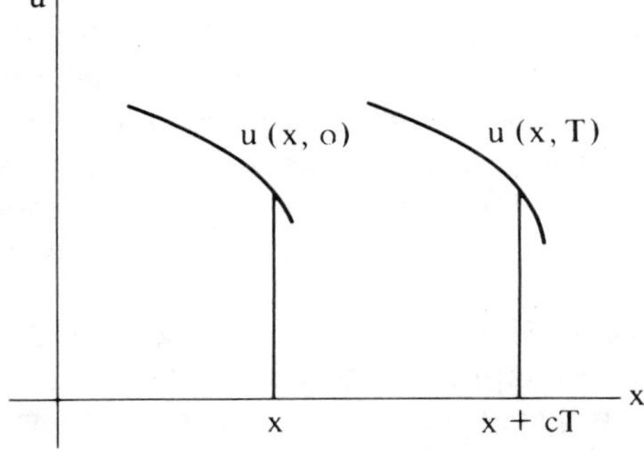

Figure 1.2

Formally this equation goes over into $v_t + cv_x = 0$ as $h, k \to 0$. We ask to what extent a solution v of (3.5) in the grid points with initial values

$$v(x,0) = f(x) \tag{3.6}$$

approximates for small h,k the solution of the initial-value problem (3.1), (3.4).

Setting $\lambda = k/h$, we write (3.5) as a recursion formula

$$v(x,t+k) = (1+\lambda c)v(x,t) - \lambda c v(x+h,t) \tag{3.7}$$

expressing v at the time $t+k$ in terms of v at the time t. Introducing the *shift operator* E defined by

$$Ef(x) = f(x+h), \tag{3.8}$$

(3.7) becomes

$$v(x,t+k) = ((1+\lambda c) - \lambda c E)v(x,t) \tag{3.8a}$$

for $t = nk$ this immediately leads by iteration to the solution of the initial-value problem for (3.5):

$$\begin{aligned} v(x,t) = v(x,nk) &= ((1+\lambda c) - \lambda c E)^n v(x,0) \\ &= \sum_{m=0}^{n} \binom{n}{m} (1+\lambda c)^m (-\lambda c E)^{n-m} f(x) \\ &= \sum_{m=0}^{n} \binom{n}{m} (1+\lambda c)^m (-\lambda c)^{n-m} f(x+(n-m)h). \end{aligned} \tag{3.9}$$

Clearly the domain of dependence for $v(x,t) = v(x,nk)$ consists of the set of points

$$x,\ x+h,\ x+2h,\ \ldots,\ x+nh \tag{3.10}$$

on the x-axis, all of which lie between x and $x+nh$. The domain of the differential equation solution consists of the point $\xi = x - ct = x - c\lambda nh$, which lies completely outside the interval $(x, x+nh)$. It is clear that v for $h,k \to 0$ cannot be expected to converge to the correct solution u of the differential equation, since in forming $v(x,t)$ we do not make use of any information on the value of $f(\xi)$, which is vital for determining $u(x,t)$, but only of more and more information on f in the interval $(x, x+(t/\lambda))$ which is irrelevant. The difference scheme fails the *Courant–Friedrichs–Lewy* test, which requires that the limit of the domain of dependence for the difference equation contains the domain of dependence for the differential equation.

That the scheme (3.5) is inappropriate also is indicated by its high degree of *instability*. In applied problems the data f are never known with perfect accuracy. Moreover, in numerical computations we cannot easily use the exact values but commit small round-off errors at every step. Now it is clear from (3.9) that errors in f of absolute value ε with the proper (alternating) sign can lead to a resulting error in $v(x,t)=v(x,nk)$ of size

$$\varepsilon \sum_{m=0}^{n} \binom{n}{m}(1+\lambda c)^m(\lambda c)^{n-m}=(1+2\lambda c)^n\varepsilon. \tag{3.11}$$

Thus for a fixed mesh ratio λ the possible resulting error in v grows exponentially with the number n of steps in the t-direction.

A more appropriate difference scheme uses "backward" difference quotients:

$$\frac{v(x,t+k)-v(x,t)}{k}+c\frac{v(x,t)-v(x-h,t)}{h}=0 \tag{3.12}$$

or symbolically

$$v(x,t+k)=((1-\lambda c)+\lambda cE^{-1})v(x,t). \tag{3.13}$$

The solution of the initial-value problem for (3.13) becomes

$$v(x,t)=v(x,nk)=\sum_{m=0}^{n}\binom{n}{m}(1-\lambda c)^m(\lambda c)^{n-m}f(x-(n-m)h). \tag{3.14}$$

In this scheme the domain of dependence for $v(x,t)$ on f consists of the points

$$x,\ x-h,\ x-2h,\ \ldots,\ x-nh=x-\frac{t}{\lambda} \tag{3.15}$$

Letting $h,k\to 0$ in such a way that the mesh ratio λ is held fixed, the set (3.15) has as its limit points the interval $[x-(t/\lambda),x]$ on the x-axis. The Courant–Friedrichs–Lewy test is satisfied, when this interval contains the point $\xi=x-ct$, that is when the mesh ratio λ satisfies

$$\lambda c\leqslant 1. \tag{3.16}$$

Stability of the scheme under the condition (3.16) is indicated by the fact that by (3.14) a maximum error of size ε in the initial function f results in a maximum possible error in the value of $v(x,t)=v(x,nk)$ of size

$$\varepsilon\sum_{m=0}^{n}\binom{n}{m}(1-\lambda c)^m(\lambda c)^{n-m}=\varepsilon((1-\lambda c)+\lambda c)^n=\varepsilon. \tag{3.17}$$

We can prove that the v represented by (3.14) actually converges to $u(x,t)=f(x-ct)$ for $h,k\to 0$ with $k/h=\lambda$ fixed, provided the stability criterion (3.16) holds and f has uniformly bounded second derivatives. For that purpose we observe that $u(x,t)$ satisfies

$$\begin{aligned}&|u(x,t+k)-(1-\lambda c)u(x,t)-\lambda cu(x-h,t)|\\ &=|f(x-ct-ck)-(1-\lambda c)f(x-ct)-\lambda cf(x-ct-h)|\leqslant Kh^2,\end{aligned} \tag{3.18}$$

where

$$K=\tfrac{1}{2}(c^2\lambda^2+\lambda c)\sup|f''|, \tag{3.19}$$

as is seen by expanding f about the point $x-ct$. Thus, setting $w=u-v$ we have

$$|w(x,t+k)-(1-\lambda c)w(x,t)-\lambda c w(x-h,t)| \leqslant Kh^2$$

and hence

$$\begin{aligned}\sup_x |w(x,t+k)| &\leqslant (1-\lambda c)\sup_x |w(x,t)| + \lambda c \sup_x |w(x-h,t)| + Kh^2 \\ &= \sup_x |w(x,t)| + Kh^2. \qquad (3.20)\end{aligned}$$

Applying (3.20) repeatedly it follows for $t=nk$ that

$$\begin{aligned}|u(x,t)-v(x,t)| &\leqslant \sup_x |w(x,nk)| \\ &\leqslant \sup_x |w(x,0)| + nKh^2 = \frac{Kth}{\lambda},\end{aligned}$$

since $w(x,0)=0$. Consequently $w(x,t)\to 0$ as $h\to 0$, that is, the solution v of the difference scheme (3.12) converges to the solution u of the differential equation.

PROBLEMS

1. Show that the solution v of (3.12) with initial data f converges to u for $h\to 0$ and a fixed $\lambda \leqslant 1/c$, under the sole assumption that f is continuous. (Hint: Use the fact that both u and v change by at most ε when we change f by at most ε.)

2. To take into account possible round-off errors we assume that instead of (3.13) v satisfies an inequality
$$|v(x,t+k)-(1-\lambda c)v(x,t)-\lambda c v(x-h,t)| < \delta.$$
Show that for a prescribed δ and for K given by (3.19) we have the estimate
$$|u(x,t)-v(x,t)| \leqslant \frac{Kth}{\lambda} + \frac{t}{\lambda h}\delta \qquad (3.21)$$
assuming that (3.16) holds and that $v(x,0)=f(x)$. Find values for λ and h based on this formula that will guarantee the smallest maximum error in computing $u(x,t)$.

3. Instability of a difference scheme under small perturbations does not exclude the possibility that in special cases the scheme converges towards the correct function, if no errors are permitted in the data or the computation. In particular let $f(x)=e^{\alpha x}$ with a complex constant α. Show that for fixed x, t and any fixed positive $\lambda = k/h$ whatsoever both the expressions (3.9) and (3.14) converge for $n\to\infty$ towards the correct limit $e^{\alpha(x-ct)}$. (This is consistent with the Courant–Friedrichs–Lewy test, since for an *analytic* f the values of f in any interval determine those at the point ξ uniquely.)

4. Quasi-linear Equations

The general first-order equation for a function $u=u(x,y,\ldots,z)$ has the form

$$f(x,y,\ldots,u,u_x,u_y,\ldots,u_z)=0. \tag{4.1}$$

Equations of this type occur naturally in the calculus of variations, in particle mechanics, and in geometrical optics. The main result is the fact that the general solution of an equation of type (4.1) can be obtained by solving systems of Ordinary Differential Equations (O.D.E.s for short). This is not true for higher-order equations or for systems of first-order equations. In what follows we shall mostly limit ourselves to the case of two independent variables x,y. The theory can be extended to more independent variables without any essential change.

We first consider the somewhat simpler case of a quasi-linear equation

$$a(x,y,u)u_x+b(x,y,u)u_y=c(x,y,u). \tag{4.2}$$

We represent the function $u(x,y)$ by a surface $z=u(x,y)$ in xyz-space. Surfaces corresponding to solutions of a P.D.E. are called *integral surfaces* of the P.D.E. The prescribed functions $a(x,y,z), b(x,y,z), c(x,y,z)$ define a field of vectors in xyz-space (or in a portion Ω of that space). Obviously only the direction of the vector, the *characteristic direction*, matters for the P.D.E. (4.2). Since $(u_x,u_y,-1)$ constitute direction numbers of the normal of the surface $z=u(x,y)$, we see that (4.2) is just the condition that the normal of an integral surface at any point is perpendicular to the direction of the vector (a,b,c) corresponding to that point. Thus integral surfaces are surfaces that at each point are tangent to the characteristic direction.

With the field of characteristic directions with direction numbers (a,b,c) we associate the family of *characteristic curves* which at each point are tangent to that direction field. Along a characteristic curve the relation

$$\frac{dx}{a(x,y,z)}=\frac{dy}{b(x,y,z)}=\frac{dz}{c(x,y,z)} \tag{4.3}$$

holds. Referring the curve to a suitable parameter t (or denoting the common ratio in (4.3) by dt) we can write the condition defining characteristic curves in the more familiar form of a system of ordinary differential equations

$$\frac{dx}{dt}=a(x,y,z), \qquad \frac{dy}{dt}=b(x,y,z), \qquad \frac{dz}{dt}=c(x,y,z). \tag{4.4}$$

The system is "autonomous" (the independent variable t does not appear explicitly). The choice of the parameter t in (4.4) is artificial. Using any other parameter along the curve amounts to replacing a,b,c by proportional quantities, which does not change the characteristic curve in xyz-space or the P.D.E. (4.2). Assuming that a,b,c are of class C^1 in a region

Ω, we know from the theory of O.D.E.s that through each point of Ω there passes exactly one characteristic curve. There is a 2-parameter family of characteristic curves in xyz-space (but a 3-parameter family of solutions $(x(t),y(t),z(t))$ of (4.4), since replacing the independent variable t by $t+c$ with a constant c changes the solution (x,y,z), but not the characteristic curve, which is its range).

If a surface S: $z=u(x,y)$ is a union of characteristic curves, then S is an integral surface. For then through any point P of S there passes a characteristic curve γ contained in S. The tangent to γ at P necessarily lies in the tangent plane of S at P. Since the tangent to γ has the characteristic direction, the normal to S at P is perpendicular to the characteristic direction, which makes S an integral surface. Conversely we can show that every integral surface S is the union of characteristic curves, or that through every point of S there passes a characteristic curve contained in S. This is a consequence of the following theorem:

Theorem. *Let the point $P=(x_0,y_0,z_0)$ lie on the integral surface S: $z=u(x,y)$. Let γ be the characteristic curve through P. Then γ lies completely on S.*

PROOF. Let γ given by $(x(t),y(t),z(t))$ be the solution of (4.4) for which $(x,y,z)=(x_0,y_0,z_0)$ for $t=t_0$. From γ and S we form the expression

$$U=z(t)-u(x(t),y(t))=U(t). \tag{4.5}$$

Obviously $U(t_0)=0$ since P lies on S. By (4.4)

$$\begin{aligned}\frac{dU}{dt}&=\frac{dz}{dt}-u_x(x(t),y(t))\frac{dx}{dt}-u_y(x(t),y(t))\frac{dy}{dt}\\&=c(x,y,z)-u_x(x,y)a(x,y,z)-u_y(x,y,z)b(x,y,z).\end{aligned}$$

This can be written as the ordinary differential equation

$$\begin{aligned}\frac{dU}{dt}&=c(x,y,U+u(x,y))-u_x(x,y)a(x,y,U+u(x,y))\\&\quad-u_y(x,y)b(x,y,U+u(x,y))\end{aligned} \tag{4.6}$$

for U, where for x,y we have to substitute the functions $x(t),y(t)$ from the description of γ. Now $U\equiv 0$ is a particular solution of (4.6), since $u(x,y)$ satisfies (4.2). By the uniqueness theorem for O.D.E.s, this is the only solution vanishing for $t=t_0$. It follows that the function $U(t)$ defined by (4.5) vanishes identically. But that just means that the whole curve γ lies on S. □

As a consequence of the theorem two integral surfaces that have a point P in common intersect along the whole characteristic curve γ through P. Conversely if the integral surfaces S_1 and S_2 intersect, without touching, along a curve γ, then γ is characteristic. For consider the tangent planes π_1,π_2 to S_1,S_2 at a point P of γ. Each of the planes has to contain the characteristic direction (a,b,c) at P. Since $\pi_1\neq\pi_2$ it follows that the intersection of π_1 and π_2 has the direction (a,b,c). Since the tangent T to γ at P also has to belong to both π_1 and π_2, it follows that T has the direction (a,b,c), and hence that γ is characteristic.

5. The Cauchy Problem for the Quasi-linear Equation

We now have a simple description for the *general solution* u of (4.2): The integral surface $z=u(x,y)$ is the union of characteristic curves. To get a better insight into the structure of the manifold of solutions it is desirable to have a definite method of generating solutions in terms of a prescribed set F of functions, called "data." Ideally we have a mapping $F\to u$ of data F onto solutions u of the P.D.E. The space of solutions is then described by the usually simpler space of data. A good deal of the theory of P.D.E.s is concerned with the "problem" of actually finding the u belonging to a given F. (Here "finding" commonly is equated with "establishing existence.")

A simple way of selecting an individual $u(x,y)$ out of the infinite set of all solutions of (4.2) consists in prescribing a curve Γ in xyz-space which is to be contained in the integral surface $z=u(x,y)$. Let Γ be represented parametrically by

$$x=f(s), \qquad y=g(s), \qquad z=h(s). \tag{5.1}$$

We are asking for a solution $u(x,y)$ of (4.2) such that the relation

$$h(s)=u(f(s),g(s)) \tag{5.2}$$

holds identically in s. Finding the function $u(x,y)$ for given data $f(s)$, $g(s)$, $h(s)$ constitutes the *Cauchy problem* for (4.2). Actually the same curve Γ has many different parametric representations (5.1) for different choices of the parameter s. Introducing a different parameter σ by a substitution $s=\phi(\sigma)$ will not change the solution $u(x,y)$ of the Cauchy problem.

We shall be satisfied here with a *local* solution u of our problem, defined for x,y near values $x_0=f(s_0)$, $y_0=g(s_0)$.

In many instances the variable y will be identified with time and x with position in space. It is then natural to pose the problem of finding a solution $u(x,y)$ from its *initial values* at the time $y=0$:

$$u(x,0)=h(x). \tag{5.3}$$

This *initial-value problem* obviously is the special Cauchy problem in which the curve Γ has the form

$$x=s, \qquad y=0, \qquad z=h(s), \tag{5.4}$$

that is, Γ lies in the xz-plane and is referred to x as parameter. We notice that in the initial-value problem we prescribe a single function $h(x)$, which in turn is determined uniquely by u, whereas in the general Cauchy problem many space curves Γ are bound to lead to the same u. An integral surface contains many curves Γ but only one intersection with the xz-plane.

Let then the functions $f(s)$, $g(s)$, $h(s)$ describing Γ be of class C^1 in a neighborhood of a value s_0. Let

$$P_0=(x_0,y_0,z_0)=(f(s_0),g(s_0),h(s_0)). \tag{5.5}$$

Assume that the coefficients $a(x,y,z)$, $b(x,y,z)$, $c(x,y,z)$ in (4.2) are of class C^1 in x,y,z near P_0. It is clear intuitively that the integral surface $z=u(x,y)$ passing through Γ will have to consist of the characteristic curves passing through the various points of Γ. Accordingly we form for each s near s_0 that solution

$$x=X(s,t), \qquad y=Y(s,t), \qquad z=Z(s,t) \tag{5.6}$$

of the characteristic differential equations (4.4) which reduces to $f(s),g(s),h(s)$ for $t=0$. The functions X,Y,Z then satisfy

$$X_t=a(X,Y,Z), \qquad Y_t=b(X,Y,Z), \qquad Z_t=c(X,Y,Z) \tag{5.7}$$

identically in s,t and also satisfy the initial conditions

$$X(s,0)=f(s), \qquad Y(s,0)=g(s), \qquad Z(s,0)=h(s). \tag{5.8}$$

From the general theorems on existence and on continuous dependence on parameters of solutions of systems of ordinary differential equations it follows that there exists a unique set of functions $X(s,t)$, $Y(s,t)$, $Z(s,t)$ of class C^1 for (s,t) near $(s_0,0)$ which satisfy (5.7), (5.8).

Equations (5.6) represent a surface Σ: $z=u(x,y)$ referred to parameters s,t if we can solve the first two equations for s,t in terms of x,y, say in the form $s=S(x,y)$, $t=T(x,y)$. Then the u defined by

$$z=u(x,y)=Z(S(x,y),T(x,y)) \tag{5.9}$$

will be the explicit representation of Σ. By (5.5), (5.8)

$$x_0=X(s_0,0), \qquad y_0=Y(s_0,0). \tag{5.10}$$

Now the implicit function theorem asserts that we can find solutions $s=S(x,y)$, $t=T(x,y)$ of

$$x=X(S(x,y),T(x,y)), \qquad y=Y(S(x,y),T(x,y)) \tag{5.11}$$

of class C^1 in a neighborhood of (x_0,y_0) and satisfying

$$s_0=S(x_0,y_0), \qquad 0=T(x_0,y_0), \tag{5.12}$$

provided the Jacobian

$$J=\begin{vmatrix} X_s(s_0,0) & Y_s(s_0,0) \\ X_t(s_0,0) & Y_t(s_0,0) \end{vmatrix} \tag{5.13}$$

does not vanish. By (5.7), (5.8) this amounts to the condition

$$J=\begin{vmatrix} f'(s_0) & g'(s_0) \\ a(x_0,y_0,z_0) & b(x_0,y_0,z_0) \end{vmatrix} \neq 0. \tag{5.14}$$

Thus (5.14) guarantees that locally (5.6) represents a surface Σ: $z=u(x,y)$. That Σ is an integral surface is clear in the parametric representation (5.6). For at any point P the quantities X_t, Y_t, Z_t give the direction of the tangent to a curve $s=$const. on Σ, which will have to lie in the tangent plane of Σ at P. Thus (5.7) shows that the tangent plane at any point contains the characteristic direction (a,b,c), and hence that Σ is an integral surface. [One can also verify analytically that the function u represented by (5.9) satisfies the differential equation (4.2) by first expressing u_x, u_y in terms of S_x, S_y, T_x, T_y, and then expressing those four quantities in terms of X_s, X_t, Y_s, Y_t using (5.11).]

This completes the local existence proof for the solution of the Cauchy problem, under the assumption (5.14). Uniqueness follows from the theorem on p. 10: Any integral surface through Γ would have to contain the characteristic curves through the points of Γ, hence would have to contain the surface represented parametrically by (5.6), and hence locally would have to be identical with the surface.

Condition (5.14) is essential for the existence of a C^1-solution $u(x,y)$ of the Cauchy problem. For if $J=0$ we would find from (5.2), (4.2) that at $s=s_0$, $x=f(s_0)$, $y=g(s_0)$ the three relations

$$bf'-ag'=0, \qquad h'=f'u_x+g'u_y, \qquad c=au_x+bu_y \tag{5.15}$$

hold. These imply that

$$bh'-cg'=0, \qquad ah'-cf'=0$$

and hence that f', g', h' are proportional to a,b,c. Hence $J=0$ is incompatible with the existence of a solution unless Γ happens to be characteristic at s_0. Incidentally the Cauchy problem will have infinitely many solutions for a characteristic curve Γ, which are obtained by passing any curve Γ^* satisfying (5.14) through a point P_0 of Γ and solving the Cauchy problem for Γ^*.

In the special case of a *linear* P.D.E. we can write (4.2) in the form

$$a(x,y)u_x+b(x,y)u_y=c(x,y)u+d(x,y). \tag{5.16}$$

Here the system of three characteristic O.D.E.s reduces to the pair

$$\frac{dx}{dt}=a(x,y), \qquad \frac{dy}{dt}=b(x,y) \tag{5.17}$$

or even to the single equivalent equation

$$\frac{dy}{dx}=\frac{b(x,y)}{a(x,y)}. \tag{5.18}$$

Equations (5.17) or (5.18) determine a system of curves in the xy-plane, called *characteristic projections*, (also, more commonly and confusingly, just "characteristics") which are the projections onto the xy-plane of the characteristic curves in xyz-space. The characteristic curve is obtained from its projection $x(t),y(t)$ by finding $z(t)$ from the linear O.D.E.

$$\frac{dz}{dt}=c(x(t),y(t))z+d(x(t),y(t)). \tag{5.19}$$

We indicate how to proceed in the more general case of a quasi-linear equation for a function $u=u(x_1,\ldots,x_n)$ of n independent variables. Such an equation has the form

$$\sum_{i=1}^{n} a_i(x_1,\ldots,x_n,u)u_{x_i}=c(x_1,\ldots,x_n,u). \tag{5.20}$$

Here the characteristic curves in $x_1\ldots x_n z$-space are given by the system of O.D.E.s

$$\frac{dx_i}{dt}=a_i(x_1,\ldots,x_n,z) \quad \text{for } i=1,\ldots,n \tag{5.21a}$$

$$\frac{dz}{dt}=c(x_1,\ldots,x_n,z). \tag{5.21b}$$

In the *Cauchy problem* we want to pass an integral surface $z=u(x_1,\ldots,x_n)$ in $\mathbb{R}^{n+1}$ through an $(n-1)$-dimensional manifold Γ given parametrically by

$$x_i=f_i(s_1,\ldots,s_{n-1}) \quad \text{for } i=1,\ldots,n \tag{5.22a}$$

$$z=h(s_1,\ldots,s_{n-1}). \tag{5.22b}$$

For that purpose we pass through each point of Γ with parameters $s_1,\ldots,s_{n-1}$ a characteristic curve (solution of (5.21a, b) reducing to $(f_1,\ldots,f_n,h)$ for $t=0$) represented by

$$x_i=X_i(s_1,\ldots,s_{n-1},t) \quad \text{for } i=1,\ldots,n \tag{5.23a}$$

$$z=Z(s_1,\ldots,s_{n-1},t). \tag{5.23b}$$

These equations form a parametric representation for the desired integral surface $z=u(x_1,\ldots,x_n)$, provided relations (5.23a) can be solved for $s_1,\ldots,s_{n-1},t$. This is the case when the Jacobian

$$J=\begin{vmatrix} \dfrac{\partial f_1}{\partial s_1} & \cdots & \dfrac{\partial f_n}{\partial s_1} \\ \vdots & & \vdots \\ \dfrac{\partial f_1}{\partial s_{n-1}} & \cdots & \dfrac{\partial f_n}{\partial s_{n-1}} \\ a_1 & \cdots & a_n \end{vmatrix} \tag{5.24}$$

does not vanish.

6. Examples

(1) (See Section 3.)

$$u_y + cu_x = 0 \qquad (c = \text{const.}) \tag{6.1a}$$

$$u(x,0) = h(x). \tag{6.1b}$$

The initial curve Γ corresponding to (6.1b) is given by

$$x = s, \qquad y = 0, \qquad z = h(s).$$

The characteristic differential equations are

$$\frac{dx}{dt} = c, \qquad \frac{dy}{dt} = 1, \qquad \frac{dz}{dt} = 0. \tag{6.2}$$

This leads to the parametric representation

$$x = X(s,t) = s + ct, \qquad y = Y(s,t) = t, \qquad z = Z(s,t) = h(s) \tag{6.3}$$

for the integral surface. Eliminating s,t we find for the solution of the initial-value problem (6.1a, b) the representation

$$z = h(x - cy) \tag{6.4}$$

in agreement with (3.3).

(2) Euler's P.D.E. for a *homogeneous* function $u(x_1,\ldots,x_n)$:

$$\sum_{k=1}^{n} x_k u_{x_k} = \alpha u \qquad (\alpha = \text{const.} \neq 0). \tag{6.5}$$

Since equation (6.5) is singular at the origin (the J defined by (5.24) cannot be different from 0) we postulate the initial-value problem

$$u(x_1,\ldots,x_{n-1},1) = h(x_1,\ldots,x_{n-1}) \tag{6.6}$$

corresponding to a curve Γ given by

$$x_i = \begin{cases} s_i & \text{for } i = 1,\ldots,n-1 \\ 1 & \text{for } i = n \end{cases} \tag{6.7}$$

$$z = h(s_1,\ldots,s_{n-1}).$$

Solving the characteristic differential equations

$$\frac{dx_i}{dt} = x_i \quad \text{for } i = 1,\ldots,n \tag{6.8}$$

$$\frac{dz}{dt} = \alpha z$$

leads to

$$x_i = \begin{cases} s_i e^t & \text{for } i = 1,\ldots,n-1 \\ e^t & \text{for } i = n \end{cases} \tag{6.9}$$

$$z = e^{\alpha t} h(s_1,\ldots,s_{n-1})$$

and thus to

$$z=u(x_1,\ldots,x_n)=x_n^{\alpha}h\left(\frac{x_1}{x_n},\ldots,\frac{x_{n-1}}{x_n}\right). \tag{6.10}$$

The solution u satisfies the functional equation

$$u(\lambda x_1,\ldots,\lambda x_n)=\lambda^{\alpha}u(x_1,\ldots,x_n) \tag{6.11}$$

for any $\lambda>0$, and thus is a homogeneous function of degree α.

For $\alpha<0$ the solutions of (6.5) generally become singular at the origin. More precisely the only solution u of (6.5) of class C^1 in a neighborhood of the origin is $u\equiv 0$. For along any ray

$$x_i=c_i t \qquad i=1,\ldots,n$$

from the origin referred to a parameter t we have by (6.5)

$$\frac{du}{dt}=\sum_{k=1}^{n}c_k u_{x_k}(c_1 t,\ldots,c_n t)=\frac{\alpha}{t}u.$$

Hence $ut^{-\alpha}$ is constant along the ray, and thus u tends to ∞ for $t\to 0$, unless u vanishes identically along the ray. We have here an example of a P.D.E. that has only a single solution if we restrict the domain of the solution to be an open set containing the origin.

(3) The solution $u=u(x,y)$ of the quasi-linear equation (see [5], [22])

$$u_y+uu_x=0 \tag{6.12}$$

can be interpreted as a velocity field on the x-axis varying with the time y. Equation (6.12) then states that every particle has zero acceleration, and hence constant velocity. Let

$$u(x,0)=h(x) \tag{6.13}$$

describe the initial velocity distribution, corresponding to the manifold Γ in xyz-space given by

$$x=s, \qquad y=0, \qquad z=h(s). \tag{6.14}$$

The characteristic differential equations

$$\frac{dx}{dt}=z, \qquad \frac{dy}{dt}=1, \qquad \frac{dz}{dt}=0 \tag{6.15}$$

combined with the initial condition (6.14) for $t=0$ lead to the parametric representation

$$x=s+zt, \qquad y=t, \qquad z=h(s) \tag{6.16}$$

for the solution $z=u(x,y)$ of (6.12), (6.13). Eliminating s,t from (6.16) yields the implicit equation

$$u=h(x-uy) \tag{6.17}$$

for u as a function of x,y. (Notice the analogy to (6.4)!)

The characteristic (projection) C_s in the xy-plane passing through the point $(s,0)$ is the line

$$x = s + h(s)y \tag{6.18a}$$

along which u has the constant value

$$u = h(s). \tag{6.18b}$$

Physically (6.18a) for a fixed s represents the path of the particle located at $x = s$ at the time $y = 0$. Now two characteristics C_{s_1} and C_{s_2} intersect at a point (x,y) with

$$y = -\frac{s_2 - s_1}{h(s_2) - h(s_1)}. \tag{6.19}$$

If $s_2 \neq s_1$ and $h(s_2) \neq h(s_1)$, the function u must take the distinct values $h(s_1)$ and $h(s_2)$ at (x,y) and hence cannot be univalued. There always exist positive y of the form (6.19), unless $h(s)$ is a nondecreasing function of s. For all other $h(s)$ the solution $u(x,y)$ becomes singular for some positive y. (Physically a particle with a higher velocity will eventually collide with one ahead of it having a lower velocity). In particular, u is bound to become singular if the initial velocity distribution h has compact support, except in the trivial case where $h(s) \equiv 0$. The nature of the singularity becomes clearer when we follow the values of the derivative $u_x(x,y)$ along the characteristic (6.18a). We find from (6.17) that

$$u_x = \frac{h'(s)}{1 + h'(s)y}. \tag{6.20}$$

Hence for $h'(s) < 0$ we find that u_x becomes infinite at the positive time

$$y = \frac{-1}{h'(s)}. \tag{6.21}$$

The smallest y for which this happens corresponds to the value $s = s_0$ at which $h'(s)$ has a minimum. At the time $T = -1/h'(s_0)$ the solution u experiences a "gradient catastrophe" or "blow-up." There cannot exist a *strict* solution u of class C^1 beyond the time T. This type of behavior is typical for a nonlinear partial differential equation.

It is possible, however, to define *weak* solutions of (6.12), (6.13) which exist beyond the time T. For that purpose (6.12) has to be given a meaning for a wider class of functions u that do not necessarily lie in C^1 or even are continuous. We can write (6.12) in *divergence form*

$$\frac{\partial R(u)}{\partial y} + \frac{\partial S(u)}{\partial x} = 0, \tag{6.22}$$

where $R(u)$, $S(u)$ are any functions for which

$$S'(u) = uR'(u). \tag{6.23}$$

Relation (6.22) implies for any a, b, y the "conservation law"

$$0 = \frac{d}{dy}\int_a^b R(u(x,y))\,dx + S(u(b,y)) - S(u(a,y)). \tag{6.24}$$

Conversely (6.22) follows from (6.24) for any $u \in C^1$. Now (6.24) makes sense for more general u and can serve to define "weak" solutions of (6.22). In particular we consider the case where u is a C^1-solution of (6.22) in each of two regions in the xy-plane separated by a curve $x = \xi(y)$, across which the value of u shall undergo a jump ("shock"). Denoting the limits of u from the left and right respectively by u^- and u^+, we find from (6.24) for $a < \xi(y) < b$

$$\begin{aligned}
0 &= S(u(b,y)) - S(u(a,y)) + \frac{d}{dy}\left(\int_a^\xi R(u)\,dx + \int_\xi^b R(u)\,dx\right)\\
&= S(u(b,y)) - S(u(a,y)) + \xi' R(u^-) - \xi' R(u^+)\\
&\quad - \int_a^\xi \frac{\partial S(u)}{\partial x}\,dx - \int_\xi^b \frac{\partial S(u)}{\partial x}\,dx\\
&= -(R(u^+) - R(u^-))\xi' - S(u^-) + S(u^+).
\end{aligned}$$

Hence we find the relation ("shock condition")

$$\frac{d\xi}{dy} = \frac{S(u^+) - S(u^-)}{R(u^+) - R(u^-)} \tag{6.25}$$

connecting the speed of propagation $d\xi/dy$ of the discontinuity with the amounts by which R and S jump. We observe that (6.25) depends not only on the original partial differential equation (6.12) but also on our choice of the functions $R(u)$, $S(u)$ satisfying (6.23). [Compare with Burgers' equation, p. 214.]

Problems

1. Solve the following initial-value problems and verify your solution:
   (a) $u_x + u_y = u^2$, $u(x,0) = h(x)$
   (b) $u_y = xuu_x$, $u(x,0) = x$
   (Answer: $x = ue^{-yu}$ implicitly.)
   (c) $xu_x + yu_y + u_z = u$, $u(x,y,0) = h(x,y)$
   (d) $xu_y - yu_x = u$, $u(x,0) = h(x)$
   (Answer: $u = h(\sqrt{x^2+y^2}\,)e^{\arctan(y/x)}$.)

2. (Picone). Let u be a solution of

$$a(x,y)u_x + b(x,y)u_y = -u$$

of class C^1 in the closed unit disk Ω in the xy-plane. Let $a(x,y)x + b(x,y)y > 0$ on the boundary of Ω. Prove that u vanishes identically. (Hint: Show that $\max_\Omega u \leqslant 0$, $\min_\Omega u \geqslant 0$, using conditions for a maximum at a boundary point.)

3. Let u be a C^1-solution of (6.12) in each of two regions separated by a curve $x=\xi(y)$. Let u be continuous, but u_x have a jump discontinuity on the curve. Prove that

$$\frac{d\xi}{dy}=u$$

and hence that the curve is a characteristic. (Hint: By (6.12)

$$(u_y^+ - u_y^-)+u(u_x^+ - u_x^-)=0.$$

Moreover $u(\xi(y),y)$ and $(d/dy)u(\xi(y),y)$ are continuous on the curve.)

4. Show that the function $u(x,y)$ defined for $y \geqslant 0$ by

$$u=-\tfrac{2}{3}\left(y+\sqrt{3x+y^2}\right) \quad \text{for } 4x+y^2>0$$

$$u=0 \quad \text{for } 4x+y^2<0$$

is a weak solution of (6.22) for the choice $R(u)=u, S(u)=\frac{1}{2}u^2$.

5. Define a *weak* solution $u(x,y)$ of (6.22) as a function for which the relation

$$\int\int (R(u)\phi_y + S(u)\phi_x)\,dx\,dy=0 \tag{6.26}$$

holds for any function $\phi(x,y)$ of class C_0^∞ (Relation (6.26) follows formally from (6.22) by integration by parts.) Show that this definition of weak solution also leads to the jump condition (6.25).

6. Show that the solution u of the quasi-linear partial differential equation

$$u_y + a(u)u_x=0 \tag{6.27}$$

with initial condition $u(x,0)=h(x)$ is given implicitly by

$$u=h(x-a(u)y) \tag{6.28}$$

Show that the solution becomes singular for some positive y, unless $a(h(s))$ is a nondecreasing function of s.

7. The General First-Order Equation for a Function of Two Variables

The general first-order partial differential equation for a function $z=u(x,y)$ has the form

$$F(x,y,z,p,q)=0, \tag{7.1}$$

where $p=u_x, q=u_y$. We assume that F where considered has continuous second derivatives with respect to its arguments x,y,z,p,q. Surprisingly enough the problem of solving the general equation (7.1) reduces to that of solving a system of ordinary differential equations. This reduction is suggested by the geometric interpretation of (7.1) as a condition on the integral surface $z=u(x,y)$ in xyz-space determined by a solution $u(x,y)$.

The geometry here is more involved than in the quasi-linear case where we were concerned principally with integral curves. We shall have to deal with more complicated geometric objects, called "strips."

Equation (7.1) can be viewed as a relation between the coordinates (x,y,z) of a point on an integral surface and the direction of the normal of the integral surface at that point, described by the direction numbers $p,q,-1$. An integral surface passing through a given point $P_0=(x_0,y_0,z_0)$ must have a tangent plane

$$z-z_0=p(x-x_0)+q(y-y_0) \tag{7.2}$$

for which the direction numbers $(p,q,-1)$ of the normal satisfy

$$F(x_0,y_0,z_0,p,q)=0. \tag{7.3}$$

Thus the differential equation restricts the possible tangent planes of an integral surface through P_0 to a one-parameter family. In general* such a one-parameter family of planes through P_0 can be expected to envelop a cone with vertex P_0, called the *Monge cone* at P_0 (Figure 1.3). Each possible tangent plane touches the Monge cone along a certain generator. In this way the partial differential equation (7.1) defines a *field of cones*. A surface $z=u(x,y)$ is an integral surface if at each of its points P_0 it touches the cone with vertex P_0. (See Figure 1.4.) In that case the generator along which the tangent plane touches the cone defines a direction on the surface. These "characteristic" directions are the key to the whole theory of integration of (7.1). In the special case of a quasi-linear equation (4.2) the Monge cone at P_0 degenerates into the line with direction (a,b,c) through P_0.

The central notion here is that of the *envelope* of a family of surfaces S_λ

$$z=G(x,y,\lambda) \tag{7.4}$$

depending on a parameter λ. We combine (7.4) with the equation

$$0=G_\lambda(x,y,\lambda). \tag{7.5}$$

For fixed λ equations (7.4), (7.5) determine a curve γ_λ. The envelope is the union of these curves. Its explicit equation is obtained by solving (7.5) for λ in the form $\lambda=g(x,y)$ and substitution into (7.4):

$$z=G(x,y,g(x,y)). \tag{7.6}$$

The envelope E touches the surface S_λ along the curve γ_λ. For in a point (x,y,z) of γ_λ we have $g(x,y)=\lambda$, so that (7.6) holds; moreover by (7.5) the direction numbers of the normal of E are $(G_x+G_\lambda g_x, G_y+G_\lambda g_y, -1)=(G_x,G_y,-1)$, the same as for S_λ. It is often advantageous to write equations (7.4), (7.5) for a fixed λ in the form of differential equations

$$dz=G_x\,dx+G_y\,dy, \qquad 0=G_{\lambda x}\,dx+G_{\lambda y}\,dy \tag{7.7}$$

satisfied along the curve γ_λ.

*For the present heuristic considerations we dispense with rigor.

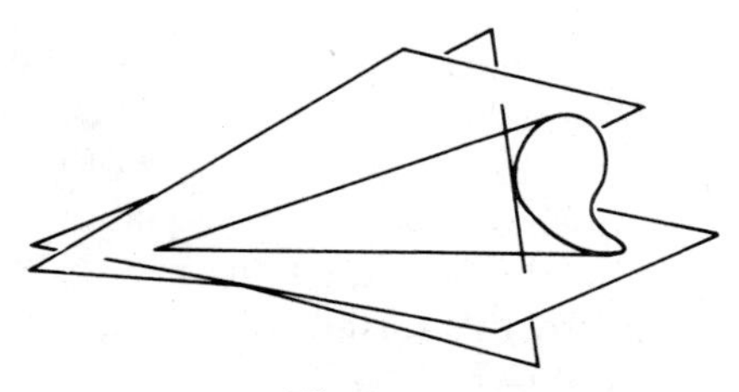

Figure 1.3

Now for fixed x_0, y_0, z_0, equation (7.2) defines a one-parameter family of planes, where we can choose p as the parameter and think of q as expressed in terms of p from (7.3). By (7.7) the generator along which the plane touches the Monge cone satisfies the equations

$$dz = p\,dx + q\,dy, \qquad 0 = dx + \frac{dq}{dp}\,dy. \tag{7.8}$$

Since by (7.3)

$$F_p + \frac{dq}{dp} F_q = 0 \tag{7.9}$$

the direction of the generator is given by

$$dz = p\,dx + q\,dy, \qquad \frac{dx}{F_p} = \frac{dy}{F_q}. \tag{7.10}$$

On a known integral surface S: $z = u(x,y)$ equations (7.10) define a direction field, since $F_p(x,y,u,u_x,u_y)$, $F_q(x,y,u,u_x,u_y)$ are then also known functions of x,y. We define the *characteristic curves* belonging to the integral surface S as those fitting the direction field. Using a suitable curve parameter t, the characteristic curves on S are given by the system of ordinary differential equations

$$\frac{dx}{dt} = F_p(x,y,z,p,q), \qquad \frac{dy}{dt} = F_q(x,y,z,p,q) \tag{7.10a}$$

$$\frac{dz}{dt} = pF_p(x,y,z,p,q) + qF_q(x,y,z,p,q), \tag{7.10b}$$

where

$$z = u(x,y), \qquad p = u_x(x,y), \qquad q = u_y(x,y). \tag{7.11}$$

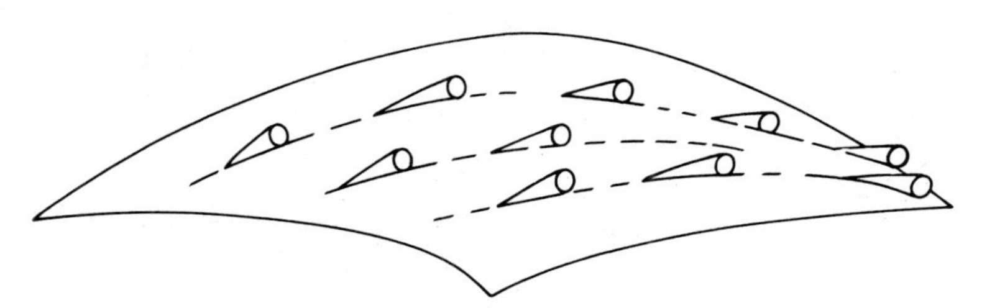

Figure 1.4

It is clear that for a quasi-linear equation (4.2) relations (7.10a,b) reduce to the characteristic differential equations (4.4). The main difference in the present more general case is that without the use of (7.11), that is, without the knowledge of the integral surface S, equations (7.10a,b) form an *underdetermined* system for the five functions x,y,z,p,q of t. However, the system is easily completed by two further equations. Partial differentiations with respect to x,y of the P.D.E.

$$F\big(x,y,u(x,y),u_x(x,y),u_y(x,y)\big)=0 \tag{7.12}$$

furnish the relations

$$F_x+u_xF_z+u_{xx}F_p+u_{xy}F_q=0 \tag{7.13a}$$

$$F_y+u_yF_z+u_{xy}F_p+u_{yy}F_q=0. \tag{7.13b}$$

Then along a characteristic curve on S we have by (7.11), (7.10a)

$$\frac{dp}{dt}=u_{xx}\frac{dx}{dt}+u_{xy}\frac{dy}{dt}=u_{xx}F_p+u_{xy}F_q=-F_x-u_xF_z$$

with a similar relation for dq/dt. Writing these relations in the form

$$\frac{dp}{dt}=-F_x(x,y,z,p,q)-pF_z(x,y,z,p,q) \tag{7.14a}$$

$$\frac{dq}{dt}=-F_y(x,y,z,p,q)-qF_z(x,y,z,p,q) \tag{7.14b}$$

relations (7.10a,b), (7.14a,b) constitute an autonomous system of five ordinary differential equations for the five functions x,y,z,p,q of t, which does not require knowledge of the integral surface $z=u(x,y)$ for its formulation.

The expression $F(x,y,z,p,q)$ is an "integral" of the system, that is, F is constant along any solution, since by (7.10a,b), (7.14a,b)

$$\begin{aligned}\frac{dF}{dt}&=F_x\frac{dx}{dt}+F_y\frac{dy}{dt}+F_z\frac{dz}{dt}+F_p\frac{dp}{dt}+F_q\frac{dq}{dt}\\&=F_xF_p+F_yF_q+F_z(pF_p+qF_q)+F_p(-F_x-pF_z)+F_q(-F_y-qF_z)=0.\end{aligned}$$

Hence along any trajectory of the system we have $F=0$ for all t, if $F=0$ for some particular t. We refer to the system of five ordinary differential equations (7.10a,b), (7.14a,b), together with the dependent relation

$$F(x,y,z,p,q)=0 \tag{7.15}$$

as the *characteristic equations*. We can base the solution of the P.D.E. on these equations without reference to their intuitive geometric significance.

A solution of the characteristic equations is a set of five functions $x(t),y(t),z(t),p(t),q(t)$. Generally, we call a quintuple (x,y,z,p,q) a *plane element* and interpret it geometrically as consisting of a point (x,y,z) combined with a plane through the point with equation

$$\zeta - z = p(\xi - x) + q(\eta - y) \tag{7.16}$$

in running coordinates ξ, η, ζ. Then $p, q, -1$ are direction numbers of the normal of the plane. An element is called *characteristic* if it satisfies (7.15). A one-parameter family of elements $(x(t), y(t), z(t), p(t), q(t))$ is called a *strip* if the elements are tangent to the curve formed by the points $(x(t), y(t), z(t))$, the *support* of the strip. For that to be the case, the *strip condition*

$$\frac{dz}{dt} = p\frac{dx}{dt} + q\frac{dy}{dt} \tag{7.17}$$

has to be satisfied. A solution of the characteristic equations will be called a *characteristic strip*. (Here the strip condition holds in consequence of relations (7.10a,b).) (See Figure 1.5.)

A surface $z = u(x,y)$ referred to parameters s,t can be thought of as consisting of a two-parameter family of elements $(x(s,t),\ y(s,t),\ z(s,t), p(s,t), q(s,t))$ formed by the points of the surface and the corresponding tangent planes. Not every two-parameter family of elements forms a surface. It is necessary again that

$$dz = p\,dx + q\,dy \tag{7.18}$$

holds along the family, i.e., that the *strip conditions*

$$\frac{\partial z}{\partial s} = p\frac{\partial x}{\partial s} + q\frac{\partial y}{\partial s}, \qquad \frac{\partial z}{\partial t} = p\frac{\partial x}{\partial t} + q\frac{\partial y}{\partial t} \tag{7.19}$$

are satisfied. Visualizing an element as a small piece of plane attached to a point, the elements of a two-parameter family belonging to a surface have to fit together smoothly somewhat like scales on a fish.

As a solution of a system of ordinary differential equations a characteristic strip is determined uniquely by any one of its elements. If that element of the strip consists of a point P of an integral surface S and of the tangent plane to S at P, then the strip is made up of the characteristic curve of S through the point P and of the tangent planes of S along that curve. If another integral surface touches S at the point P then that surface will also touch S all along the characteristic curve.

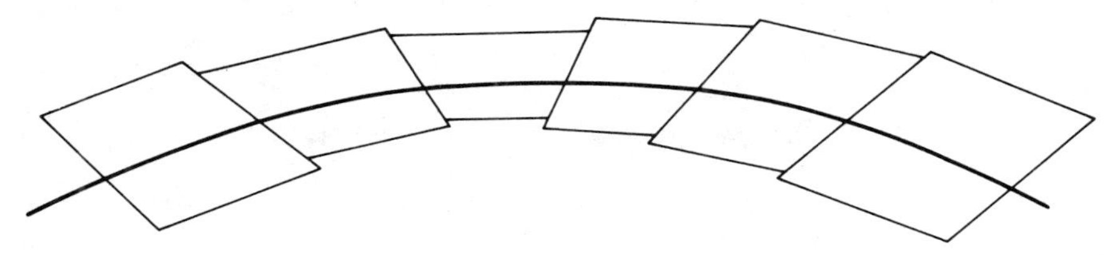

Figure 1.5

Equations (7.10a,b), (7.14a,b) describe a "law of propagation" of characteristic elements or of tangent planes on an integral surface along characteristic curves. A lower-dimensional geometric interpretation often is useful. We associate with (x,y,z,p,q) a "line element" in the xy-plane consisting of the point (x,y) and the line

$$0=p(\xi-x)+q(\eta-y) \tag{7.20}$$

in running coordinates ξ,η. The line (7.20) is the level line $\zeta=z$ of the plane (7.16). The geometric line element does not determine x,y,p,q uniquely, since p and q can be replaced by any proportional numbers. A characteristic strip gives rise to a family of line elements. Using z instead of t as a parameter, the line elements "propagate" for varying z in a definite way described by the equations

$$\frac{dx}{dz}=\frac{F_p}{pF_p+qF_q}, \qquad \frac{dy}{dz}=\frac{F_q}{pF_p+qF_q} \tag{7.21a}$$

$$\frac{dp}{dz}=-\frac{F_x+pF_z}{pF_p+qF_q}, \qquad \frac{dq}{dz}=-\frac{F_y+qF_z}{pF_p+qF_q}. \tag{7.21b}$$

We can start out with an initial line element (x_0,y_0,p_0,q_0) for $z=z_0$ (in which p_0,q_0 are to be replaced by proportional quantities so that

$$F(x_0,y_0,z_0,p_0,q_0)=0 \tag{7.22}$$

holds). With changing z the line element moves along the curve $x=x(t)$, $y=y(t)$ (called a "ray"). The line elements are not tangent to the ray (no "strip condition"), since generally

$$p\frac{dx}{dt}+q\frac{dy}{dt}=pF_p+qF_q\neq 0.$$

8. The Cauchy Problem

The Cauchy problem for (7.1) consists in passing an integral surface through an "arbitrary" initial curve Γ given parametrically in the form

$$x=f(s), \qquad y=g(s), \qquad z=h(s). \tag{8.1}$$

This will be achieved by passing suitable characteristic strips through Γ. We assume that f,g,h are of class C^1 for s near a value s_0 corresponding to a point

$$P_0=(x_0,y_0,z_0)=(f(s_0),g(s_0),h(s_0)). \tag{8.1a}$$

We first have to complete Γ into a strip consisting of characteristic elements. That is, we have to find functions

$$p=\phi(s), \qquad q=\psi(s) \tag{8.2}$$

such that

$$h'(s) = \phi(s) f'(s) + \psi(s) g'(s) \tag{8.3a}$$

$$F(f(s), g(s), h(s), \phi(s), \psi(s)) = 0. \tag{8.3b}$$

Since equation (8.3b) is nonlinear there may be one, or several, or no solution (ϕ, ψ) of (8.3a,b). We assume that we are given a special solution p_0, q_0 of

$$h'(s_0) = p_0 f'(s_0) + q_0 g'(s_0), \qquad F(x_0, y_0, z_0, p_0, q_0) = 0 \tag{8.4}$$

such that

$$\Delta = f'(s_0) F_q(x_0, y_0, z_0, p_0, q_0) - g'(s_0) F_p(x_0, y_0, z_0, p_0, q_0) \neq 0. \tag{8.5}$$

That is, we have found a characteristic plane tangent to Γ at P_0 such that the generator of contact between the plane and the Monge cone at P_0 has a different projection onto the xy-plane than the tangent to Γ at P_0. By the implicit function theorem there exist then unique functions $\phi(s), \psi(s)$ of class C^1 near s_0, satisfying (8.3a,b) and reducing to p_0, q_0 for $s = s_0$.

Through each element $(f(s), g(s), h(s), \phi(s), \psi(s))$ we now pass the characteristic strip reducing to that element for $t = 0$. In this way we find five functions

$$x = X(s,t), \qquad y = Y(s,t), \qquad z = Z(s,t), \qquad p = P(s,t), \qquad q = Q(s,t) \tag{8.6}$$

defined for $|s - s_0|$ and $|t|$ sufficiently small, which satisfy for fixed s the differential equations (7.10a,b), (7.14a,b) as functions of t, and for $t = 0$ reduce respectively to $f(s), g(s), h(s), \phi(s), \psi(s)$. The relation

$$F(X, Y, Z, P, Q) = 0 \tag{8.7}$$

holds identically in s, t, since it holds for $t = 0$ by (8.3b).

From what preceded it is clear that if there exists an integral surface S through Γ containing the element $(x_0, y_0, z_0, p_0, q_0)$, then that surface must be the union of the supports of the characteristic strips we constructed. In particular, the first three equations in (8.6) must constitute a parametric representation for S.

Conversely we shall prove that (8.6) represents a solution of our Cauchy problem in parametric form in a neighborhood of the point P_0. First of all we can solve the first two equations (8.6) for s, t in terms of x, y for (x, y) near (x_0, y_0). For by (8.1a)

$$x_0 = f(s_0) = X(s_0, 0), \qquad y_0 = g(s_0) = Y(s_0, 0)$$

and we have a nonvanishing Jacobian at $s = s_0$, $t = 0$:

$$\frac{\partial(x, y)}{\partial(s, t)} = \begin{vmatrix} X_s & Y_s \\ X_t & Y_t \end{vmatrix} = \begin{vmatrix} f' & g' \\ F_p & F_q \end{vmatrix} = \Delta \neq 0. \tag{8.7a}$$

Substituting into the third equation (8.6) yields an explicit equation $z = u(x,y)$ for a surface passing through Γ. By (8.7) this will be an integral surface, if we can show that the p,q defined by the last two equations (8.6) are identical with u_x and u_y. Now from the first three equations in (8.6) we can determine u_x, u_y uniquely in terms of s,t by the chain rule which yields the relations

$$Z_s = u_x X_s + u_y Y_s, \qquad Z_t = u_x X_t + u_y Y_t \tag{8.8}$$

We have proved that

$$u_x = P(s,t), \qquad u_y = Q(s,t), \tag{8.9}$$

if we can verify the identities

$$Z_s = PX_s + QY_s, \qquad Z_t = PX_t + QY_t. \tag{8.10}$$

These equations just express that the two-parameter family of elements (8.6) belongs to a surface. The second equation in (8.10) is a consequence of the characteristic differential equations (7.10a,b) satisfied by X, Y, Z, P, Q, as functions of t. To verify the first equation we introduce the expression

$$A(s,t) = Z_s - PX_s - QY_s. \tag{8.11}$$

Here

$$A(s,0) = h' - \phi f' - \psi g' = 0 \tag{8.12}$$

by (8.3a). Moreover, making use of the characteristic equations for X, Y, Z, P, Q, we have

$$\begin{aligned} A_t &= Z_{st} - P_t X_s - Q_t Y_s - PX_{st} - QY_{st} \\ &= \frac{\partial}{\partial s}(Z_t - PX_t - QY_t) + P_s X_t + Q_s Y_t - Q_t Y_s - P_t X_s \\ &= P_s F_p + Q_s F_q + X_s(F_x + F_z P) + Y_s(F_y + F_z Q) \\ &= \frac{\partial F}{\partial s} - F_z(Z_s - PX_s - QY_s) = -F_z A. \end{aligned} \tag{8.13}$$

But then, by integration,

$$A(s,t) = A(s,0)\exp\left(-\int_0^t F_z\, dt\right) = 0$$

because of (8.12). This completes the local existence proof for a solution of the Cauchy problem under the assumption that we have a solution p_0, q_0 of (8.4), (8.5).

The same methods apply to first-order equations in more independent variables. The general equation for a function $u(x_1,\ldots,x_n)$ has the form

$$F(x_1,\ldots,x_n,z,p_1,\ldots,p_n)=0, \tag{8.14}$$

where $z=u$, $p_i=u_{x_i}$. The Cauchy problem here consists in finding an integral surface in $x_1\ldots x_n z$-space that passes through an $(n-1)$-dimensional manifold Γ given parametrically by

$$z=h(s_1,\ldots,s_{n-1}), \qquad x_i=f_i(s_1,\ldots,s_{n-1}) \quad \text{for } i=1,\ldots,n.$$

This is achieved as before by passing through each point P of Γ a characteristic strip tangent to Γ at P. We first complete Γ into a strip by finding functions $p_i=\phi_i(s_1,\ldots,s_{n-1})$ for which

$$\frac{\partial h}{\partial s_i}=\sum_{k=1}^{n}\phi_k\frac{\partial f_k}{\partial s_i} \quad \text{for } i=1,\ldots,n-1 \tag{8.14a}$$

$$F(f_1,\ldots,f_n,h,\phi_1,\ldots,\phi_n)=0 \tag{8.14b}$$

in analogy to (8.3a,b). A characteristic strip here is a set of "elements" $(x_1,\ldots,x_n,z,p_1,\ldots,p_n)$ depending on a parameter t that satisfies in addition to (8.14) the system of ordinary differential equations

$$\frac{dx_i}{dt}=F_{p_i}, \qquad \frac{dp_i}{dt}=-F_{x_i}-F_z p_i \quad \text{for } i=1,\ldots,n \tag{8.15a}$$

$$\frac{dz}{dt}=\sum_{i=1}^{n}p_i F_{p_i}. \tag{8.15b}$$

For existence assumption (8.7a) here has to be replaced by

$$\Delta=\begin{vmatrix} \dfrac{\partial f_1}{\partial s_1} & \cdots & \dfrac{\partial f_n}{\partial s_1} \\ \vdots & & \vdots \\ \dfrac{\partial f_1}{\partial s_{n-1}} & \cdots & \dfrac{\partial f_n}{\partial s_{n-1}} \\ F_{p_1} & & F_{p_n} \end{vmatrix}\neq 0 \tag{8.15c}$$

at one point of Γ.

An instructive example is given by the equation

$$c^2(p^2+q^2)=1 \tag{8.16}$$

($c=\text{const.}>0$), which arises in geometric optics. There the level lines of a solution u are interpreted as "wave fronts," marking the location to which light has spread. (We shall obtain the same lines as lines of discontinuity for solutions of the wave equation.) Geometrically (8.16) states that tangent planes of an integral surface make a fixed angle $\theta=\arctan c$ with the

z-axis. The Monge cone enveloped by possible tangent planes through a point is then a circular cone with opening angle 2θ.

For convenience we write (8.16) as

$$F=\tfrac{1}{2}(c^2p^2+c^2q^2-1)=0 \tag{8.17}$$

leading to the characteristic equations

$$\frac{dx}{dt}=c^2p, \qquad \frac{dy}{dt}=c^2q,$$

$$\frac{dz}{dt}=c^2(p^2+q^2)=1, \qquad \frac{dp}{dt}=\frac{dq}{dt}=0. \tag{8.18}$$

A given initial curve

$$\Gamma: \qquad x=f(s), \qquad y=g(s), \qquad z=h(s) \tag{8.19}$$

is completed into a strip by choosing $p=\phi(s), q=\psi(s)$ according to the equations

$$h'(s)=\phi(s)f'(s)+\psi(s)g'(s), \qquad \phi^2(s)+\psi^2(s)=c^{-2}. \tag{8.20}$$

These equations have no real solution when $f'^2+g'^2<c^2h'^2$, that is, when Γ forms an angle less than θ with the z-axis (Γ is "timelike"). For a "spacelike" Γ with $f'^2+g'^2>c^2h'^2$ we can solve (8.20) in two different ways, giving rise to two different solutions of the Cauchy problem.

Of special interest is the case where Γ is the intersection

$$x=f(s), \qquad y=g(s), \qquad z=0 \tag{8.21}$$

of the integral surface $z=u(x,y)$ with the xy-plane. For this space-like Γ there are two solutions $(\phi(s),\psi(s))$ of

$$0=\phi f'+\psi g', \qquad \phi^2+\psi^2=c^{-2}, \tag{8.22}$$

differing only in sign. The characteristic strips with initial elements on Γ are given by

$$x=f(s)+c^2t\phi(s), \qquad y=g(s)+c^2t\psi(s), \tag{8.23a}$$

$$z=t, \qquad p=\phi(s), \qquad q=\psi(s). \tag{8.23b}$$

We represent the integral surface by its level lines in the xy-plane

$$\gamma_t: \qquad u(x,y)=z=\text{const.}=t \tag{8.24}$$

interpreted as a "wave front" moving with the "time" t. Equations (8.23a) for fixed t give the curve γ_t referred to the parameter s. We think of γ_t as made up of line elements (x,y,p,q), i.e., of points (x,y) and corresponding tangents

$$p(\xi-x)+q(\eta-y)=0. \tag{8.25}$$

The elements propagate individually according to (8.23a,b) for fixed s. The point (x,y) moves along the *ray* (8.23a) with constant speed

$$\sqrt{\left(\frac{dx}{dt}\right)^2+\left(\frac{dy}{dt}\right)^2}=c^2\sqrt{\phi^2+\psi^2}=c. \tag{8.26}$$

The ray is the straight line of direction $(\phi,\psi)=(p,q)$, hence coincides at each point with the normal to the level line γ_t. We see that the wave fronts γ_t form a family of curves with common normals ("parallel" curves). Here γ_t can also be obtained from γ_0 by laying off a fixed distance ct along each normal of γ_0. Thus γ_t is also the curve of constant "normal distance" ct from γ_0. (The choice between "interior" or "exterior" normal directions corresponds to the choice between the two solutions of the Cauchy problem.)

9. Solutions Generated as Envelopes

The envelope S of a family of integral surfaces S_λ of (7.1) with equation

$$z=G(x,y,\lambda) \tag{9.1}$$

is again an integral surface. This is obvious from the geometric interpretation of (7.1) and the fact that every tangent plane of the envelope is a tangent plane of one of the surfaces of the family. Here S touches S_λ along the curve given by the two equations

$$z=G(x,y,\lambda), \qquad 0=G_\lambda(x,y,\lambda). \tag{9.2}$$

Since integral surfaces can touch only along characteristic strips, we see that completing equations (9.2) for fixed λ by

$$p=G_x(x,y,\lambda), \qquad q=G_y(x,y,\lambda) \tag{9.3}$$

describes a characteristic strip. We obtain its parametric representation with (x,y,z,p,q) as functions of some parameter t by solving the equation $0 = G_\lambda(x,y,\lambda)$ in some parametric form $x = x(t,\lambda), y = y(t,\lambda)$ and substituting into the other equations. In this way a one-parameter family of solutions of the partial differential equation (7.1) yields a further solution and a one-parameter family of characteristic strips, purely by elimination.

Using this observation we can obtain the "general" solution of (7.1) and the "general" characteristic strip from any special *two-parameter* family

$$u=G(x,y,\lambda,\mu) \tag{9.4}$$

of solutions of (7.1). Many different one-parameter families are obtained from (9.4) by substituting for μ any function $\mu=w(\lambda)$. Here for a particular λ and a suitable function w the values of $\mu=w(\lambda)$ and $\nu=w'(\lambda)$ are arbitrary. It follows that for general λ,μ,ν the equations

$$z=G(x,y,\lambda,\mu), \qquad 0=G_\lambda(x,y,\lambda,\mu)+\nu G_\mu(x,y,\lambda,\mu) \tag{9.4a}$$

$$p=G_x(x,y,\lambda,\mu), \qquad q=G_y(x,y,\lambda,\mu) \tag{9.4b}$$

describe a characteristic strip. We obtain in this way the "general" characteristic strip containing an arbitrary characteristic element $(x_0, y_0, z_0, p_0, q_0)$. We only have to determine λ, μ, ν in such a way that equations (9.4a, b) hold for x, y, z, p, q replaced by x_0, y_0, z_0, p_0, q_0. These are essentially only 3 conditions for λ, μ, ν since the equation $F(x,y,z,p,q)=0$ holds for all elements determined from (9.4a, b) as well as for the prescribed characteristic element.

The Cauchy problem of finding an integral surface of (7.1) passing through a curve Γ can be solved directly with a knowledge of the two-parameter family (9.4). We only have to find a one-parameter subfamily of integral surfaces that touch Γ and then form their envelope. For Γ given by (8.1) this amounts to finding functions $\lambda = \lambda(s)$, $\mu = \mu(s)$ such that the relations

$$h = G(f, g, \lambda, \mu), \qquad h' = G_x(f, g, \lambda, \mu) f' + G_y(f, g, \lambda, \mu) g' \tag{9.5}$$

hold identically in s. The solution is then determined by eliminating s from the equations

$$u = G(x, y, \lambda, \mu), \qquad 0 = G_\lambda(x, y, \lambda, \mu)\lambda' + G_\mu(x, y, \lambda, \mu)\mu'. \tag{9.6}$$

Of special interest is the limiting case, when Γ degenerates into a point (x_0, y_0, z_0). Taking the solutions (9.4) for which λ, μ are such that

$$z_0 = G(x_0, y_0, \lambda, \mu)$$

and forming their envelope we obtain an integral surface with a conical singularity at (x_0, y_0, z_0), called the "conoid" solution. The solution of the general Cauchy problem can be obtained by taking the envelope of the conoids that have their singularities on the prescribed curve Γ.

Any two-parameter family of functions (9.4) determines a first-order partial differential equation (7.1) of which these functions are solutions. We only have to eliminate the parameters λ, μ between the equations

$$u = G(x, y, \lambda, \mu), \qquad u_x = G_x(x, y, \lambda, \mu), \qquad u_y = G_y(x, y, \lambda, \mu). \tag{9.7}$$

For the resulting partial differential equation we can immediately find all characteristic strips and solve the Cauchy problem.

An example is furnished by the equation

$$c^2\left(u_x^2 + u_y^2\right) = 1 \tag{9.8}$$

considered earlier. The equation has the special two-parameter family of linear solutions

$$u = G(x, y, \lambda, \mu) = c^{-1}(x\cos\lambda + y\sin\lambda) + \mu \tag{9.9}$$

representing planes in xyz-space forming an angle $\theta = \arctan c$ with the z-axis. The general integral surface $z = u(x,y)$ is the envelope of a one-parameter subfamily of the planes (9.9). The conoid with "vertex" or singular point (x_0, y_0, z_0), here identical with the Monge cone, is the envelope of the planes (9.9) passing through (x_0, y_0, z_0), i.e., of the planes

$$z - z_0 = c^{-1}((x - x_0)\cos\lambda + (y - y_0)\sin\lambda). \tag{9.10}$$

The conoid thus is the circular cone

$$c^2(z - z_0)^2 = (x - x_0)^2 + (y - y_0)^2. \tag{9.11}$$

We can find the solution (9.8) passing through an initial curve γ_0 in the xy-plane by forming the envelope of the conoids with vertex on γ_0. Since the level line z = const. = t of the conoid with vertex $(x_0, y_0, 0)$ is a circle of radius ct and center (x_0, y_0), we obtain the level line γ_t of the envelope by forming the envelopes of the circles of radius ct with centers on γ_0. This agrees with the earlier representation of γ_t as curve of normal distance ct from γ_0. It corresponds to Huygens's generation of the wave front γ_t as the envelope of "circular" waves issuing from the original wave front γ_0.

PROBLEMS

1. For the equation

$$u_x^2 + u_y^2 = u^2$$

find (a) the characteristic strips; (b) the integral surfaces passing through the circle

$$x = \cos s, \qquad y = \sin s, \qquad z = 1$$

[Answer: $z = \exp[\pm(1 - \sqrt{x^2 + y^2})]$]; (c) the integral surfaces through the line $x = s$, $y = 0$, $z = 1$ [Answer: $u = \exp(\pm y)$].

2. For the equation

$$u_y = u_x^3$$

(a) find the solution with $u(x,0) = 2x^{3/2}$ [Answer: $u = 2x^{3/2}(1 - 27y)^{-1/2}$]; (b) prove that every solution regular for all x, y is linear.

3. For the equation

$$u = xu_x + yu_y + \tfrac{1}{2}(u_x^2 + u_y^2)$$

find a solution with $u(x,0) = \frac{1}{2}(1 - x^2)$.

4. Given the family of spheres of radius 1 with centers in the xy-plane

$$u = G(x,y,\lambda,\mu) = \sqrt{1 - (x - \lambda)^2 - (y - \mu)^2}\,,$$

find the first-order partial differential equation they satisfy. Find all characteristic strips and give a geometric description. Find the conoid solution with vertex $(0, 0, \frac{1}{2})$. Find the integral surfaces through the line $x = s$, $y = 0$, $z = \frac{1}{2}$.

5. (Characteristics as extremals of a variational problem). Consider for a given function $H(x_1, \ldots, x_n, t, p_1, \ldots, p_n)$ the partial differential equation

$$F = \frac{\partial u}{\partial t} + H\left(x_1, \ldots, x_n, t, \frac{\partial u}{\partial x_1}, \ldots, \frac{\partial u}{\partial x_n}\right) = 0$$

(Hamilton–Jacobi equation) for $u = u(x_1, \ldots, x_n, t)$. Obtain the characteristic equations in the form

$$\frac{dx_i}{dt} = H_{p_i}, \qquad \frac{du}{dt} = \sum_i p_i H_{p_i} - H, \qquad \frac{dp_i}{dt} = -H_{x_i}.$$

Setting $dx_i/dt = v_i$, $du/dt = L$ we use the first $n+1$ equations to express L as a function of $x_1, \ldots, x_n, t, v_1, \ldots, v_n$. Show then that

$$L_{v_i} = p_i, \qquad L_{x_i} = -H_{x_i},$$

which implies that

$$\frac{d}{dt} L_{v_i} - L_{x_i} = 0.$$

These are the Euler–Lagrange equations for an extremal of the functional $\int L(x_1, \ldots, x_n, t, dx_1/dt, \ldots, dx_n/dt)\, dt$.

6. Find all solutions $u(x, y)$ of $u_y = F(u_x)$ for which $u(x, 0) = h(x)$. [Answer:

$$u = (F(p) - pF'(p))y + h(x + yF'(p)) \tag{9.12}$$

where p is the solution of the implicit equation

$$p = h'(x + yF'(p)). \tag{9.13}$$
]

Second-Order Equations: Hyperbolic Equations for Functions of Two Independent Variables* 2

1. Characteristics for Linear and Quasi-linear Second-Order Equations†

We start with the general quasi-linear second-order equation for a function $u(x,y)$:

$$au_{xx} + 2bu_{xy} + cu_{yy} = d, \tag{1.1}$$

where a,b,c,d depend on x,y,u,u_x,u_y. Here the *Cauchy problem* consists in finding a solution u of (1.1) with given (compatible) values of u, u_x, u_y on a curve γ in the xy-plane. Thus, for γ given parametrically by

$$x = f(s), \qquad y = g(s), \tag{1.2}$$

we prescribe on γ the *Cauchy data*

$$u = h(s), \qquad u_x = \phi(s), \qquad u_y = \psi(s). \tag{1.3}$$

The values of any function $v(x,y)$ and of its first derivatives $v_x(x,y)$, $v_y(x,y)$ along the curve γ are connected by the compatibility condition ("strip condition")

$$\frac{dv}{ds} = v_x f'(s) + v_y g'(s) \tag{1.4}$$

which follows by differentiating $v(f(s), g(s))$ with respect to s. Applied to the solution u of the Cauchy problem this implies the identity

$$h'(s) = \phi(s) f'(s) + \psi(s) g'(s) \tag{1.5}$$

between the Cauchy data. Thus no more than two of the functions h, ϕ, ψ

* ([5], [6], [13], [15])
† ([26])

can be prescribed arbitrarily. Instead we might give on γ the values of u and of its *normal derivative*:

$$u=h(s), \qquad \frac{-u_x g' + u_y f'}{\sqrt{f'^2+g'^2}} = \chi(s). \tag{1.6}$$

Compatibility conditions also hold for the higher partial derivatives of any function on γ. Thus taking $v=u_x$ or $v=u_y$ we find that

$$\frac{du_x}{ds} = u_{xx} f'(s) + u_{xy} g'(s), \qquad \frac{du_y}{ds} = u_{xy} f'(s) + u_{yy} g'(s). \tag{1.7}$$

Similar relations are valid for the s-derivatives of u_{xx}, u_{xy}, u_{yy}, etc.

If now $u(x,y)$ is a solution of (1.1), (1.3) we have the three linear equations

$$au_{xx} + 2bu_{xy} + cu_{yy} = d \tag{1.8a}$$

$$f'u_{xx} + g'u_{xy} = \phi' \tag{1.8b}$$

$$f'u_{xy} + g'u_{yy} = \psi' \tag{1.8c}$$

for the values of u_{xx}, u_{xy}, u_{yy} along γ, with coefficients that are known functions of s. These determine u_{xx}, u_{xy}, u_{yy} uniquely unless

$$\Delta = \begin{vmatrix} f' & g' & 0 \\ 0 & f' & g' \\ a & 2b & c \end{vmatrix} = ag'^2 - 2bf'g' + cf'^2 = 0. \tag{1.9}$$

We call the "initial" curve γ *characteristic* (with respect to the differential equation and data), if $\Delta=0$ along γ, noncharacteristic if $\Delta \neq 0$ along γ. Along a noncharacteristic curve the Cauchy data uniquely determine the second derivatives of u on γ. As a matter of fact, we can then also find successively the values of all higher derivatives of u on γ, as far as they exist. We obtain, e.g., three linear equations with determinant Δ for $u_{xxx}, u_{xxy}, u_{xyy}$ by differentiating (1.1) partially with respect to x, and using the two equations obtained from (1.4) for $v=u_{xx}$ and $v=u_{xy}$; the coefficients in the three equations only involve the values of u and its first and second derivatives, known already. Obtaining in this way the values of all derivatives of u in some particular point (x_0,y_0) on γ, we could write down a *formal* power series for the solution of the Cauchy problem in terms of powers of $x-x_0, y-y_0$. It would be an actual representation of the solution u in a neighborhood of (x_0,y_0), if u were known to be analytic. This procedure will be legitimized in the case of analytic Cauchy data, when we discuss the Cauchy–Kowalevski theorem below.

In the case of a *characteristic* initial curve γ, equations (1.8a, b, c) are inconsistent, unless additional identities are satisfied by the data. Hence the Cauchy problem with Cauchy data prescribed on a characteristic curve generally has no solution. Write condition (1.9) for a characteristic curve as

$$a\,dy^2 - 2b\,dx\,dy + c\,dx^2 = 0. \tag{1.10}$$

We can solve (1.10) for dy/dx in the form

$$\frac{dy}{dx} = \frac{b \pm \sqrt{b^2 - ac}}{a}. \tag{1.11}$$

When the characteristic curve γ is given implicitly by an equation $\phi(x,y) =$ const., we have $\phi_x\,dx + \phi_y\,dy = 0$ along γ so that (1.10) reduces to the equation

$$a\phi_x^2 + 2b\phi_x\phi_y + c\phi_y^2 = 0. \tag{1.12}$$

Relation (1.11) is an ordinary differential equation for γ provided a,b,c are known functions of x,y. This is the case when either a fixed solution $u = u(x,y)$ of (1.1) is considered, or when equation (1.1) is linear, that is,

$$\begin{gathered} a = a(x,y), \qquad b = b(x,y), \qquad c = c(x,y), \\ d = -2D(x,y)u_x - 2E(x,y)u_y - F(x,y)u - G(x,y). \end{gathered} \tag{1.13}$$

The equation (1.1) is called *elliptic* if $ac - b^2 > 0$, *hyperbolic* if $ac - b^2 < 0$, and *parabolic* if $ac - b^2 = 0$. Restricting ourselves to the case of real variables, we observe that corresponding to the choice of $\pm$ in (1.11) there are two families of characteristic curves in the hyperbolic case, one in the parabolic case, and none in the elliptic case. We note, though, that in the nonlinear case, "type" (elliptic, parabolic, hyperbolic) is not determined by the differential equation, but can depend on the individual solution, and even for a linear equation might be different in different regions of the plane.

2. Propagation of Singularities

The characteristic curves are closely associated with the propagation of certain types of singularities. Along a noncharacteristic curve the Cauchy data uniquely determine the second derivatives of a solution. One approach to defining "generalized" solutions of (1.1), not necessarily of class C^2, consists in considering solutions of class C^1 with second derivatives that have jump discontinuities along a curve γ. More precisely, we assume that we have a certain region in the xy-plane divided by a curve γ into two portions I and II. There shall be two solutions u^{I} and u^{II} of (1.1) respectively defined and of class C^2 in the closed regions I and II. u^{I} and u^{II} together define a function u in the union of I and II with discontinuities along γ. The resulting u, pieced together ordinarily cannot be considered a "generalized" solution of (1.1), unless (1.1) in some generalized sense still holds on γ. This requires certain *transition conditions* along γ. The simplest case arises when the resulting u is required to be of class C^1, and hence the functions u^{I} and u^{II} as well as their first derivatives coincide along γ. If also the second derivatives coincide, u actually is a "strict" solution of class C^2. Of interest, therefore, is the case where the second derivatives of u^{I} and u^{II} are not the same on γ. Since however, by assumption u^{I} and u^{II} have the same Cauchy data along γ, a discontinuity in the second derivatives of u can occur only if γ is a characteristic curve.

We analyze the situation in more detail for a linear second-order equation

$$0=Lu=a(x,y)u_{xx}+2b(x,y)u_{xy}+c(x,y)u_{yy}$$
$$+2d(x,y)u_x+2e(x,y)u_y+f(x,y)u \tag{2.1}$$

with regular coefficients a,b,c,d,e,f.

Let Ω be an open set in the xy-plane and γ be an arc in Ω such that $\Omega - \gamma$ consists of two open disjoint sets I and II. Let γ be given by an equation $x = \phi(y)$. (Precise regularity assumptions justifying what follows can easily be supplied by the reader. These are not relevant for the present discussion.) We say that a function u defined in Ω has a *jump discontinuity* along γ, if $u = u^{\mathrm{I}}$ in I, $u = u^{\mathrm{II}}$ in II, where u^{I} is continuous in I + γ and u^{II} is continuous in II + γ. We denote by

$$[u]=u^{\mathrm{II}}(\phi(y),y)-u^{\mathrm{I}}(\phi(y),y) \tag{2.2}$$

the *jump* of u at a point $(\phi(y),y)$ of γ. Then along γ

$$\frac{d}{dy}[u]=(u_x^{\mathrm{II}}(\phi(y),y)-u_x^{\mathrm{I}}(\phi(y),y))\phi'(y)+u_y^{\mathrm{II}}(\phi(y),y)-u_y^{\mathrm{I}}(\phi(y),y)$$
$$=[u_x]\phi'+[u_y]. \tag{2.3}$$

In particular $[u]=0$ when u is continuous along γ, and thus

$$0=[u_x]\phi'+[u_y]. \tag{2.4}$$

We consider now the case where $u^{\mathrm{I}}(x,y)$, $u^{\mathrm{II}}(x,y)$ are of class C^3 and satisfy (2.1) respectively in I + γ and II + γ, defining a function u of class C^1 in Ω. Then

$$[u]=[u_x]=[u_y]=0. \tag{2.5}$$

Moreover, subtracting (2.1) formed for u^{I} and u^{II} along γ yields

$$a[u_{xx}]+2b[u_{xy}]+c[u_{yy}]=0, \tag{2.6}$$

since a,b,c,d,e,f are continuous. From (2.5) we find that

$$[u_{xx}]\phi'+[u_{xy}]=0, \qquad [u_{xy}]\phi'+[u_{yy}]=0. \tag{2.7}$$

Hence for a C^1-function u the jumps in the second derivatives are not independent. Knowing, for example, the jump of u_{xx}, those of u_{xy}, u_{yy} are uniquely determined. Setting $[u_{xx}]=\lambda$, we have

$$[u_{xx}]=\lambda, \qquad [u_{xy}]=-\phi'\lambda, \qquad [u_{yy}]=\phi'^2\lambda. \tag{2.8}$$

For a solution of the differential equation it follows now from (2.6) that

$$a-2b\phi'+c\phi'^2=0 \tag{2.9}$$

unless $\lambda=0$, that is, unless $u\in C^2$. Recalling that $\phi'=dx/dy$, equation (2.9) asserts that γ is a characteristic curve. According to (2.8) the quantity λ measures the "intensity" of the jumps in the second derivatives.

We interpret y as the time, and ϕ as the point $x=\phi(y)$ moving along the x-axis. For every y we have in $u(x,y)$ a function of x whose second derivative is discontinuous at the moving point $\phi(y)$. The speed dx/dy of "propagation of the discontinuity" is determined by (2.9).

It is remarkable that the jumps in different points of γ are related to each other. There is a definite law according to which the *intensity* λ of the jump propagates along γ. To find it we derive from (2.8) the relations

$$\lambda'=\frac{d\lambda}{dy}=[u_{xxx}]\phi'+[u_{xxy}],$$
$$-(\phi'\lambda)'=[u_{xxy}]\phi'+[u_{xyy}]. \tag{2.10}$$

In addition, differentiating (2.1) formed for u^{I} and u^{II} with respect to x and forming the jumps on γ, we find that

$$0=a[u_{xxx}]+2b[u_{xxy}]+c[u_{xyy}]+a_x[u_{xx}]$$
$$+2b_x[u_{xy}]+c_x[u_{yy}]+2d[u_{xx}]+2e[u_{xy}]$$

Eliminating the third derivatives, using (2.8), (2.9), and (2.10), one arrives at the relation

$$0=2(b-c\phi')\lambda'+\left(a_x-2b_x\phi'+c_x\phi'^2+2d-2e\phi'-c\phi''\right)\lambda. \tag{2.11}$$

This is an ordinary differential equation for the jump intensity λ, which regulates its growth during propagation. If, for example, $\lambda=0$ in one point of γ, it follows that $\lambda=0$ all along γ, so that no jump at all occurs.

3. The Linear Second-Order Equation

We analyze in more detail the linear second-order equation

$$au_{xx}+2bu_{xy}+cu_{yy}+2du_x+2eu_y+fu=0 \tag{3.1}$$

with coefficients a,b,c,d,e,f depending on x,y.

Introducing new independent variables ξ,η by the substitution,

$$\xi=\phi(x,y), \qquad \eta=\psi(x,y), \tag{3.2}$$

we transform this linear equation into one of the same type,

$$Lu=A(\xi,\eta)u_{\xi\xi}+2B(\xi,\eta)u_{\xi\eta}+\cdots=0, \tag{3.3}$$

where,

$$A = a\phi_x^2 + 2b\phi_x\phi_y + c\phi_y^2$$
$$B = a\phi_x\psi_x + b(\phi_x\psi_y + \phi_y\psi_x) + c\phi_y\psi_y \tag{3.4}$$
$$C = a\psi_x^2 + 2b\psi_x\psi_y + c\psi_y^2$$
etc.

This suggests, in the hyperbolic case, that one can simplify the differential equation by introducing the characteristics as new coordinate lines. Let

$$\xi = \phi(x,y) = \text{const.}, \qquad \eta = \psi(x,y) = \text{const.}$$

be the two families of characteristics in the xy-plane, so that both ϕ and ψ satisfy (1.12). By (3.4) this implies that $A=0$ and $C=0$, and the hyperbolic equation reduces, after division by B, to the *normal* form,

$$u_{\xi\eta} + 2Du_\xi + 2Eu_\eta + Fu = 0. \tag{3.5}$$

The new equation has the lines $\xi=$const. and $\eta=$const. as characteristic curves. By a further linear transformation,

$$x' = \xi + \eta, \qquad y' = \xi - \eta, \tag{3.6}$$

one can also transform (3.5) into the alternate form

$$u_{y'y'} - u_{x'x'} + 2D'u_{x'} + 2E'u_{y'} + F'u = 0. \tag{3.7}$$

In the elliptic case, where $ac - b^2 > 0$, there exist no real characteristics. We can attempt to find a real transformation (3.2) taking (3.1) into an equation of the form,

$$u_{\xi\xi} + u_{\eta\eta} + 2Du_\xi + 2Eu_\eta + Fu = 0. \tag{3.8}$$

This means we want to choose $\phi(x,y), \psi(x,y)$, so that $A = C$, $B = 0$. This can be achieved by taking for ϕ, ψ solutions of the system of equations,

$$\phi_x = \frac{b\psi_x + c\psi_y}{W}, \qquad \phi_y = -\frac{a\psi_x + b\psi_y}{W}, \tag{3.9}$$

where $W = \sqrt{ac - b^2}$. Eliminating ϕ from (3.9) we see that ψ has to be a solution of the *Beltrami equation*,

$$\left(\frac{a\psi_x + b\psi_y}{W}\right)_x + \left(\frac{b\psi_x + c\psi_y}{W}\right)_y = 0. \tag{3.10}$$

Example (*The Tricomi Equation*).

$$u_{yy} - yu_{xx} = 0. \tag{3.11}$$

For this equation, $ac - b^2 = -y$. Hence for $y < 0$, $ac - b^2 > 0$, and the equation is elliptic. For $y > 0$, $ac - b^2$ is negative and the equation is hyperbolic. On the x-axis, it is parabolic. (See Figure 2.1.) Here, the characteristic equation (1.10) reduces to $-y\,dy^2 + dx^2 = 0$ or

$$dx \pm \sqrt{y}\,dy = 0 \quad \text{for } y > 0. \tag{3.12}$$

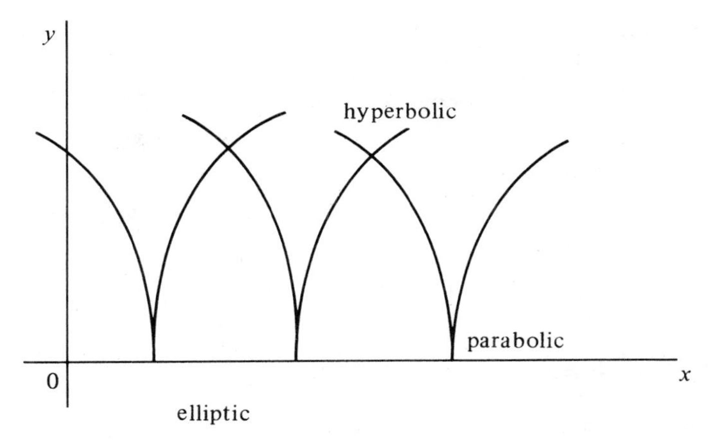

Figure 2.1

The characteristic curves in the half plane $y>0$ are therefore

$$3x \pm 2y^{3/2} = \text{const.} \tag{3.13}$$

The transformation,

$$\xi = 3x - 2y^{3/2}, \qquad \eta = 3x + 2y^{3/2}, \tag{3.14}$$

reduces the equation to the normal form:

$$u_{\xi\eta} - \frac{1}{6}\frac{u_\xi - u_\eta}{\xi - \eta} = 0. \tag{3.15}$$

The curves

$$3x - 2y^{3/2} = \text{const.}$$

are the branches of cubic curves having positive slope and the curves

$$3x + 2y^{3/2} = \text{const.}$$

are the symmetric curves with negative slopes. On $y=0$, the curves have cusps with a vertical tangent.

Problems

1. For the equation of minimal surfaces (2.12), p. 3 of Chapter 1, find
(a) all minimal surfaces of revolution about the z-axis (i.e., $u = f(\sqrt{x^2+y^2}\,)$)
(b) the differential equations for the (imaginary) characteristic curves.

2. Find by power series expansion with respect to y the solution of the initial-value problem

$$u_{yy} = u_{xx} + u$$
$$u(x,0) = e^x, \qquad u_y(x,0) = 0.$$

3. (Legendre transformation, Hodograph method). Let $u(x,y)$ be a solution of a quasi-linear equation of the form

$$a(u_x,u_y)u_{xx}+2b(u_x,u_y)u_{xy}+c(u_x,u_y)u_{yy}=0.$$

Introduce new independent variables ξ,η and a new unknown function ϕ by

$$\xi=u_x(x,y), \qquad \eta=u_y(x,y), \qquad \phi=xu_x+yu_y-u. \tag{3.16}$$

Prove that ϕ as a function of ξ,η satisfies $x=\phi_\xi$, $y=\phi_\eta$ and the *linear* differential equation

$$a(\xi,\eta)\phi_{\eta\eta}-2b(\xi,\eta)\phi_{\xi\eta}+c(\xi,\eta)\phi_{\xi\xi}=0. \tag{3.17}$$

4. The One-Dimensional Wave Equation

The simplest of all hyperbolic differential equations is the one-dimensional wave equation

$$Lu=u_{tt}-c^2u_{xx}=0, \tag{4.1}$$

where u is a function of two independent variables x and t. The variable x is commonly identified with "position" and t with "time"; c is a given positive constant. Physically u can represent the normal displacement of the particles of a vibrating string. Here the characteristics are the two families of lines $x\pm ct=$ constant in the xt-plane. Introducing them as coordinates by putting

$$x+ct=\xi, \qquad x-ct=\eta, \tag{4.2}$$

(4.1) becomes

$$u_{\xi\eta}=0. \tag{4.3}$$

Assume that the domain of u, as a function of x,t or, equivalently, as a function of ξ,η is convex. Since $(u_\xi)_\eta=0$ it follows that u_ξ is independent of η, say, $u_\xi=f'(\xi)$, and then $u=\int f'(\xi)\,d\xi+G(\eta)$. That is,

$$u=F(\xi)+G(\eta). \tag{4.4}$$

In the original variables we find that u is of the form

$$u=F(x+ct)+G(x-ct). \tag{4.5}$$

Here $u\in C^2$ if and only if $F,G,\in C^2$. Thus the general solution of (4.1) is obtained by superposition of a solution $F(x+ct)=v$ of $v_t-cv_x=0$ and of a solution $G(x-ct)=w$ of $w_t+cw_x=0$. This corresponds to the fact that the differential operator

$$L=\frac{\partial^2}{\partial t^2}-c^2\frac{\partial^2}{\partial x^2} \tag{4.6}$$

can be decomposed into

$$L=\left(\frac{\partial}{\partial t}-c\frac{\partial}{\partial x}\right)\left(\frac{\partial}{\partial t}+c\frac{\partial}{\partial x}\right). \tag{4.7}$$

Thus the graph of $u(x,t)$ in the xu-plane consists of two waves propagating without change of shape with velocity c in opposite directions along the x-axis. (See p. 5.)

We impose the initial conditions

$$u(x,0)=f(x), \qquad u_t(x,0)=g(x). \tag{4.8}$$

For u of the form (4.5), we have at $t=0$

$$u(x,0)=F(x)+G(x)=f(x), \tag{4.9}$$

$$u_t(x,0)=cF'(x)-cG'(x)=g(x). \tag{4.10}$$

Differentiating (4.9) with respect to x and solving the two linear equations for F' and G' we obtain

$$F'(x)=\frac{cf'(x)+g(x)}{2c}, \qquad G'(x)=\frac{cf'(x)-g(x)}{2c} \tag{4.11}$$

or,

$$F(x)=\frac{f(x)}{2}+\frac{1}{2c}\int_0^x g(\xi)\,d\xi+\delta, \tag{4.12}$$

$$G(x)=\frac{f(x)}{2}-\frac{1}{2c}\int_0^x g(\xi)\,d\xi+\varepsilon,$$

with suitable constants δ,ε. Here $\delta+\varepsilon=0$ by (4.9). Hence

$$\begin{aligned}u(x,t)&=F(x+ct)+G(x-ct)\\&=\frac{1}{2}(f(x+ct)+f(x-ct))+\frac{1}{2c}\int_{x-ct}^{x+ct}g(\xi)\,d\xi.\end{aligned} \tag{4.13}$$

For $f\in C^2$ and $g\in C^1$ this actually represents a solution $u\in C^2$ of the initial-value problem (4.1), (4.8).

We see from (4.13) that $u(x,t)$ is determined uniquely by the values of the initial functions f,g in the interval $[x-ct,x+ct]$ of the x-axis whose end points are cut out by the characteristics through the point (x,t). This interval represents the *domain of dependence* for the solution at the point (x,t) as shown in Figure 2.2.

Conversely, the initial values at a point $(\xi,0)$ of the x-axis *influence* $u(x,t)$ at points (x,t) in the wedge-shaped region bounded by the characteristics through $(\xi,0)$, i.e., for $\xi-ct<x<\xi+ct$. This indicates that for our equation "disturbances" or "signals" only travel with speed c as shown in Figure 2.3.

We saw that formula (4.5) represents a solution $u\in C^2(\mathbb{R}^2)$ of (4.1) for any $f,g,\in C^2(\mathbb{R})$. One is tempted to consider any u of the form (4.5) for "general" $f,g,$ as a *generalized* or *weak solution* of (4.1) even though u may not have derivatives in the ordinary sense. One easily verifies that any function u of the form (4.5) satisfies the functional equation

$$\begin{aligned}u(x,t)-u(x+c\zeta,t+\zeta)-u(x-c\eta,t+\eta)&\\+u(x+c\zeta-c\eta,t+\zeta+\eta)&=0.\end{aligned} \tag{4.14}$$

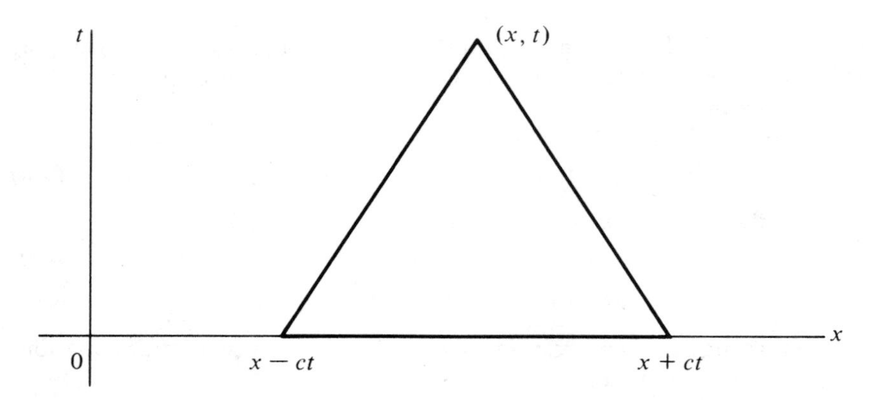

Figure 2.2

Geometrically, for any parallelogram $ABCD$ in the xt-plane bounded by four characteristic lines, (see Figure 2.4), the sums of the values of u in opposite vertices are equal, that is,

$$u(A)+u(C)=u(B)+u(D). \tag{4.15}$$

Every solution of (4.1) is of the form (4.5) with $F, G, \in C^2$ and thus satisfies (4.14). Conversely, using Taylor expansions for small ζ,η, every C^2-solution of (4.14) satisfies (4.1). Thus (4.14) can be viewed as a weak formulation of equation (4.1).

We use (4.15) to solve an "initial-boundary-value" problem for (4.1). Assume the wave equation to be satisfied only in a fixed x-interval $0<x<L$ for all $t>0$. Then we can prescribe, in addition to the initial data

$$u=f(x), \qquad u_t=g(x) \quad \text{for } 0<x<L,\ t=0 \tag{4.16}$$

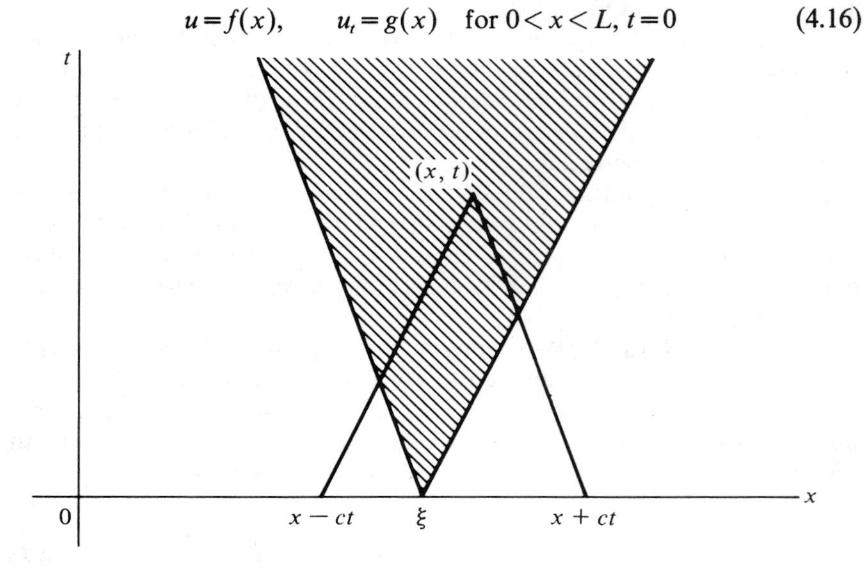

Figure 2.3

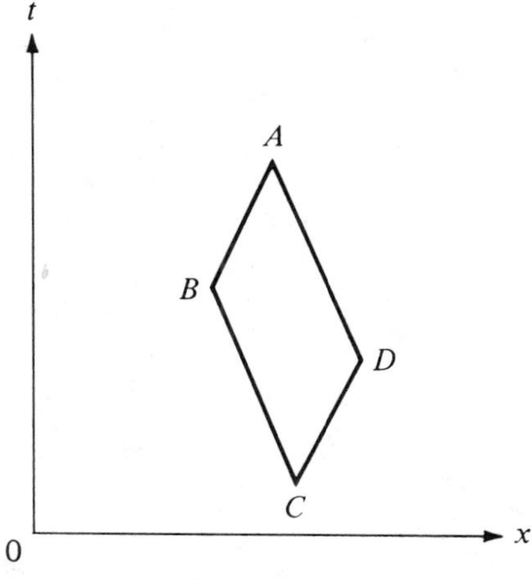

Figure 2.4

certain "boundary" conditions, for example

$$u=\alpha(t) \quad \text{for } x=0 \qquad 0<t, \tag{4.17a}$$

$$u=\beta(t) \quad \text{for } x=L \qquad 0<t. \tag{4.17b}$$

We are interested in the solution of (4.1) in the strip

$$0<x<L, \qquad 0<t. \tag{4.18}$$

We divide the strip into a number of regions by the characteristics through the corners and through the points of intersections of the characteristics with the boundaries, etc. as shown in Figure 2.5.

In region I the solution u is determined by the formula (4.13) from the initial data alone. In a point $A=(x,t)$ of region II we form the characteristic parallelogram with vertices A, B, C, D and get $u(A)$ from (4.15) as

$$u(A)=-u(C)+u(B)+u(D), \tag{4.19}$$

with $u(B)$ known from boundary condition (4.17a) and $u(C), u(D)$ known since C, D lie in I. Similarly, we get u successively in all points of the regions III, IV, V, ... If we want the solution u of this "mixed" problem to be regular (e.g., to be of class C^2) in the closure of the strip, the data f, g, α, β have to *fit together* in the corners so that u and its first and second derivatives come out to be the same when computed either from f, g or from α, β. We clearly need the *compatibility conditions*,

$$\begin{aligned} &\alpha(0)=f(0), \qquad \alpha'(0)=g(0), \qquad \alpha''(0)=c^2 f''(0) \\ &\beta(0)=f(L), \qquad \beta'(0)=g(L), \qquad \beta''(0)=c^2 f''(L). \end{aligned} \tag{4.20}$$

These actually are also sufficient for $u \in C^2$ when $f, \alpha, \beta \in C^2$ and $g \in C^1$. For example, for $A \in$ II and $u(A)$ given by (4.19) we take the limit as $A \to D$ for D fixed. Then $u(B) \to \alpha(0)$, $u(C) \to f(0)$ and $u(A) \to -f(0)+u(D)+\alpha(0)$ $=u(D)$, if (4.20) holds. If, instead, $\alpha(0) \neq f(0)$, we would find that u has a jump all along the line $x=ct$.

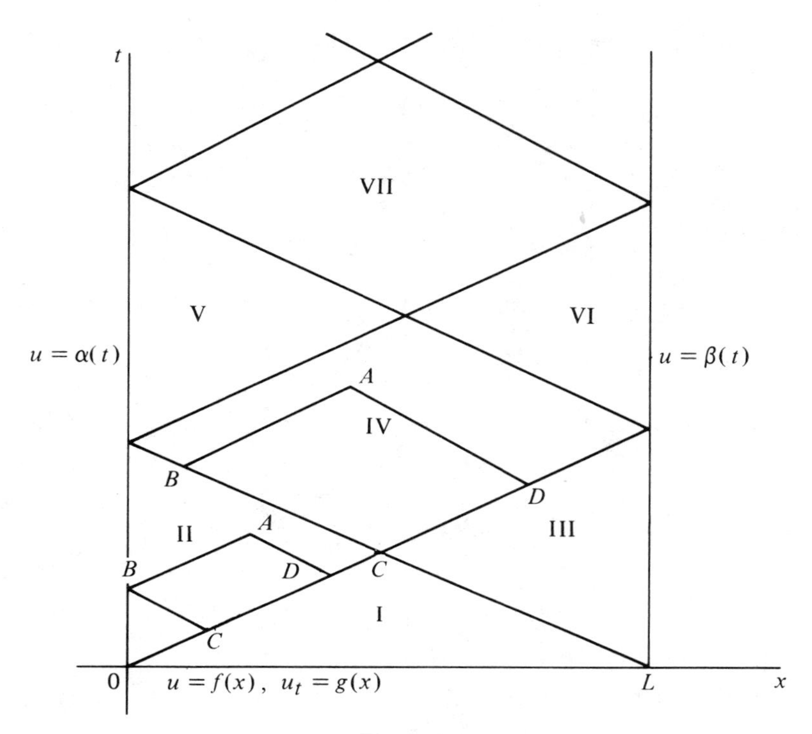

Figure 2.5

An alternative method ("separation of variables") gives the solution at one stroke by *expansion into eigenfunctions*. For simplicity, take the case where $L=\pi$ and $\alpha=\beta=0$ in (4.17), so that we have *homogeneous* boundary conditions. Then, for each t, $u(x,t)$ can be expanded into a Fourier sine series

$$u=\sum_{n=1}^{\infty} a_n(t)\sin nx. \tag{4.21}$$

Substituting into (4.1) we find that $a_n(t)$ satisfies the ordinary differential equation $a_n''+n^2c^2a_n=0$, hence that a_n is of the form

$$a_n(t)=c_n\cos(nct)+d_n\sin(nct). \tag{4.22}$$

Here the constants c_n, d_n can be found from the initial conditions (4.16) which require that

$$f(x)=\sum_{n=1}^{\infty} c_n\sin nx, \qquad g(x)=c\sum_{n=1}^{\infty} nd_n\sin nx, \tag{4.23}$$

so that by Fourier

$$c_n=\frac{2}{\pi}\int_0^{\pi} f(x)\sin nx\,dx, \qquad d_n=\frac{2}{cn\pi}\int_0^{\pi} g(x)\sin nx\,dx. \tag{4.24}$$

In applications we usually deal only with a *bounded* domain for our solutions, and are led to initial-boundary-value problems rather than to "pure" initial-value problems. An example is the normal displacement $u(x,t)$ of a vibrating string of length π. Then u satisfies the boundary conditions $u(x,0)=u(x,\pi)=0$, if the ends of the string are held fixed, and initial conditions of type (4.16) if the initial normal displacement and normal velocity of each particle of the string are prescribed.

PROBLEMS

1. Let $f(x),g(x)$ have compact support (i.e., vanish for all sufficiently large $|x|$). Show that the solution $u(x,t)$ of (4.1), (4.8) has compact support in x for each fixed t. Show that the functions F,G in the decomposition (4.5) for u can be of compact support only when $\int_{-\infty}^{\infty} g(\xi)\,d\xi=0$.

2. Let α be a constant $\neq -c$. Find the solution $u(x,t)$ of (4.1) in the quadrant $x>0, t>0$, for which

$$u=f(x), \qquad u_t=g(x) \quad \text{for } t=0,\ x>0$$
$$u_t=\alpha u_x \quad \text{for } x=0,\ t>0,$$

where f and g are of class C^2 for $x>0$ and vanish near $x=0$. (Hint: Use (4.5)). Show that generally no solution exists when $\alpha=-c$.

3. Solve the initial-boundary-value problem ("mixed problem")

$$u_{tt}=u_{xx} \quad \text{for } 0<x<\pi,\ 0<t,$$
$$u=0 \quad \text{for } x=0,\pi;\ 0<t,$$
$$u=1, \qquad u_t=0 \quad \text{for } 0<x<\pi,\ t=0,$$

by (a) piecing together, using (4.15), and (b) Fourier series. Check that the solutions agree.

4. Let the operators L_1, L_2 be defined by

$$L_1u=au_x+bu_y+cu, \qquad L_2u=du_x+eu_y+fu,$$

where a,b,c,d,e,f are constants with $ae-bd\neq 0$. Prove that

(a) the equations $L_1u=w_1$, $L_2u=w_2$ have a common solution u, if $L_1w_2=L_2w_1$. (Hint: By linear transformation reduce to the case $a=e=1$, $b=d=0$.)

(b) The general solution of $L_1L_2u=0$ has the form $u=u_1+u_2$, where $L_1u_1=0$, $L_2u_2=0$.

5. Solve

$$u_{tt}-c^2u_{xx}=x^2 \quad \text{for } 0<t \quad \text{and all } x$$
$$u=x, \qquad u_t=0 \quad \text{for } t=0.$$

(Hint: First find a special time independent solution of the P.D.E.)

6. Find a solution of

$$u_{tt}-c^2u_{xx}=\lambda^2u$$

(λ constant) of the form $u=f(x^2-c^2t^2)=f(s)$, when $f(0)=1$, in form of a power series in s.

7. With L defined by (4.1) prove that
(a) $Lu=0$, $Lv=0$ implies $L(u_tv_t+c^2u_xv_x)=0$,
(b) Prove that for $Lu = 0$, $Lv = 0$ for $a < x < b$, $t > 0$ and $u = 0$, $v = 0$ for $x = a,b$; $t > 0$,

$$\frac{d}{dt}\int_a^b \frac{1}{2}(u_tv_t+c^2u_xv_x)\,dx=0.$$

8. For the solution (4.21), (4.22) of the wave equation express the "energy"

$$\int_0^\pi \frac{1}{2}(u_t^2+u_x^2)\,dx,$$

in terms of c_n, d_n.

9. Find the Fourier series solution (4.21) for the case

$$u(x,0)=\frac{\pi}{2}-\left|\frac{\pi}{2}-x\right|, \qquad u_t(x,0)=0 \quad \text{for } 0<x<\pi,$$

(vibration of string plucked at center) and calculate its energy.

10. Find in closed form the solution $u(x,t)$ of

$$Lu=u_{tt}-c^2u_{xx}=0 \quad \text{for } 0<x,\ 0<t,$$
$$u(0,t)=h(t) \quad \text{for } 0<t,$$
$$u(x,0)=f(x) \qquad u_t(x,0)=g(x) \quad \text{for } 0<x,$$

with $f,g,h\in C^2$ for nonnegative arguments and satisfying

$$h(0)=f(0), \qquad h'(0)=g(0), \qquad h''(0)=c^2f''(0).$$

Verify that the u obtained has continuous second derivatives even on the characteristic line $x=ct$.

5. Systems of First-Order Equations

It is convenient to treat general second-order equations as part of a still more general theory, that of first-order systems, to which in principle all higher-order single equations can be reduced. Thus the linear equation (3.1) for $u(x,y)$ is reduced to a first-order system by introducing the new dependent variables

$$u=u_1, \qquad u_x=u_2, \qquad u_y=u_3, \tag{5.1}$$

yielding the equations

$$\frac{\partial u_1}{\partial y}=u_3, \qquad \frac{\partial u_2}{\partial y}=\frac{\partial u_3}{\partial x} \tag{5.2}$$

$$a\frac{\partial u_2}{\partial x}+b\left(\frac{\partial u_2}{\partial y}+\frac{\partial u_3}{\partial x}\right)+c\frac{\partial u_3}{\partial y}+2du_2+2eu_3+fu_1=0. \tag{5.3}$$

More generally, writing t for y, we consider an N-vector (column vector)

$$u=\begin{bmatrix} u_1 \\ \vdots \\ u_N \end{bmatrix}=u(x,t), \tag{5.4}$$

satisfying an equation

$$A(x,t)\frac{\partial u}{\partial t}+B(x,t)\frac{\partial u}{\partial x}=C(x,t)u+D(x,t), \tag{5.5}$$

for given square matrices A,B,C of order N and column vector D. The Cauchy problem for (5.5) prescribes the value of u on a curve $t=\phi(x)$ in the xt-plane

$$u=u(x,\phi(x))=f(x). \tag{5.6}$$

The curve is *characteristic* if we cannot find the derivatives of u from the data on the curve. Now (5.5), (5.6) imply on $t=\phi(x)$ that

$$Au_t+Bu_x=Cf+D, \qquad u_x+\phi'(x)u_t=f', \tag{5.7}$$

or

$$(A-\phi'B)u_t=Cf+D-Bf'. \tag{5.8}$$

Thus, characteristic curves are those for which the matrix $A-\phi'B$ is singular, or for which the Nth-degree differential equation

$$\det(A\,dx-B\,dt)=0 \tag{5.9}$$

holds.

For the *initial-value problem* to be treated here we prescribe the values of u on the x-axis, assumed to be non-characteristic:

$$u(x,0)=f(x), \tag{5.10}$$

$$\det A\neq 0 \quad \text{for } t=0. \tag{5.11}$$

For small t we can then solve for u_t in (5.5) getting a new equation

$$u_t+B(x,t)u_x=Cu+D \tag{5.12}$$

(with *new* matrices B,C and vector D.) The characteristic differential equation (5.9) now takes the form

$$\det\left(\frac{dx}{dt}I-B\right)=0 \tag{5.13}$$

(I = unit matrix) which factors into conditions of the form

$$\frac{dx}{dt}=\lambda_i(x,t), \tag{5.14}$$

where $\lambda_i(x,t)$ denotes the ith eigenvalue of the matrix $B(x,t)$. We assume these eigenvalues to be real, so that for each i (5.14) is satisfied by a one-parameter family of characteristic curves C_i.

More precisely, we assume the system (5.5) to be *hyperbolic* in the sense that there exists a complete set of real eigenvectors $\xi^1,\ldots,\xi^N$ of B such that

$$B\xi^k=\lambda_k\xi^k, \tag{5.15}$$

where the ξ^k are linearly independent and depend "regularly" on x and t. (This certainly holds where the eigenvalues of B are real and *distinct*.) The column vectors ξ^k then form the columns of a nondegenerate matrix $\Gamma=\Gamma(x,t)$ for which

$$B\Gamma=\Gamma\Lambda, \tag{5.16}$$

where Λ is the diagonal matrix whose diagonal elements are the eigenvalues λ_k.

Introducing a new unknown vector v by $u=\Gamma v$, we find from (5.12) that v satisfies

$$v_t+\Lambda v_x=cv+d, \tag{5.17}$$

with new coefficients

$$c=\Gamma^{-1}C\Gamma-\Gamma^{-1}\Gamma_t-\Gamma^{-1}B\Gamma_x, \qquad d=\Gamma^{-1}D$$

and initial conditions

$$v=\Gamma^{-1}f=g(x) \quad \text{for } t=0. \tag{5.18}$$

We have reduced (5.5) to "canonical form" where $A=I$ and B is diagonal. If v has components v_i, $c=(c_{ik})$, and d has components d_i we find from (5.17), along a characteristic C_i of the ith family,

$$\frac{dv_i}{dt}=\frac{\partial v_i}{\partial t}+\frac{dx}{dt}\frac{\partial v_i}{\partial x}=\frac{\partial v_i}{\partial t}+\lambda_i\frac{\partial v_i}{\partial x}=\sum_k c_{ik}v_k+d_i. \tag{5.19}$$

The ith backward characteristic C_i through a point (X,T) has an equation

$$x=\alpha_i(t,X,T) \tag{5.20}$$

(obtained by solving the ordinary differential equation (5.14)) with $x = X$ for $t = T$. Then by (5.18), (5.19), integrating along C_i (see Figure 2.6)

$$v_i(X,T)=g_i(\alpha_i(0,X,T))+\int_0^T\left(\sum_k c_{ik}v_k+d_i\right)dt, \tag{5.21}$$

where in the integrand x has to be replaced by $\alpha_i(t,X,T)$. Formula (5.21) resembles a system of integral equations except that the domain of integration is different for each component of v. We write (5.21) symbolically as

$$v=W+Sv, \tag{5.22}$$

where W is the vector (considered known) with components,

$$W_i(X,T)=g_i(\alpha_i(0,X,T))+\int_0^T d_i(\alpha_i(t,X,T),t)\,dt \tag{5.23}$$

and S is the linear operator taking a vector v with components $v_k(x,t)$ into

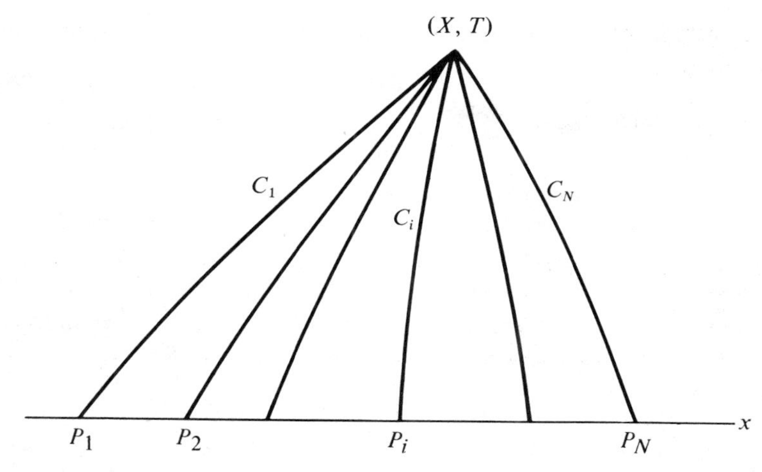

Figure 2.6

a vector $w = Sv$ with components

$$w_i(X,T) = \int_0^T \sum_k c_{ik}(\alpha_i(t,X,T),t)\,v_k(\alpha_i(t,X,T),t)\,dt. \tag{5.24}$$

Given sufficient regularity of our data, the mapping S: $C \to C$ is continuous in the space C of continuous bounded vectors $v(x,t)$ with domain in the strip $0 \leqslant t \leqslant \tau$, using in C the "maximum norm":

$$\|v\| = \sup_{\substack{k=1,\ldots,N \\ x;\ 0 \leqslant t \leqslant \tau}} |v_k(x,t)|. \tag{5.25}$$

The norm $\|S\|$ of the operator S (the supremum of $\|Sv\|$ for $\|v\| = 1$) obviously is bounded* by the constant

$$q = \tau \sup_{\substack{i,x \\ 0 \leqslant t \leqslant \tau}} \sum_k |c_{ik}(x,t)|. \tag{5.26}$$

For $q < 1$, i.e., for τ sufficiently small, the mapping S is *contractive*. Then (5.22) has a unique solution $v \in C$ obtainable by the process of iteration

$$v^{n+1} = W + Sv^n, \qquad v^0 = 0. \tag{5.27}$$

(Convergence follows by comparison with a geometric series.)

The resulting v will satisfy the "integral equations" (5.21), but it is not certain that these imply the differential equations (5.19) or (5.17) until we have established existence of continuous partial derivatives for v. For that purpose we shall work in the narrower Banach space C^1 of vectors $v(x,t)$ for which v and v_x are continuous and bounded for $0 \leqslant t \leqslant \tau$ and all x, in which we choose as norm

$$|||v||| = \max(\|v\|, \|v_x\|). \tag{5.28}$$

*We assume, for convenience, that the $c_{ik}(x,t)$ and their first derivatives are bounded uniformly for $-\infty < x < \infty$, $0 \leqslant t \leqslant \tau$.

The restriction of S to C^1 maps C^1 into itself and has its norm for C^1 bounded by

$$q^* = \tau\left(\sum_k \left(\sup_{x,t,i} (|c_{ikx}(x,t)| + |c_{ik}(x,t)|)\right) \sup_{t,X,T} |\alpha_{iX}(t,X,T)|\right)$$
$$+ \tau \sum_k \sup_{x,t,i} |c_{ik}(x,t)|. \tag{5.29}$$

For $q^* < 1$ equation (5.22) has a solution in C^1, provided $W \in C^1$. Obviously Sv for $v \in C^1$ can be differentiated with respect to T as well. Thus, convergence of the v^n and v_x^n implies convergence of v_t^n. Hence in the limit we obtain a solution v of the original initial-value problem for our partial differential equations. The condition $q^* < 1$ will again be satisfied for τ sufficiently small. In this way, we obtain an existence theorem for a sufficiently small t-interval. Here v can be continued as long as our ordinary differential equations for the characteristics lead to sufficiently regular solutions permitting us to define functions $\alpha_i(t,X,T)$ describing characteristics from (X,T) backwards to the x-axis. Further extensions to *quasi-linear* systems were given by Courant and Lax. (See [6].)

We have solved the pure initial-value problem in the linear case, assuming that the initial data are known on the whole x-axis. The method described extends to mixed initial-boundary-value problems as well. For simplicity take as domain for the vector v the first quadrant $x > 0$, $t > 0$. We then need, in addition to initial conditions,

$$v = g(x) \quad \text{for } 0 < x,\ t = 0, \tag{5.30}$$

boundary conditions on the half line $x = 0$, $0 < t$. The number of these conditions depends on the number of *positive* eigenvalues λ_k of the matrix B. Let, say, the number r be such that

$$\lambda_1 > \lambda_2 \ldots > \lambda_r > 0 > \lambda_{r+1} > \ldots > \lambda_N. \tag{5.31}$$

The characteristics C_i of the ith family then have positive slopes for $i = 1, \ldots, r$, negative ones for $i = r+1, \ldots, N$. A backward characteristic

$$x = \alpha_i(t, X, T) \tag{5.32}$$

through a point (X,T) with $0 < X$, $0 < T$ for $i = r+1, \ldots, N$ will hit the positive x-axis in a point $P_i(X,T)$. For $i = 1, \ldots, r$, it will either hit the positive x-axis or positive t-axis in a point, again denoted by $P_i(X,T)$. On the positive t-axis we prescribe the values of r linear combinations of the components of v, which we assume to have been brought into the form

$$v_i - \sum_{k=r+1}^{n} \gamma_{ik}(t) v_k = h_i(t) \quad \text{for } 0 < t,\ x = 0;\ i = 1, \ldots, r. \tag{5.33}$$

We redefine the operator S acting on vectors v. Let, for a vector $v(x,t)$, the component $w_i(X,T)$ of $w = Sv$ be defined by (5.24) if $P_i(X,T)$ lies on the

x-axis, in particular, whenever $i=r+1,\ldots,N$. If, on the other hand, $P_i(X,T)$ lies on the t-axis, say $P_i=(0,t_0)$, we take

$$w_i(X,T)=\sum_{k=r+1}^{N}\gamma_{ik}(t_0)\int_0^{t_0}\sum_{s=1}^{N}c_{ks}(\alpha_k(t,0,t_0),t)v_s(\alpha_k(t,0,t_0),t)\,dt$$
$$+\sum_{k=1}^{N}\int_{t_0}^{T}c_{ik}(\alpha_i(t,X,T),t)v_k(\alpha_i(t,X,T),t)\,dt. \tag{5.34}$$

(Here the second sum corresponds to the expression for $v_i(X,T)$ in terms of $v_i(P_i)$ and integrals along C_i, the first sum corresponds to $v_i(P_i)$ expressed by the $v_k(P_i)$ for $k>r$, which, in turn, are expressed by integrals over backward characteristics through P_i [See Figure 2.7].) We again get a set of equations formally describable by (5.22) with a known W, which can be solved for $0\leqslant t<\tau$, with τ sufficiently small. Solutions of problems in a finite interval $0\leqslant x\leqslant L$ can be reduced to pure initial-value problems, and to problems of the type just discussed, by breaking up the domain $0\leqslant x\leqslant L$, $0<t\leqslant\tau$ into suitable portions by characteristics, as was done earlier for the one-dimensional wave equation. The number of boundary conditions on $x=0$ equals the number of positive λ_k, that on $x=L$ equals the number of negative λ_k.

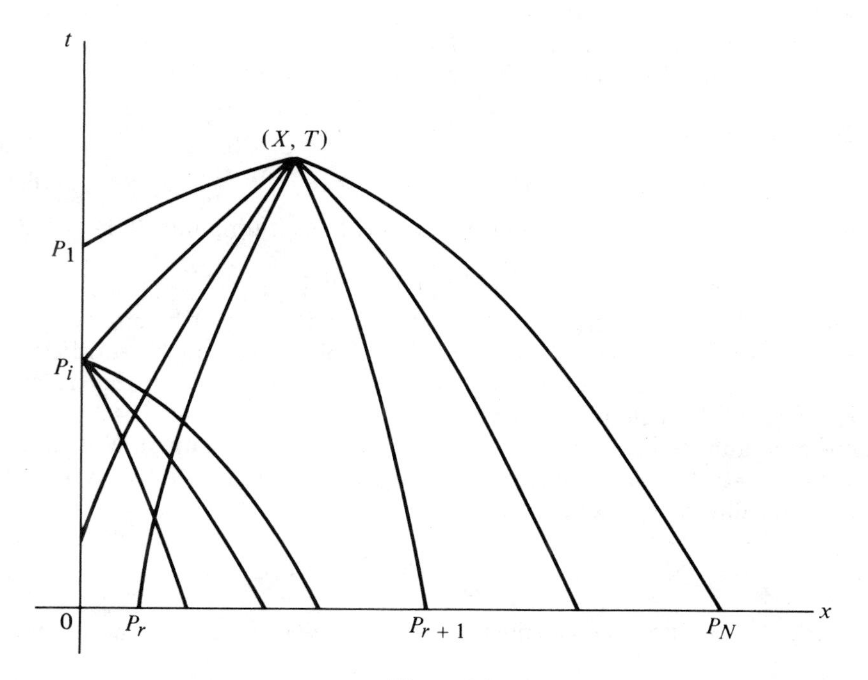

Figure 2.7

PROBLEM

Write

$$u_{tt}=c^2u_{xx}, \qquad u(x,0)=f(x), \qquad u_t(x,0)=g(x)$$

as an initial-value problem for the vector $(u_1,u_2)=(u_t,u_x)$. Reduce the system to the canonical form (5.17) and solve the problem. Solve the mixed problem

$$u_{tt}-c^2u_{xx}=0 \quad \text{for } x>0, \quad t>0$$

$$u(x,0)=f(x), \qquad u_t(x,0)=g(x) \quad \text{for } x>0$$

$$u_t(0,t)+au_x(0,t)=h(t) \quad \text{for } t>0$$

with $a=$ const.

6. A Quasi-linear System and Simple Waves*

Consider a quasi-linear system of equations for functions $v_1(x,t),\ldots,$ $v_N(x,y)$ forming a column vector $v(x,t)$, of the special form

$$v_t+B(v)v_x=0 \tag{6.1}$$

with a square matrix $B(v)$ depending on the dependent variables. We assume that (6.1) is hyperbolic in the sense that $B(v)$ has real distinct eigenvalues for all v in question. The general solution $v(x,t)$ can be visualized as forming a two-dimensional surface in N-dimensional v-space, referred to the two parameters x,t. Special explicit solutions, "simple waves," can be obtained by requiring that this surface degenerates into a curve, that is, that the range of the solution v is one-dimensional. Such v can be represented in the form

$$v=F(\theta), \tag{6.2}$$

where the scalar θ is a function of x and t. Substituting into (6.1) yields

$$F'(\theta)\theta_t+B(F)F'(\theta)\theta_x=0. \tag{6.3}$$

Thus $F'(\theta)$ must be an eigenvector of the matrix $B(F)$ belonging to an eigenvalue λ such that $\theta_t+\lambda\theta_x=0$. These conditions can be satisfied by thinking of $B(v)$ as constituting a field of matrices in N-dimensional v-space. With each point v we associate an eigenvector $V=V(v)$ and corresponding eigenvalue $\lambda=\lambda(v)$, varying smoothly with v. We take a vector $F=F(\theta)$ corresponding to a particular solution of the system of ordinary differential equations

$$\frac{dF}{d\theta}=V(F). \tag{6.4}$$

Along this solution the eigenvalue $\lambda(v)=\lambda(F(\theta))$ becomes a known func-

*([5])

tion $c(\theta)$. Taking for $\theta=\theta(x,t)$ any solution of the scalar equation

$$\theta_t + c(\theta)\theta_x = 0 \tag{6.5}$$

we have in $v=F(\theta(x,t))$ a solution of (6.1) called a *simple wave*.

The solution of (6.5) with initial values

$$\theta=\phi(x) \quad \text{for } t=0 \tag{6.6}$$

is given by the implicit equation [See Chapter 1, (6.28)].

$$\theta=\phi(x-c(\theta)t).$$

Problem

Find all simple wave solutions of the equation

$$u_{tt} = (1+u_x)^2 u_{xx} \tag{6.7}$$

with $u(x,0)=h(x)$. [Hint: Write (6.7) as a first order system for the vector $v=(u_x, u_t)$ and find solutions with $u_x=\theta$, $u_t=F(\theta)$. Use Problem 6., p. 19.

[Answer:

$$u = \mp\tfrac{1}{2}\theta^2 + h(x \pm (1+\theta)t) + ct \tag{6.8}$$

where $c=$ const. and θ is the solution of

$$\theta = h'(x \pm (1+\theta)t)]. \tag{6.9}$$

3 Characteristic Manifolds and the Cauchy Problem*

1. Notation of Laurent Schwartz†

This multi-index notation is extremely convenient for partial differential equations, keeping us from drowning in a flood of subscripts. We consider here vectors $x = (x_1, x_2, \ldots, x_n)$ with n (usually) real components x_k or equivalently points x in $\mathbb{R}^n$. The vectors $\alpha = (\alpha_1, \alpha_2, \ldots, \alpha_n)$ whose components are non-negative integers α_k are called *multi-indices* and form a subset Z^n of $\mathbb{R}^n$. Letters $x, y, \ldots, \xi, \eta, \ldots$ will be used for vectors, and $\alpha, \beta, \gamma, \ldots$ for multi-indices. The dimension n varies and has to be inferred from the context. Components are always indicated by adding a subscript to the vector symbol: ξ has components ξ_k. In $\mathbb{R}^n$ we introduce the null-vector 0 and the one-vector 1 (both belonging to Z^n) by

$$0 = (0,0,\ldots,0); \qquad 1 = (1,1,\ldots,1). \tag{1.1a}$$

For the points in $\mathbb{R}^n$ we apply the usual rules for operations in a linear vector space over the reals. Following L. Schwartz we define for $\alpha \in Z^n$ the scalars

$$|\alpha| = \alpha_1 + \alpha_2 + \cdots + \alpha_n; \qquad \alpha! = \alpha_1!\alpha_2!\cdots\alpha_n!, \tag{1.1b}$$

and for $x \in \mathbb{R}^n$, $\alpha \in Z^n$ the monomial

$$x^\alpha = x_1^{\alpha_1} x_2^{\alpha_2} \cdots x_n^{\alpha_n}. \tag{1.1c}$$

In Z^n we define a partial ordering by writing

$$\alpha \geqslant \beta \quad \text{whenever } \alpha_i \geqslant \beta_i \quad \text{for } i = 1, \ldots, n. \tag{1.2}$$

*([6])
†([12], [16])

By C_α we generally denote a coefficient depending on n nonnegative integers $\alpha_1, \ldots, \alpha_n$:

$$C_\alpha = C_{\alpha_1 \cdots \alpha_n}. \tag{1.3}$$

The C_α may be real numbers, or vectors in a space $\mathbb{R}^m$. The general m-th degree polynomial in $x_1, \ldots, x_n$ is then of the form

$$P(x) = \sum_{|\alpha| \leq m} C_\alpha x^\alpha \tag{1.4}$$

with $\alpha \in Z^n$, $x \in \mathbb{R}^n$, $C_\alpha \in \mathbb{R}$.

Using the Cauchy differentiation symbol $D_k = \partial/\partial x_k$, we introduce the "gradient vector" $D = (D_1, \ldots, D_n)$, and define the gradient of a function $u(x_1, \ldots, x_n)$ as the vector

$$Du = (D_1 u, \ldots, D_n u). \tag{1.5}$$

The general partial differentiation operator of order m is then

$$D^\alpha = D_1^{\alpha_1} D_2^{\alpha_2} \cdots D_n^{\alpha_n} = \frac{\partial^m}{\partial x_1^{\alpha_1} \cdots \partial x_n^{\alpha_n}} \tag{1.6}$$

where $|\alpha| = m$.

Problems

Let x,y denote vectors and α,β multi-indices with n components. Prove the following relations:

1. the binomial theorem

$$(x+y)^\alpha = \sum_{\substack{\beta,\gamma \\ \beta+\gamma=\alpha}} \frac{\alpha!}{\beta!\gamma!} x^\beta y^\gamma. \tag{1.7}$$

2. Taylor expansion* for a polynomial $f(x)$ of degree m

$$f(x) = \sum_{|\alpha| \leq m} \frac{1}{\alpha!} (D^\alpha f(0)) x^\alpha. \tag{1.8}$$

3. Multinomial theorem. For an integer $m \geq 0$ and $x = (x_1, \ldots, x_n)$

$$(x_1 + \cdots + x_n)^m = \sum_{|\alpha|=m} \frac{m!}{\alpha!} x^\alpha. \tag{1.9}$$

4. For a multi-index $\alpha = (\alpha_1, \ldots, \alpha_n)$

$$\alpha! \leq |\alpha|! \leq n^{|\alpha|} \alpha!. \tag{1.10}$$

5. Leibnitz's rule for scalar valued functions $f(x)$, $g(x)$ of a vector x

$$D^\alpha(fg) = \sum_{\substack{\beta,\gamma \\ \beta+\gamma=\alpha}} \frac{\alpha!}{\beta!\gamma!} (D^\beta f)(D^\gamma g). \tag{1.11}$$

* Quite generally we write $D^\alpha f(z)$ for $D^\alpha f$ taken for the argument z.

6.
$$D^\beta x^\alpha = \frac{\alpha!}{(\alpha-\beta)!} x^{\alpha-\beta} \quad \text{if } \alpha \geqslant \beta \tag{1.12a}$$
$$D^\beta x^\alpha = 0 \quad \text{otherwise.} \tag{1.12b}$$

7. Higher order directional derivatives. Let f be a function from $\mathbb{R}^n$ to $\mathbb{R}$; then for vectors x,y and a scalar t
$$\frac{d^m}{dt^m} f(x+ty) = \sum_{|\alpha|=m} \frac{|\alpha|!}{\alpha!} (D^\alpha f(x+ty)) y^\alpha. \tag{1.13}$$
(Hint: Induction over m or (1.9) with x_i replaced by $y_i D_i$.)

2. The Cauchy Problem*

In the Schwartz notation the general mth-order linear differential equation for a function $u(x) = u(x_1, \dots, x_n)$ takes the simple form
$$Lu = \sum_{|\alpha| \leqslant m} A_\alpha(x) D^\alpha u = B(x). \tag{2.1}$$
The same formula describes the general mth-order *system* of N differential equations in N unknowns if we interpret u and B as column vectors with N components and the A_α as $N \times N$ square matrices. Similarly the general mth-order quasi-linear equation (respectively system of such equations) is
$$Lu = \sum_{|\alpha|=m} A_\alpha D^\alpha u + C = 0, \tag{2.2}$$
where now the A and C are functions of the independent variables x_k and of the derivatives $D^\beta u$ of the unknown u of orders $|\beta| \leqslant m-1$. More general nonlinear equations or systems
$$F(x, D^\alpha u) = 0 \tag{2.3}$$
can be reduced formally to quasi-linear ones by applying a first-order differential operator to (2.3). On the other hand, an mth-order quasi-linear system (2.2) can be reduced to a (larger) first-order one, by introducing all derivatives $D^\beta u$ with $|\beta| \leqslant m-1$ as new dependent variables, and making use of suitable compatibility conditions for the $D^\beta u$.

The *Cauchy Problem* consists of finding a solution u of (2.2) or (2.1) having prescribed *Cauchy data* on a hyper-surface $S \subset \mathbb{R}^n$ given by
$$\phi(x_1, \dots, x_n) = 0. \tag{2.4}$$
Here ϕ shall have m continuous derivatives and the surface should be regular in the sense that
$$D\phi = (\phi_{x_1}, \dots, \phi_{x_n}) \neq 0. \tag{2.5}$$
The *Cauchy data* on S for an mth-order equation consist of the derivatives

*([27])

of u of orders less than or equal to $m-1$. They cannot be given arbitrarily but have to satisfy the compatibility conditions valid on S for all functions regular near S (instead *normal* derivatives of order less than m can be given independently from each other). We are to find a solution u near S which has these Cauchy data on S. We call S *noncharacteristic* if we can get all $D^\alpha u$ for $|\alpha|=m$ on S from the linear algebraic system of equations consisting of the compatibility conditions for the data and the partial differential equation (2.2) taken on S. We call S *characteristic* if at each point x of S the surface S is not noncharacteristic. Characteristic surfaces naturally occur in connection with *singular solutions* of a certain type. If u is a (generalized) solution of (2.1) of class C^{m-1} which has jump discontinuities in its mth-order derivatives along a surface S, then S must be a characteristic surface. (See Chapter 2, p. 36.)

To get an algebraic criterion for characteristic surfaces we first consider the special case where the hyper-surface S is the coordinate plane $x_n=0$. The Cauchy data then consist of the $D^\beta u$ with $|\beta|<m$ taken for $x_n=0$. Singling out the "normal" derivatives on S of orders $\leqslant m-1$:

$$D_n^k u=\psi_k(x_1,\ldots,x_{n-1}) \quad \text{for } k=0,\ldots,m-1 \text{ and } x_n=0 \tag{2.6}$$

we have on S

$$D^\beta u=D_1^{\beta_1}D_2^{\beta_2}\ldots D_{n-1}^{\beta_{n-1}}\psi_{\beta_n} \tag{2.7}$$

provided that $\beta_n<m$. In particular for $|\beta|\leqslant m-1$ we have here the compatibility conditions expressing all Cauchy data in terms of normal derivatives on S. Let α^* denote the multi-index

$$\alpha^*=(0,\ldots,0,m). \tag{2.8}$$

In the differential equation (2.1) or (2.2) taken on S it is only the term with $\alpha=\alpha^*$ that is not expressible by (2.7) in terms of $\psi_0,\ldots,\psi_{m-1}$ and hence in terms of the Cauchy data. All others contain derivatives $D^\alpha u$ with $\alpha_n\leqslant m-1$. Thus $D^{\alpha^*}u$, and hence all $D^\alpha u$ with $|\alpha|\leqslant m$, are determined uniquely on S, if we can solve the differential equation for the term $D^{\alpha^*}u$. This is always possible in a unique way if and only if the matrix A_{α^*} is nondegenerate, i.e., $\det(A_{\alpha^*})\neq 0$. For a single scalar differential equation this condition reduces to $A_{\alpha^*}\neq 0$. In the linear case the validity of the condition

$$\det(A_{\alpha^*})\neq 0 \tag{2.9}$$

does not depend on the Cauchy data on S; in the quasi-linear case, however, where the A_α depend on the $D^\beta u$ with $|\beta|\leqslant m-1$ and on x, one has to know the ψ_k in order to decide if S is noncharacteristic.

Condition (2.9) involves coefficients of mth-order derivatives. We define the *principal part* L_{pr} of L (both in (2.2) and (2.1)) as consisting of the highest order terms of L:

$$L_{\text{pr}}=\sum_{|\alpha|=m}A_\alpha D^\alpha. \tag{2.10}$$

The "symbol" of this differential operator is the matrix form ("characteristic matrix" of L):

$$\Lambda(\zeta) = \sum_{|\alpha|=m} A_\alpha \zeta^\alpha. \tag{2.11}$$

Here the $N \times N$ matrix $\Lambda(\zeta)$ has elements that are mth-degree forms in the components of the vector $\zeta = (\zeta_1, \ldots, \zeta_n)$. In particular, the multiplier of D_n^m in L_{pr} is $A_{\alpha^*} = \Lambda(\eta)$, where

$$\eta = (0, \ldots, 0, 1) = D\phi. \tag{2.12}$$

is the unit normal to the surface $\phi = x_n = 0$. The condition for the plane $\phi = x_n = 0$ to be noncharacteristic is then

$$Q(D\phi) \neq 0, \tag{2.13}$$

where $Q = Q(\zeta)$ is the *characteristic form* defined by

$$Q(\zeta) = det(\Lambda(\zeta)) \tag{2.14}$$

for any vector ζ. (In the case of a scalar equation ($N = 1$) the characteristic form $Q(\zeta)$ coincides with the polynomial $\Lambda(\zeta)$.) We shall see that quite generally (2.13) is the condition for a surface $\phi = 0$ to be noncharacteristic.

Take now a general S described by (2.4). By assumption (2.5) the first derivatives of ϕ do not vanish simultaneously. Suppose that in a neighborhood of a given point of S, the condition $\phi_{x_n} \neq 0$ holds. The transformation

$$y_i = \begin{cases} x_i & \text{for } i = 1, \ldots, n-1 \\ \phi(x_1, \ldots, x_n) & \text{for } i = n \end{cases} \tag{2.15}$$

is then locally regular and invertible. By the chain rule,

$$\frac{\partial u}{\partial x_i} = \sum_k C_{ik} \frac{\partial u}{\partial y_k}, \tag{2.16}$$

where the

$$C_{ik} = \frac{\partial y_k}{\partial x_i} \tag{2.17}$$

are functions of x or of y. Denoting by C the matrix of the C_{ik} and introducing the gradient operator d with respect to y with components

$$d_i = \frac{\partial}{\partial y_i}, \tag{2.18}$$

we can write (2.16) symbolically, as

$$D = Cd, \tag{2.19}$$

taking D and d to be column vectors. Generally, then for $|\alpha| = m$

$$D^\alpha = (Cd)^\alpha + R_\alpha, \tag{2.20}$$

where R_α is a linear differential operator involving only derivatives of orders $\leqslant m-1$, (arising from the dependence of C on x) and $(Cd)^\alpha$ is formed as if C were a constant matrix, i.e., not applying differentiations to the elements of C. Then the principal part of the operator L in (2.2) or (2.1) transformed to y-coordinates is given by

$$L_{\mathrm{pr}} = \sum_{|\alpha|=m} A_\alpha (Cd)^\alpha = l_{\mathrm{pr}} \tag{2.21}$$

and its symbol, the characteristic matrix of l, by

$$\lambda(\eta) = \sum_{|\alpha|=m} A_\alpha (C\eta)^\alpha. \tag{2.22}$$

For the regular mapping (2.15) x-derivatives of orders less than or equal to r are linear combinations of y-derivatives of orders less than or equal to r, and conversely. Hence noncharacteristic behavior of S is preserved under the transformation. Thus S is noncharacteristic for L if the plane $y_n = 0$ is noncharacteristic with respect to the operator L transformed to y-coordinates, i.e., if

$$\det(\lambda(\eta)) = \det\left(\sum_{|\alpha|=m} A_\alpha (C\eta)^\alpha \right) \neq 0, \tag{2.23}$$

for the column vector η with components $(0,\ldots,0,1)$. But then $\zeta = C\eta$ is just the column vector with components $\partial y_n / \partial x_i = D_i \phi$, that is, the vector $D\phi$. Thus, the condition for noncharacteristic behavior of S can again be written as (2.13).

If u in (2.2) stands for a vector with N components, the condition for S to be a characteristic surface

$$Q(D\phi) = \det\left(\sum_{|\alpha|=m} A_\alpha (D\phi)^\alpha \right) = 0 \tag{2.24}$$

signifies the vanishing on S of a form of degree Nm in the components of $D\phi$. In the linear case (where the coefficients of that form only depend on x and not on u) we can consider a one-parameter family $\phi(x_1,\ldots,x_n) =$ const. $= c$ of characteristic surfaces. Then (2.24) becomes a first-order partial differential equation of ϕ, homogeneous of degree Nm in the first derivatives of ϕ, from which ϕ can be determined by the methods of solution for single first-order equations. For example in the case of a linear first-order system

$$Lu = \sum_{i=1}^{n} A_i(x) \frac{\partial u}{\partial x_i} + B(x)u = w(x), \tag{2.25}$$

(u and w being N-vectors and the A_i and B, $N \times N$ matrices) the condition for a characteristic surface is

$$\det\left(\sum_{i=1}^{n} \frac{\partial \phi}{\partial x_i} A_i \right) = 0. \tag{2.26}$$

Alternately a single characteristic surface S gives rise to a partial differential equation, when described by an *explicit* equation

$$\phi(x_1,\ldots,x_n)=x_n-\psi(x_1,\ldots,x_{n-1})=0. \tag{2.27}$$

Equation (2.24) in the linear case becomes a first-order partial differential equation for the function ψ. Take, e.g., the wave equation

$$u_{tt}=c^2(u_{x_1x_1}+u_{x_2x_2}) \tag{2.28a}$$

for $u=u(x_1,x_2,t)$. A characteristic surface $t=\psi(x_1,x_2)$ then satisfies the equation

$$1=c^2\left(\psi_{x_1}^2+\psi_{x_2}^2\right) \tag{2.28b}$$

already encountered in Chapter 1, p. 27. Thus equation (2.28b) regulates the propagation of singularities for equation (2.28a).

The characteristic form $Q(\zeta)$ of the operator L defined by (2.14), (2.11) generally depends on ζ and on the arguments of the A_α, that is, on x and the $D^\beta u$ with $|\beta|\leqslant m-1$. A hypersurface is characteristic for the operator L (and in the nonlinear case for given Cauchy data $D^\beta u$) if $Q(\zeta)=0$ for the normal vector ζ of the surface. We call L elliptic if $Q(\zeta)\neq 0$ for all real $\zeta\neq 0$. In that case there exist no real characteristic hypersurfaces. In the case of an operator L with *real* coefficients, ellipticity of L is then equivalent to *definiteness* of the form $Q(\zeta)$ (at least for $n>1$): the form $Q(\zeta)$ is of constant sign for $\zeta\neq 0$. This, of course, can only happen when the degree mN of Q is even. The standard example for an elliptic L is the Laplace operator $\Delta=D_1^2+\cdots+D_n^2$ for which

$$Q(\zeta)=\sum_{i=1}^{n}\zeta_i^2$$

is positive definite. The extension of the notion of "hyperbolicity" to general quasi-linear systems is more complicated.

We can define characteristic manifolds for more general nonlinear equations as well. Take, for example, an mth-order scalar equation for a function $u=u(x_1,\ldots,x_n)$, which we write in the form

$$F(x,p_\alpha)=0, \tag{2.29a}$$

where $p_\alpha=D^\alpha u$ for $|\alpha|\leqslant m$. Suppose that on the hypersurface S given by (2.4) we have $\phi_{x_n}\neq 0$. Differentiating (2.29a) with respect to x_n we obtain the equation

$$0=\frac{\partial F}{\partial x_n}+\sum_\alpha \frac{\partial F}{\partial p_\alpha}D_nD^\alpha u. \tag{2.29b}$$

Since $\zeta_n\neq 0$ for $\zeta=D\phi$, the condition for S to be characteristic with respect to (2.29b) is simply

$$\sum_{|\alpha|=m}\frac{\partial F}{\partial p_\alpha}(D\phi)^\alpha=0. \tag{2.30}$$

We use (2.30) to define what is meant by S to be characteristic with respect to (2.29a). Observe that this condition generally involves not only the Cauchy data but all derivatives of u of orders $\leqslant m$ on S. In the example of the first-order equation (7.1) of Chapter 1, condition (2.30) for a characteristic (projection) $\phi(x,y)=\text{const.}$ would read

$$F_p\phi_x + F_q\phi_y = 0.$$

Since here $\phi_x\,dx + \phi_y\,dy = 0$ along S, the condition is equivalent to the second equation in (7.10) of Chapter 1.

PROBLEMS

1. Identify the special cases of "characteristic curves" in Chapter 2 (1.12) and (5.13) with the general formula (2.13).

2. Let $u(x)=u(x_1,\ldots,x_n)$ and its derivatives of orders $<m$ vanish on the hypersurface S given by (2.4). Show that on S

$$D^\alpha u = \mu(D\phi)^\alpha \quad \text{for } |\alpha| = m, \tag{2.31}$$

where the factor of proportionality μ depends on u but not on α. Show that in particular for u of the form $u=\phi^m v(x)$ with $v\in C^m$

$$D^\alpha u = m!(D\phi)^\alpha v. \tag{2.32}$$

(Hint: Transform S by (2.15) into the plane $y_n=0$.)

3. Real Analytic Functions and the Cauchy–Kowalevski Theorem*

This theorem quite generally asserts the local existence of solutions of the non-characteristic Cauchy problem, provided the equation and data are *analytic*. Hence we need to consider the class of analytic functions, which plays a significant role in many questions concerning partial differential equations. Global behavior of analytic functions can only be understood when we consider such functions for *complex* arguments, or as solutions of systems of Cauchy–Riemann equations. However, for most purposes in differential equations it is sufficient to consider the restrictions of analytic functions to real arguments, the so-called *real analytic* functions. These form a curious subclass C^ω of C^∞, that can either be characterized by the property of local representability by power series, or by the way their partial derivatives increase with increasing order.

* The name of the second person connected with this theorem is given variously as Kowalewski, Sophie von Kowalevsky, Sonja Kovalevsky, Sofya Kovalevskaya. In her own German and French papers in the *Acta Mathematica* she calls herself (rightly or wrongly) Sophie Kowalevski, and this is the spelling adopted here.

(a) *Multiple infinite series*

We shall consider multiple infinite series of the form

$$\sum_{\alpha} C_{\alpha} \tag{3.1a}$$

where C_α are real numbers defined for all $\alpha \in Z^n$. (Summation is always over all of Z^n, unless some limitation is indicated.) The term *convergence* will be used exclusively to denote *absolute convergence.* Thus (3.1) converges, iff

$$\sum_{\alpha} |C_{\alpha}| < \infty. \tag{3.1b}$$

The sum of a convergent series does not depend on the order of summation. We shall also have occasion to consider series (3.1a) whose terms C_α are vectors in $\mathbb{R}^m$. Convergence then shall mean *component-wise* absolute convergence.

EXAMPLES

(1) For $x \in \mathbb{R}^n$, $\alpha \in \hat{Z^n}$ (see (3.1a))

$$\sum_{\alpha} x^{\alpha} = \prod_{i=1}^{n} \left(\sum_{\alpha_i=0}^{\infty} x_i^{\alpha_i} \right) = \frac{1}{(1-x_1)(1-x_2)\cdots(1-x_n)} = \frac{1}{(1-x)^1} \tag{3.2a}$$

provided

$$|x_i| < 1 \qquad \text{for all } i. \tag{3.2b}$$

(2) For $x \in \mathbb{R}^n$, $\alpha \in Z^n$ (see (1.9))

$$\sum_{\alpha} \frac{|\alpha|!}{\alpha!} x^{\alpha} = \sum_{j=0}^{\infty} \sum_{|\alpha|=j} \frac{|\alpha|!}{\alpha!} x^{\alpha} = \sum_{j=0}^{\infty} (x_1 + \cdots + x_n)^j$$

$$= \frac{1}{1-(x_1+\cdots+x_n)} \tag{3.3a}$$

provided

$$|x_1| + |x_2| + \cdots + |x_n| < 1. \tag{3.3b}$$

Convergence of infinite series of scalar functions $C_\alpha(x)$ defined and continuous for x in a set $S \subset \mathbb{R}^n$ often will be established by *comparison* with a series of constants c_α. If

$$|C_{\alpha}(x)| \leqslant c_{\alpha} \quad \text{for all } \alpha \in Z^n, \quad x \in S \tag{3.4a}$$

and if

$$\sum_{\alpha} c_{\alpha} < \infty, \tag{3.4b}$$

then

$$\sum_\alpha C_\alpha(x) \tag{3.4c}$$

converges uniformly for $x \in S$ and represents a continuous function. If S is open and all the $C_\alpha(x)$ are of class $C^j(S)$, and if the formally differentiated series

$$\sum_\infty D^\beta C_\alpha(x) \tag{3.4d}$$

converge uniformly for $x \in S$ and each β with $|\beta| \leqslant j$, then the sum of the series (3.4c) belongs to $C^j(S)$ and

$$D^\beta \sum_\alpha C_\alpha(x) = \sum_\alpha D^\beta C_\alpha(x) \quad \text{for } x \in S, \quad |\beta| \leqslant j. \tag{3.4e}$$

PROBLEMS

1. Show that the series in (3.2a), resp. (3.3a) do not converge (that is do not converge absolutely) unless conditions (3.2b), resp. (3.3b) are satisfied.

2. Show that for $\alpha, \beta \in Z^n$, $x \in \mathbb{R}^n$, $|x_i| < 1$ for all i

$$\sum_{\substack{\alpha \\ \alpha \geqslant \beta}} \frac{\alpha!}{(\alpha-\beta)!} x^{\alpha-\beta} = D^\beta \frac{1}{(1-x)^1} = \frac{\beta!}{(1-x)^{1+\beta}}. \tag{3.5}$$

3. Show that for $|x_1| + \cdots + |x_n| < 1$

$$\sum_{\substack{\alpha \\ \alpha \geqslant \beta}} \frac{|\alpha|!}{(\alpha-\beta)!} x^{\alpha-\beta} = D^\beta \frac{1}{1 - x_1 - \cdots - x_n} = \frac{|\beta|!}{(1 - x_1 - \cdots - x_n)^{1+|\beta|}}. \tag{3.6}$$

Of particular importance are the power series

$$\sum_\alpha c_\alpha x^\alpha \tag{3.7a}$$

where $x \in \mathbb{R}^n$, $\alpha \in Z^n$, $c_\alpha \in \mathbb{R}$. Assume that the series converges for a certain z, and hence that

$$\mu = \sum_\alpha |c_\alpha||z^\alpha| < \infty. \tag{3.7b}$$

Then (3.7a) converges uniformly for all x with

$$|x_i| \leqslant |z_i| \quad \text{for all } i. \tag{3.7c}$$

Hence

$$f(x) = \sum_\alpha c_\alpha x^\alpha \tag{3.7d}$$

defines a continuous function f in the set (3.7c). We show next that all series obtained by formal differentiation of (3.7a) converge in the interior of the set (3.7c), and even uniformly in every compact subset of the interior. Indeed the interior of (3.7c) is given by

$$|x_i| < |z_i| \quad \text{for all } i. \tag{3.7e}$$

It is non-empty only if $z_i \neq 0$ for all i. For all x in a compact subset of (3.7e) we have

$$|x_i| \leqslant q|z_i| \quad \text{for all } i \tag{3.7f}$$

with a certain q between 0 and 1. Then, using (3.5) with x replaced by the vector $(q, \ldots, q) = q1$,

$$\begin{aligned}\sum_\alpha |D^\beta c_\alpha x^\alpha| &\leqslant \sum_{\alpha \geqslant \beta} \frac{\alpha!}{(\alpha-\beta)!} |c_\alpha| q^{|\alpha-\beta|} |z^{\alpha-\beta}| \\ &\leqslant \frac{\mu}{|z^\beta|} \sum_{\alpha \geqslant \beta} \frac{\alpha!}{(\alpha-\beta)!} q^{|\alpha-\beta|} \\ &= \frac{\mu}{|z^\beta|} \frac{\beta!}{(1-q)^{n+|\beta|}}.\end{aligned} \tag{3.7g}$$

Consequently, the function $f(x)$ defined by (3.7d) is of class C^∞ in the set (3.7e). In any compact subset of the set (3.7e) f satisfies an inequality of the form

$$|D^\beta f(x)| \leqslant M|\beta|!r^{-|\beta|} \tag{3.7h}$$

for all $\beta \in Z^n$, where

$$M = \frac{\mu}{(1-q)^n}, \qquad r = (1-q) \operatorname*{Min}_i |z_i|. \tag{3.7i}$$

Termwise differentiation of (3.7d) at the origin also shows that the coefficients c_α of the power series are given by

$$c_\alpha = \frac{1}{\alpha!} D^\alpha f(0). \tag{3.7j}$$

(b) *Real analytic functions*

Definition. Let $f(x)$ be a function whose domain is an open set Ω in $\mathbb{R}^n$ and whose range lies in $\mathbb{R}$. For $y \in \Omega$ we call f *real analytic at* y if there exist $c_\alpha \in \mathbb{R}$ and a neighbourhood N of y (all depending on y) such that

$$f(x) = \sum_\alpha c_\alpha (x-y)^\alpha \tag{3.8}$$

for all x in N. We say f is *real analytic in* Ω (in symbols: $f \in C^\omega(\Omega)$), if f is real analytic at each $y \in \Omega$. A *vector* $f(x) = (f_1(x), \ldots, f_m(x))$ defined in Ω is called real analytic, if each of its components is real analytic.

Our previous results immediately yield the following theorem:

Theorem. *If $f = (f_1, \ldots, f_m) \in C^{\omega}(\Omega)$, then $f \in C^{\infty}(\Omega)$. Moreover, for any $y \in \Omega$ there exists a neighborhood N of y and positive numbers M, r such that for all $x \in N$*

$$f(x) = \sum_{\alpha} \frac{1}{\alpha!} (D^{\alpha} f(y))(x - y)^{\alpha} \tag{3.9a}$$

$$|D^{\beta} f_k(x)| \leqslant M |\beta|! r^{-|\beta|} \quad \text{for all } \beta \in Z^n \quad \text{and all } k. \tag{3.9b}$$

Real analytic f enjoy the property of *unique continuation* expressed by the following theorem:

Theorem. *Let $f \in C^{\omega}(\Omega)$ where Ω is a* connected *open set* in $\mathbb{R}^n$. Let $z \in \Omega$.*

Then f is determined uniquely in Ω if we know the $D^{\alpha} f(z)$ for all $\alpha \in Z^n$. In particular f is determined uniquely in Ω by its values in any non-empty open subset of Ω.

PROOF. Let $g, h \in C^{\omega}(\Omega)$ and let $D^{\alpha} g(z) = D^{\alpha} h(z)$ for all $\alpha \in Z^n$. Write $f = g - h$ and decompose Ω into

$$\Omega_1 = \{x \mid x \in \Omega;\ D^{\alpha} f(x) = 0 \quad \text{for all } \alpha \in Z^n\}$$

$$\Omega_2 = \{x \mid x \in \Omega;\ D^{\alpha} f(x) \neq 0 \quad \text{for some } \alpha \in Z^n\}.$$

Then Ω_2 is open by continuity of f. The set Ω_1 also is open; for if $y \in \Omega_1$, we have $f(x) = 0$ in a neighborhood of y by (3.9a). Since $z \in \Omega_1$ and Ω is connected, Ω_2 must be empty. □

Local power series representations are not the only way to characterize real analytic functions. The class C^{ω} can be obtained by growth restrictions on the higher derivatives in the real or by restricting differentiable functions in a complex domain to real arguments.

Definition. Let $f(x) = (f_1(x), \ldots, f_m(x))$ be defined in an open set Ω of $\mathbb{R}^n$. If $y \in \Omega$ and M, r are positive real numbers, we say that $f \in C_{M,r}(y)$, if $f \in C^{\infty}$ in a neighborhood of y, and

$$|D^{\beta} f_k(y)| \leqslant M |\beta|! r^{-|\beta|} \quad \text{for all } \beta \in Z^n \quad \text{and } k = 1, \ldots, m. \tag{3.10}$$

Theorem. *Let f be defined in the open set Ω. Then necessary and sufficient for $f \in C^{\omega}(\Omega)$ is that $f \in C^{\infty}(\Omega)$ and that for every compact $S \subset \Omega$ we can find M, r such that $f \in C_{M,r}(y)$ for all $y \in S$.*

PROOF. Let $f \in C^{\omega}(\Omega)$. Then $f \in C^{\infty}(\Omega)$. Moreover, by (3.9b) we can find for every $z \in \Omega$ positive numbers $M = M(z)$, $r = r(z)$ and a neighborhood $N = N(z)$ such that

$$|D^{\beta} f_k(x)| \leqslant M |\beta|! r^{-|\beta|} \tag{3.11}$$

* An open set is connected if it cannot be decomposed into two disjoint non-empty open sets.

for all $\beta \in Z^n$, $x \in N(z)$ and k. A finite number of the $N(z)$, say $N(z^1), \ldots, N(z^j)$, covers the compact S. Then (3.10) holds for all $y \in S$ with $M = \operatorname{Max} M(z^i)$, $r = \operatorname{Min} r(z^i)$. Conversely, let $f \in C^\infty(\Omega)$ and let there exist for every compact $S \subset \Omega$ numbers M,r such that (3.10) holds for all $y \in S$. Take any $y \in \Omega$ and take for S the compact set

$$S = \{x \mid |x - y| \leqslant s\}$$

where the positive number s is chosen so small that $S \subset \Omega$. Let M,r be the corresponding values such that (3.11) holds for all $\beta \in Z^n$, $x \in S$. We shall prove that

$$f_k(x) = \sum_\alpha \frac{1}{\alpha!}(D^\alpha f_k(y))(x - y)^\alpha \tag{3.12}$$

for any x with

$$d = |x_1 - y_1| + |x_2 - y_2| + \cdots + |x_n - y_n| \leqslant \operatorname{Min}(r,s). \tag{3.13}$$

Introduce the function

$$\phi(t) = f_k(y + t(x - y))$$

of the scalar argument t. Here $y + t(x - y) \in S$ for $0 \leqslant t \leqslant 1$, and consequently by (1.13), (3.11), (1.9) for any integer $j \geqslant 0$

$$\begin{aligned}\left|\frac{1}{j!}\frac{d^j}{dt^j}\phi(t)\right| &= \left|\sum_{|\alpha|=j}\frac{1}{\alpha!}(D^\alpha f_k(y + t(x - y))(x - y)^\alpha\right| \\ &\leqslant \sum_{|\alpha|=j} M\frac{|\alpha|!}{\alpha!}r^{-|\alpha|}|(x - y)^\alpha| = Mr^{-j}d^j.\end{aligned}$$

By Taylor's theorem

$$f_k(x) = \phi(1) = \sum_{k=0}^{j-1}\frac{1}{k!}\phi^{(k)}(0) + r_j$$

where

$$|r_j| = \left|\frac{1}{(j-1)!}\int_0^1 (1 - t)^{j-1}\phi^{(j)}(t)\,dt\right| \leqslant Mr^{-j}d^j.$$

Since here $d < r$, relation (3.12) follows for $k \to \infty$. □

PROBLEMS

1. A set $\Omega \subset \mathbb{R}$ is *convex* if for $x,y \in \Omega$, $\theta \in \mathbb{R}$, $0 < \theta < 1$ also $\theta x + (1 - \theta)y \in \Omega$. Show that an open convex set is connected.

2. Let Ω_1,Ω_2 be open convex sets in $\mathbb{R}^n$, and let $f_1 \in C^\omega(\Omega_1)$, $f_2 \in C^\omega(\Omega_1)$. Show that $f_1(x) = f_2(x)$ for all $x \in \Omega_1 \cap \Omega_2$, if $D^\alpha f_1(y) = D^\alpha f_2(y)$ for all α at a single point y of $\Omega_1 \cap \Omega_2$. Show by an example that this may not be true if the Ω_k are just *connected* instead of *convex*.

3. Let the power series (3.7a) converge for a certain z with $z_i \neq 0$ for all i. Then the series represents a real analytic function in the set (3.7e).

4. Let f be real analytic in an open *convex* set Ω. Let x,z be points of Ω for which the series

$$\sum_\alpha \frac{1}{\alpha!}(D^\alpha f(z))(x-z)^\alpha \tag{3.14}$$

converges. Then the series has the sum $f(x)$.

The classical proof of the Cauchy–Kowalevski theorem works with the *method of majorants*, in which problems are compared with others that in some ways behave worse but can be solved explicitly. In this way complicated estimates are replaced by simple explicit manipulations.

Definition. Let f,F be functions with domain in $\mathbb{R}^n$ and range in $\mathbb{R}^m$, of class C^∞ in a neighborhood of the origin. We say f is *majorised* by F (in symbols: $f \ll F$), if

$$|D^\alpha f_k(0)| \leqslant D^\alpha F_k(0) \quad \text{for all } \alpha \in Z^n \quad \text{and all } k. \tag{3.15}$$

For f that are analytic at 0, and more generally, for f belonging to some class $C_{M,r}$, we can always find simple rational majorants:

Theorem. *The vector $f(x)$ belongs to $C_{M,r}(0)$ if and only if*

$$f \ll (\phi,\ldots,\phi) = \phi 1 \tag{3.16a}$$

where ϕ is the scalar

$$\phi(x) = \frac{Mr}{r - x_1 - x_2 - \cdots - x_n}. \tag{3.16b}$$

Moreover, $f \in C_{M,r}(0)$ and $f(0) = 0$ is equivalent to

$$f \ll (\phi - M,\ldots,\phi - M) = (\phi - M)1 \tag{3.16c}$$

where

$$\phi - M = \frac{M(x_1 + \cdots + x_n)}{r - x_1 - \cdots - x_n}. \tag{3.16d}$$

The theorem is obvious since

$$D^\alpha\phi(0) = M|\alpha|!r^{-|\alpha|}. \tag{3.16e}$$

Certain operations preserve majorisation. Thus, trivially $f \ll F$ implies $D^\alpha f \ll D^\alpha F$ for all α. More interesting is the case of compound functions:

Theorem. *Let $f(x)$, $F(x)$ be C^∞-maps of a neighborhood of the origin of $\mathbb{R}^n$ into $\mathbb{R}^m$, and let $g(u)$, $G(u)$ be C^∞-maps of a neighborhood of the origin of $\mathbb{R}^m$ into $\mathbb{R}^p$. Let $f(0) = F(0) = 0$, and $f \ll F, g \ll G$. Then*

$$g(f(x)) \ll G(F(x)). \tag{3.17}$$

PROOF. It is clear that $g(f(x)) \in C^\infty$ near $x = 0$. By repeated application of the chain rule of differentiation we can express any derivative of $g(f(x))$ in terms of derivatives of $g(u)$ with respect to u at $u = f(x)$ and derivatives of $f(x)$ with respect to x. More precisely, setting $h(x) = g(f(x))$, $H(x) = G(F(x))$, we have for each $\alpha \in Z^n$ and $k = 1,2,\ldots,p$ a formula of the type

$$D^\alpha h_k(0) = P_\alpha(\delta^\beta g_k(0), D^\gamma f_j(0)).$$

Here P_α is a polynomial in its arguments whose coefficients are non-negative integers; δ stands for $(\partial/\partial u_1,\ldots,\partial/\partial u_m)$; $j = 1,\ldots,m$; $\beta \in Z^m, \gamma \in Z^n$; $|\beta| \leqslant |\alpha|$, $|\gamma| \leqslant |\alpha|$. By assumption then

$$\begin{aligned} |D^\alpha h_k(0)| &\leqslant P_\alpha(|\delta^\beta g_k(0)|, |D^\gamma f_j(0)|) \\ &\leqslant P_\alpha(\delta^\beta G_k(0), D^\gamma F_j(0)) = D^\alpha H_k(0) \end{aligned}$$

which was to be proved. □

We can apply this theorem to obtain estimates for the derivatives of a compound function. Let $f(x)$ map a neighborhood of a point $y \in \mathbb{R}^n$ into $\mathbb{R}^m$, and $g(u)$ map a neighborhood of $v = f(y)$ into $\mathbb{R}^p$. Let

$$f \in C_{M,r}(y), g \in C_{\mu,\rho}(v) \tag{3.18a}$$

Then

$$h(x) = g(f(x)) \in C_{\mu,\rho r/(mM+\rho)}(y). \tag{3.18b}$$

Indeed

$$h(y + x) = g(v + f(y + x) - f(y)) = g^*(f^*(x)),$$

where

$$g^*(u) = g(v + u) \in C_{\mu,\rho}(0) \tag{3.18c}$$

$$f^*(x) = f(y + x) - f(y) \in C_{M,r}(0); \qquad f^*(0) = 0. \tag{3.18d}$$

Thus $f^*(x) \ll (\phi - M,\ldots,\phi - M)$ with $\phi - M$ given by (3.16d), while $g^*(u) \ll (\psi,\ldots,\psi)$ with

$$\psi(u) = \frac{\mu\rho}{\rho - u_1 - \cdots - u_m}.$$

It follows that

$$h(y + x) \ll (\chi(x),\ldots,\chi(x)),$$

where

$$\begin{aligned} \chi(x) &= \frac{\mu\rho}{\rho - m(\phi(x) - M)} = \frac{\mu\rho(r - x_1 - \cdots - x_n)}{\rho r - (\rho + mM)(x_1 + \cdots + x_n)} \\ &\ll \frac{\mu\rho r}{\rho r - (\rho + mM)(x_1 + \cdots + x_n)}. \end{aligned}$$

This implies (3.18b).

Since functions are real analytic, if they belong to some class $C_{M,r}$ uniformly for all points in a neighborhood of a given point, we have immediately:

Theorem. *If $f(x)$ and $g(u)$ are real analytic, then $g(f(x))$ is real analytic at all x for which $f(x)$ lies in the domain of g.*

PROBLEMS

1. Show that the functions $f(x) = 1/x$ and $f(x) = \sqrt{x}$ are real analytic for real $x > 0$ by showing that for $y > 0$ we have $f \in C_{M,r}(y)$ for appropriate M,r.

2. Show that a scalar rational function of $x = (x_1, \ldots, x_n)$ is real analytic wherever the denominator does not vanish.

3. Let p denote a positive real number. Show that the function $f: \mathbb{R} \to \mathbb{R}$ defined by

$$f(x) = \exp(-x^{-p}) \quad \text{for } x > 0 \tag{3.19a}$$

$$f(x) = 0 \quad \text{for } x \leqslant 0 \tag{3.19b}$$

belongs to $C^\infty(\mathbb{R})$, but is not a real analytic at $x = 0$.

4. (a) Show that the function

$$f(x) = \sum_{n=1}^{\infty} \frac{\cos(n!x)}{(n!)^n}$$

belongs to $C^\infty(\mathbb{R})$ and has period 2π.

(b) Show that f is not real analytic at $x = 0$. [Hint: $f \notin C_{M,r}(0)$.]

(c) Show that f is not real analytic at any real x. [Hint: $f(x + 2\pi n/m) - f(x)$ is analytic for any integers n,m with $m \neq 0$.]

5. Let f,g be real analytic in $x = (x_1, \ldots, x_n)$ at the origin, and represented by their respective power series in x for

$$|x_1| < r, \ldots, |x_n| < r. \tag{3.19c}$$

Let $f \ll g$. Show that

$$|f(x_1, \ldots, x_n)| \leqslant g(|x_1|, \ldots, |x_n|) \tag{3.19d}$$

for all x satisfying (3.19c).

6. (Implicit function theorem for real analytic functions). Let

$$f(x,y) = (f_1, \ldots, f_m) = f(x_1, \ldots, x_n, y_1, \ldots, y_m)$$

be real analytic at the point (x^0,y^0) of $\mathbb{R}^n \times \mathbb{R}^m$. Let

$$f(x^0,y^0) = 0; \quad \det(f_y(x^0,y^0)) \neq 0$$

where f_y is the square matrix formed by the $\partial f_i/\partial y_k$. Show that $f(x,y) = 0$ has a solution $y = y(x)$ which is real analytic at x^0 with $y^0 = y(x^0)$. [Hint: (a) By a linear substitution in x,y reduce the equation to the form $y = g(x,y)$ where $g(x,y)$ is real analytic at (0,0), and $g(0,0) = 0$, $g_y(0,0) = 0$. (b) Show that necessary and sufficient for a function $y = h(x)$ analytic at 0 to satisfy $y = g(x,y)$ are a set of recursion relations

$$c_\alpha = P(D^\beta \delta^\gamma g(0,0), c_\varepsilon); \qquad c_0 = 0 \tag{3.20}$$

where $c_\alpha = D^\alpha h(0)$, $|\beta + \gamma| \leqslant |\alpha|$, $|\varepsilon| < |\alpha|$, and δ denotes y-differentiation, and P is a polynomial with non-negative coefficients in its arguments. (c) Let $g(x,y) \ll G(x,y)$ where G is real analytic at (0,0), $G(0,0) = 0$, $G_y(0,0) = 0$. Let $y = H(x)$ be a solution of $y = G(x,y)$ analytic at 0. Show that $|c_\alpha| \leqslant D^\alpha H(0)$ for the c_α obtained from (3.20) by recursion, and show that $\sum c_\alpha x^\alpha$ defines an analytic solution of $y = g(x,y)$. (d) Use a $G(x,y) = (G_1, \ldots, G_m)$ of the form

$$G_k(x,y) = \frac{Mr}{r - x_1 - \cdots - x_n - y_1 - \cdots - y_m} - M - \frac{M}{r}(y_1 + \cdots + y_m)].$$

7. Let the scalar $f(x)$ for $x \in \mathbb{R}$ with $|x| < 1$ be defined by

$$f(x) = c \sum_{k=1}^{\infty} \frac{x^k}{k^2}$$

where c is a positive constant. Show that $f^2 \ll f$ if c is sufficiently small.

(c) *Analytic and real analytic functions*

The domain of an analytic function lies in the space $\mathbb{C}^n$ of vectors $x = (x_1, \ldots, x_n)$ with n *complex* components x_k. For $x \in \mathbb{C}^n$, $\alpha \in Z^n$ we again define x^α by (1.1c). We then define analytic functions of a complex vector $x \in \mathbb{C}^n$ with values in a space $\mathbb{C}^m$ in complete analogy with real analytic functions. Thus for f whose domain is an open set $\Sigma \subset \mathbb{C}^n$, we call f analytic in Σ (in symbols: $f \in C^a(\Sigma)$), if for each $y \in \Sigma$ we can represent f in the form (3.8) in a neighborhood of y. Here $c_\alpha \in \mathbb{C}^m$. It follows again that f can be differentiated arbitrarily often and that the coefficients c_α are given by

$$c_\alpha = \frac{1}{\alpha!} D^\alpha f(y). \tag{3.21}$$

Analytic functions also share with real analytic ones the unique continuation property expressed by the theorem on p. 65, as long as the domain Σ is connected. There is, however, a fundamental difference expressed by the following theorem.

Theorem. *A function $f(x)$ with range in $\mathbb{C}^m$ and domain in $\mathbb{C}^n$ is analytic in the open set Σ iff f is continuous and has continuous first derivatives with respect to the independent variables.**

PROOF. Let $y \in \Sigma$ and let the closure of the "polydisc"

$$\sigma = \{x \,|\in \mathbb{C}^n; |x_k - y_k| < r \quad \text{for } k = 1, \ldots, n\} \tag{3.22a}$$

lie in Σ. By repeatedly applying the ordinary Cauchy formula for analytic functions of a single variable we find that for any $x \in \sigma$

$$f(x) = (2\pi i)^{-n} \int_{\gamma_1} \frac{dz_1}{z_1 - x_1} \int_{\gamma_2} \frac{dz_2}{z_2 - x_2} \cdots \int_{\gamma_n} \frac{dz_n}{z_n - x_n} f(z_1, \ldots, z_n) \tag{3.22b}$$

* The assumption of continuity of f and its first derivatives is unnecessary by Hartog's theorem. See [17].

where the γ_k are the positively oriented circles $|z_k - y_k| = r$. For short (see (3.2a))

$$f(x) = (2\pi i)^{-n} \int_\Gamma \frac{f(z)}{(z-x)^1}\, dz_1 \cdots dz_n \tag{3.22c}$$

with $\Gamma = \gamma_1 \times \gamma_2 \times \cdots \times \gamma_n$. Since by (3.2a) with x_k replaced by $\zeta_k = (x_k - y_k)/(z_k - y_k)$

$$\frac{1}{(z-x)^1} = \frac{1}{(z-y)^1(1-\zeta)^1} = \sum_\alpha \frac{(x-y)^\alpha}{(z-y)^{1+\alpha}} \tag{3.22d}$$

uniformly for a fixed $x \in \sigma$ for all $z \in \Gamma$, we find that for $x \in \sigma$

$$f(x) = \sum_\alpha c_\alpha (x-y)^\alpha \tag{3.22e}$$

with

$$c_\alpha = (2\pi i)^{-n} \int_\Gamma \frac{f(z)}{(z-y)^{1+\alpha}}\, dz_1 \cdots dz_n \tag{3.22f}$$

which shows that f is analytic in Σ. □

Let $|f_k(x)| \leqslant M$ for $x \in \Gamma$, $k = 1,\ldots,m$. From (3.22c) we get the *Cauchy derivative estimate*

$$\begin{aligned} |D^\alpha f_k(y)| &= \left| \frac{\alpha!}{(2\pi i)^n} \int_\Gamma \frac{f_k(z)}{(z-y)^{1+\alpha}}\, dz_1 \cdots dz_n \right| \\ &\leqslant M\alpha!\, r^{-|\alpha|} \leqslant M|\alpha|!\, r^{-|\alpha|}. \end{aligned} \tag{3.23}$$

Another trivial consequence of the Cauchy representation (3.22c) is that *the limit of a sequence of analytic functions that converges uniformly in an open set is itself analytic.*

It is obvious that the real part of an analytic f is real analytic for real arguments. More precisely, if $f(x)$ for a certain $y \in \mathbb{R}^n$ is analytic in the polydisc

$$\Sigma = \{x \mid x \in \mathbb{C}^n;\ |x_k - y_k| < r \quad \text{for } k = 1,\ldots,n\}, \tag{3.24a}$$

then $\operatorname{Re}(f)$ is real analytic in the set

$$S = \{x \mid x \in \mathbb{R}^n;\ |x_k - y_k| < r \quad \text{for } k = 1,\ldots,n\}. \tag{3.24b}$$

If, moreover, $|f_k(x)| \leqslant M$ for $x \in \Sigma$ and all k, then the restriction of f to real arguments belongs to $C_{M,r}$, as a consequence of (3.23).

Conversely, if f is real analytic at a point y and $f \in C_{M,r}(y)$, then the expansion of f in powers of $x - y$ converges and defines an analytic function in the set of x with (see (3.3b))

$$s = |x_1 - y_1| + \cdots + |x_n - y_n| < r. \tag{3.25a}$$

Moreover, in that set

$$|f_k(x)| \leqslant M \sum_\alpha \frac{|\alpha|!}{\alpha!} r^{-|\alpha|} |(x-y)^\alpha| = \frac{Mr}{r-s}. \tag{3.25b}$$

Thus for the real analytic functions belonging to $C_{M,r}(y)$ the parameter r indicates how far the functions can be extended into the complex, while M gives a measure for the size of the extended functions.

More generally, let $f \in C^\omega(\Omega)$, where Ω is an open set in $\mathbb{R}^n$, and let S be a compact subset of Ω. Then there are M,r such that $f \in C_{M,r}(y)$ for all $y \in S$. We can assume here that r is so small that the set (3.25a) lies in Ω for real x. In the complex set (3.25a) then the series

$$\sum_\alpha \frac{1}{\alpha!} (D^\alpha f(y))(x-y)^\alpha \tag{3.26a}$$

defines an analytic function $F^y(x)$ with $F^y(x) = f(x)$ when x is a *real* point of the set (3.25a). (See problem 4, p. 67). The union U of a suitable finite number of the sets (3.25a) corresponding to points $y^1, \ldots, y^N$ contains S. Then the various functions $F^y(x)$ corresponding to $y = y^1, \ldots, y^N$, (denoted by $F^1, \ldots, F^n$) uniquely define an analytic function $F(x)$ in the set U. For if the convex sets

$$\sum_k |x_k - y_k^i| < r \quad \text{and} \quad \sum_k |x_k - y_k^j| < r \tag{3.26b}$$

in $\mathbb{C}^n$ have a non-empty intersection σ, then σ also contains a real point z. At that point $D^\alpha F^i(z) = D^\alpha F^j(z) = D^\alpha f(z)$ for all α. Since σ is convex and connected it follows that $F^i(x) = F^j(x)$ throughout σ. Since U contains a complex neighborhood of the real set S, we have:

Theorem. *If $f \in C^\omega(\Omega)$ with $\Omega \subset \mathbb{R}^n$, then for every compact subset S of Ω there exists a neighborhood Σ of S in $\mathbb{C}^n$ and a function $F \in C^a(\Sigma)$, such that $F(x) = f(x)$ for all $x \in S$.*

PROBLEMS

1. Let f be a scalar analytic function of $x \in \mathbb{C}^n$. Put $x = y + iz$ with $y,z \in \mathbb{R}^n$ and $f = u + iv$ with $u,v \in \mathbb{R}$.

(a) Show that u,v satisfy the system of Cauchy–Riemann equations

$$\frac{\partial u}{\partial y_k} = \frac{\partial v}{\partial z_k}, \frac{\partial u}{\partial z_k} = -\frac{\partial v}{\partial y_k}; \qquad k = 1, \ldots, n. \tag{3.27}$$

(b) Show that the function $u(y_1, \ldots, y_n, z_1, \ldots, z_n)$ is real analytic in its arguments. (Hint: Use $|a| + |b| \leqslant \sqrt{2}|a + ib|$ for $a, b \in \mathbb{R}$.)

2. Let $u(y_1, y_2, z_1, z_2)$ be a scalar real-valued C^2-function of its arguments in a ball in $\mathbb{R}^4$. Let u satisfy the system of equations

$$u_{y_1y_1} + u_{z_1z_2} = u_{z_1y_2} - u_{z_2y_1} = u_{y_1y_2} + u_{z_1z_2} = u_{y_2y_2} + u_{z_1z_2} = 0. \tag{3.28}$$

Show that u is the real part of an analytic function of $(y_1 + iz_1, y_2 + iz_2)$. [Hint: find v from its differential dv in terms of derivatives of u.]

3. (See problem 3, p. 69.) Let p denote a positive real number and $g\colon \mathbb{R} \to \mathbb{R}$ be the function defined by

$$g(x) = \exp(-x^{-p}) \quad \text{for } x > 0; \qquad g(x) = 0 \quad \text{for } x \leqslant 0. \tag{3.29a}$$

(a) Show that we can find a $\theta = \theta(p)$ with $0 < \theta < 1$, such that

$$|g^{(k)}(y)| \leqslant \frac{k!}{(\theta y)^k} \exp(-\tfrac{1}{2} y^{-p}) \quad \text{for } y > 0 \tag{3.29b}$$

(Hint: Find bounds for $g^{(k)}(y)$ from Cauchy's derivative formula (3.23) applied to $g = \exp(-x^{-p})$ in the complex x-plane, using as path of integration a circle γ with center y and radius θy, with θ so small that $\operatorname{Re}(x^{-p}) > \frac{1}{2} y^{-p}$ on γ.)

(b) Show that there exist M, r only depending on p, such that

$$|g^{(k)}(y)| \leqslant M(k!)^{1+1/p} r^{-k} \quad \text{for all } k \tag{3.29c}$$

for all $x \in \mathbb{R}$. (Functions g that satisfy (3.29c) for all x in an interval form the *Gevrey class* $1 + 1/p$. Real analytic functions are of class 1.)

4. A function $f\colon \mathbb{R}^n \to \mathbb{R}$ is *entire* if it is represented by a power series (3.7a) for all $x \in \mathbb{R}^n$.

(a) Show that for an entire f and any $x, y \in \mathbb{R}^n$ formula (3.9a) holds.

(b) Show that $f \in C^\omega(\mathbb{R}^n)$ is entire if and only if for every $y \in \mathbb{R}^n$ and $r > 0$ there exists an M such that

$$f \in C_{M,r}(y).$$

(c) For $x, \xi \in \mathbb{R}^n, \alpha \in Z^n, r > 0$ find an M such that $f \in C_{M,r}(0)$ when $f(x) = x^\alpha$ and when $f(x) = \cos(\sum_k \xi_k x_k)$.

5. Let $\sum$ be an open set in $\mathbb{C}^n$ and S a compact set in $\mathbb{R}^m$. Let $f(x,y)$ be defined for $x \in \sum$, $y \in S$, and let f be continuous as a function of $(x,y) \in \sum \times S$, and belong to $C^a(\sum)$ as a function of x for any fixed $y \in S$. Show that

$$F(x) = \int_S f(x,y)\, dy_1 \cdots dy_m$$

belongs to $C^a(\sum)$. (Hint: Represent F as uniform limit of analytic functions.)

(d) *The proof of the Cauchy–Kowalevski theorem*

The theorem concerns the existence of a real analytic solution of the Cauchy problem for the case of real analytic data and equations. We restrict ourselves to quasi-linear systems of type (2.2), since more general nonlinear systems can be reduced to quasi-linear ones by differentiation. We assume that the initial surface S is real analytic in a neighborhood of one of its points x^0, that is near x^0 the surface S is given by an equation $\phi(x) = 0$, where ϕ is real analytic at x^0 and $D\phi \neq 0$ at x^0, say, $D_n \phi \neq 0$. On S we prescribe compatible Cauchy data $D^\beta u$ for $|\beta| < m$ which shall be real analytic at x^0, that is represented locally by power series in $x_1 - x_1^0, \ldots, x_{n-1} - x_{n-1}^0$. The coefficients A_α, C shall be real analytic functions of their arguments x, $D^\beta u$ at the point x^0, $D^\beta u^0$, that is given by convergent power series in $x - x^0$, $D^\beta u - D^\beta u^0$

near x^0, $D^\beta u^0$, where $D^\beta u^0$ is the value of $D^\beta u$ corresponding to the Cauchy data at x^0. We assume, moreover, S to be non-characteristic at x^0 (and hence in a neighborhood of x^0) in the sense that $Q(D\phi) \neq 0$, with Q defined by (2.14), (2.11). Then the Cauchy–Kowalevski theorem asserts that there exists a unique solution which is real analytic at x^0.

The proof of this general theorem consists in showing that all coefficients of the expansion of a prospective solution u in powers of $x - x^0$ can be obtained uniquely by successive differentiation from the differential equation and the Cauchy data, and that the resulting series actually converges to a solution.

The proof becomes easier, if we reduce the problem before constructing the power series. First of all, one transforms x^0 into the origin and S locally by an analytic transformation into a neighborhood of the origin in the plane $x_n = 0$. Then by introducing derivatives of orders less than m as new dependent variables one reduces the system to one of first order. We make use here of the fact that the set of real analytic functions is closed under differentiation and composition. One arrives at a first order system in which the coefficient matrix of the term with $\partial u/\partial x_n$ is non-generate because S is non-characteristic. Hence we can solve for $\partial u/\partial x_n$, obtaining a system in the *standard form*

$$\frac{\partial u}{\partial x_n} = \sum_{i=1}^{n-1} a^i(x,u) \frac{\partial u}{\partial x_i} + b(x,u) \tag{3.30}$$

where the $a^i(x,u)$ are square matrices (a^i_{jk}), and $b(x,u)$ is a column vector with components b_i. On $x_n = 0$, near 0, we have prescribed initial values $u = f(x_1,\ldots,x_{n-1})$. Here we can assume that $f = 0$, introducing $u - f$ as the new unknown function. We can add x_n as an additional dependent variable u^* or component of u satisfying the equation $\partial u^*/\partial x_n = 1$, and the initial condition $u^* = 0$. This has the effect that the a^i and b in (3.30) do not depend on x_n. Writing (3.30) componentwise, we only have to prove the following version of the Cauchy–Kowalevski theorem:

Theorem. *Let the a^i_{jk} and b_j be real analytic functions of $z = (x_1,\ldots,x_{n-1}, u_1,\ldots,u_N)$ at the origin of $\mathbb{R}^{N+n-1}$. Then the system of differential equations*

$$\frac{\partial u_j}{\partial x_n} = \sum_{i=1}^{n-1} \sum_{k=1}^{N} a^i_{jk}(z) \frac{\partial u_k}{\partial x_i} + b_j(z) \quad \text{for } j = 1,\ldots,N \tag{3.31a}$$

with initial conditions

$$u_j = 0 \quad \text{for } x_n = 0; \qquad j = 1,\ldots,N \tag{3.31b}$$

has a unique system of solutions $u_j(x_1, \ldots, x_n)$ that is real analytic at the origin.*

* i.e., unique among the real analytic u_j.

PROOF. For any solutions $u_j(x)$ of (3.31a) that are real analytic at the origin and any $\alpha \in Z^n$ we obtain by applying D^α and setting $x = 0$ relations of the form

$$D_n D^\alpha u_j(0) = P_\alpha(d^\beta a^i_{jk}(0), d^\gamma b_j(0), D^\delta u_k(0)). \tag{3.32a}$$

Here d is the gradient operator with respect to z:

$$d = \left(\frac{\partial}{\partial z_1}, \ldots, \frac{\partial}{\partial z_{N+n-1}}\right); \tag{3.32b}$$

$\beta, \gamma, \delta, i, k$ range over

$$\beta, \gamma \in Z^{N+n-1}; \qquad |\beta|, |\gamma| \leqslant |\alpha|; \qquad \delta \in Z^n; \qquad |\delta| \leqslant |\alpha| + 1; \qquad \delta_n \leqslant \alpha_n \tag{3.32c}$$

$$i = 1, \ldots, n-1; \qquad j,k = 1, \ldots, N \tag{3.32d}$$

and P_α is a polynomial in its arguments, whose coefficients are non-negative integers (neither the chain rule for differentiation nor the rule for differentiating a product can lead to anything else). In addition, from (3.31b)

$$D^\alpha u_j(0) = 0 \quad \text{for } \alpha_n = 0. \tag{3.32e}$$

By induction over α_n we obtain from (3.32a,e) all $D^\alpha u_j(0)$ in terms of the $d^\beta a^i_{jk}(0)$, $d^\gamma b_j(0)$ alone. Hence the $u_j(x)$ are determined uniquely by (3.31a,b) (as far as they are represented by their power series). Conversely, if we calculate recursively quantities c^α_j from (3.32a,e) replacing everywhere the $D^\alpha u_j(0)$ by c^α_j, and if the series

$$\sum_\alpha \frac{1}{\alpha!} c^\alpha_j x^\alpha$$

converge and represent functions $u_j(x)$ near 0, then the $u_j(x)$ form a solution of (3.31a,b) real analytic at 0. Indeed then $c_j = D^\alpha u_j(0)$ is satisfied. Moreover, for real analytic $u_k(x)$ both sides of equation (3.31a) will be real analytic at 0, and (3.32a) just guarantees that the coefficients in the power series for both sides are identical.

Thus all that remains to show is that the formal power series for the $u_j(x)$, with coefficients $D^\alpha u_j(0)/\alpha!$ obtained from 3.32a,e), converge near $x = 0$. This is easily achieved by the method of majorants. Let

$$a^i_{jk}(z) \ll A^i_{jk}(z); \qquad b_j(z) \ll B_j(z) \tag{3.33a}$$

and let the $U_j(x)$ form a solution of the "majorising problem"

$$\frac{\partial U_j}{\partial x_n} = \sum_{i=1}^{n-1} \sum_{k=1}^{N} A^i_{jk}(z) \frac{\partial U_k}{\partial x_i} + B_j(z) \quad \text{for } j = 1, \ldots, N \tag{3.33b}$$

$$U_j = 0 \quad \text{for } x_n = 0; \qquad j = 1, \ldots, N \tag{3.33c}$$

which is real analytic at 0. Then clearly

$$|D^\alpha u_j(0)| \leqslant D^\alpha U_j(0) \tag{3.33d}$$

holds for the $D^\alpha u_j(0)$ computed from (3.32a,e). Hence the formal power series based on those $D^\alpha u_j(0)$ actually converges and represents a solution of our Cauchy problem in a neighborhood of the origin.

It remains to produce a majorising system whose solution is real analytic at 0. Assume that all the $a^i_{jk}(z), b_j(z)$ belong to $C_{M,r}(0)$, and hence are majorised by the function

$$\frac{Mr}{r - z_1 - \cdots - z_{N+n-1}}. \tag{3.34a}$$

Then a majorising Cauchy problem is provided by

$$\frac{\partial U_j}{\partial x_n} = \frac{Mr}{r - x_1 - \cdots - x_{n-1} - U_1 - \cdots - U_N}\left(1 + \sum_{i=1}^{n-1}\sum_{k=1}^{N}\frac{\partial U_k}{\partial x_i}\right)$$

$$U_j = 0 \quad \text{for } x_n = 0; \qquad j = 1,\ldots,N.$$

This problem has solutions of the form

$$U_j(x_1,\ldots,x_n) = V(x_1 + \cdots + x_{n-1},x_n) \quad \text{for } j = 1,\ldots,N \tag{3.34b}$$

where $V(s,t)$ as a function of $s = x_1 + \cdots + x_{n-1}, t = x_n$ is a solution of the scalar first order Cauchy problem

$$V_t = \frac{Mr}{r - s - NV}(1 + N(n-1)V_s) \tag{3.34c}$$

$$V(s,0) = 0 \tag{3.34d}$$

Here V can be found explicitly by the methods of Chapter I to be the function

$$V(s,t) = \frac{1}{Nn}\left(r - s - \sqrt{(r-s)^2 - 2nMNrt}\right), \tag{3.34e}$$

which obviously is real analytic in (s,t) at the origin.

The function $V(s,t)$ in (3.34e) depends on the parameters M,r and the integers n,N. Thus it belongs to a certain class $C_{\mu,\rho}$, with certain μ,ρ only depending on M,r,N,n. Its expansion in terms of powers of s and t converges then for $|s| + |t| < \rho$. Hence the power series for the solution of the Cauchy problem (3.31a,b) converges for

$$|x_1| + |x_2| + \cdots + |x_n| < \rho$$

where ρ only depends on the numbers of dependent and independent variables and on the class $C_{M,r}(0)$ to which the coefficients $a^i_{jk}(z)$ and $b_j(z)$ belong. □

The theorem of Cauchy and Kowalevski is *local* in character, and applies only to analytic solutions of analytic Cauchy problems. It does not guarantee *global* existence of solutions; it does not exclude the possibility that other non-analytic solutions exist for the same Cauchy problem, nor the possibility that an analytic solution becomes *non-analytic* some distance away from the initial surface. Nor is a similar general existence theorem valid, if we just restrict the data to belong to some class C^s or even to C^∞, instead of C^ω.*

Many physical problems lead to analytic partial differential equations. But the restriction to analytic Cauchy data and solutions is unrealistic. It implies that the solution is determined globally by local conditions near one point, and makes it impossible to describe phenomena that do not interact instantaneously with the whole universe (hence it excludes boundary-initial value problems). The Cauchy problem is "ill-posed" except for hyperbolic equations. Real analytic solutions, however, are useful in discussing more general solutions. In addition, some equations (the elliptic analytic ones) have the property that all solutions are analytic in the interior of their domain of existence.

The Cauchy–Kowalevski theorem gives only local existence near one point of the initial surface. However, the solution can easily be seen to exist everywhere near the whole initial surface. More precisely, let Ω be an open set in $\mathbb{R}^{n-1}$, and let the coefficients $a^i_{jk}(z)$ and $b_j(z)$ in (3.31a) be real analytic in $z = (z_1, \ldots, z_{n+N-1})$ at all points z with

$$(z_1, \ldots, z_{n-1}) \in \Omega;\ z_n = z_{n+1} = \cdots = z_{n+N-1} = 0.$$

Identify Ω with the set of points $y \in \mathbb{R}^n$ for which

$$(y_1, \ldots, y_{n-1}) \in \Omega, \qquad y_n = 0. \tag{3.35a}$$

For any $y \in \Omega$ there exists then a solution $u^y(x)$ of (3.31a,b) represented by a convergent series in terms of powers of $x - y$ in a set

$$\sigma_y = \{x \mid x \in \mathbb{R}^n;\ |x_1 - y_1| + \cdots + |x_{n-1} - y_{n-1}| + |x_n| < \rho(y)\}. \tag{3.35b}$$

Here $\rho(y)$ can be assumed to be so small that the intersection of σ_y with the plane $x_n = 0$ lies in Ω. The various $u^y(x)$ for different y in Ω uniquely define a solution $u(x)$ of (3.31a,b) in the set

$$U = \bigcup_{y \in \Omega} \sigma_y \tag{3.35c}$$

(why?). Here U is an open set in $\mathbb{R}^n$ containing Ω. In particular, any compact subset of Ω has a neighborhood in $\mathbb{R}^n$ contained in U, and a solution of the Cauchy problem contained in that neighborhood.

How far the domain of existence of the Cauchy problem extends in the x_n-direction depends on the class $C_{M,r}$ to which the coefficients a^i_{jk} and b^i_j belong.

* In the case of *linear* equations with analytic coefficients uniqueness holds for non-analytic u (see Holmgren's theorem below). Local existence holds if the coefficients are at least analytic in the variables $x_1, \ldots, x_{n-1}$ (see [34]).

PROBLEMS

1. Let $u(x) = u(x_1, \ldots, x_n)$ denote a solution of the wave equations

$$u_{x_n x_n} = \sum_{i=1}^{n-1} u_{x_i x_i}. \tag{3.36a}$$

Consider the Cauchy problem for (3.36a) with data

$$u = f(x_1, \ldots, x_{n-1}); \qquad u_{x_n} = g(x_1, \ldots, x_{n-1}) \tag{3.36b}$$

prescribed on a hypersurface S with equation

$$x_n = \phi(x_1, \ldots, x_{n-1}). \tag{3.36c}$$

Assume that x^0 is a point of S, and that the functions f, g, ϕ are real analytic at $(x_1^0, \ldots, x_{n-1}^0)$, and that S is non-characteristic at x^0:

$$\sum_{i=1}^{n-1} \phi_{x_i}^2 \neq 1. \tag{3.36d}$$

Reduce this Cauchy problem explicitly to the standard form (3.31a,b).

2. Find explicitly a value $\rho = \rho(n,N,M,r)$ such that a solution of (3.31a,b) exists and is represented by a power series for

$$|x_1| + \cdots + |x_{n-1}| < \rho, |x_n| < \rho \tag{3.37}$$

for any real analytic $a_{jk}^i(z)$, $b_k^i(z)$ belonging to $C_{M,r}(0)$. [Hint: Find ρ such that the function $V(s,t)$ in (3.34e) is analytic for complex s,t with $|s| < \rho, |t| < \rho$].

3. Let $u = (u_1, \ldots, u_N) = u(x_1, \ldots, x_n)$. Consider the non-linear differential equations

$$\frac{\partial u_j}{\partial x_n} = G_j\left(x_p, u_q, \frac{\partial u_i}{\partial x_k}\right) \quad \text{for } j = 1, \ldots, N \tag{3.38a}$$

$(p,k = 1, \ldots, n-1;\ i,q = 1, \ldots, N)$ with initial conditions

$$u_j(x_j, \ldots, x_{n-1}, 0) = 0 \quad \text{for } j = 1, \ldots, N \tag{3.38b}$$

where the G_j are real analytic at the origin in their arguments. Show *directly*, by the method of majorants (without reduction to the quasi-linear case), that this Cauchy problem has a solution which is real analytic at the origin.

4. Let $u(x_1,x_2)$ denote a solution of the Laplace equation $u_{x_1x_1} + u_{x_2x_2} = 0$, referred to polar coordinates $x_1 = r \cos \theta$, $x_2 = r \sin \theta$. Prescribe Cauchy data on the unit circle

$$u = f(\theta), \qquad \frac{\partial u}{\partial r} = g(\theta) \quad \text{for } r = 1,$$

where f,g are real analytic and of period 2π for all real θ.

(a) Show that a real analytic solution u exists for all real θ and $|r - 1|$ sufficiently small.

(b) Show that in the case where f,g are trigonometric polynomials in θ, the solution exists in the whole x_1x_2-plane with the origin deleted. [Hint: Use special solutions $e^{in\theta}r^{\pm n}$].

5. Let $f(z)C^a(\sum)$ where $\sum$ is the unit disc $|z| < 1$ in $\mathbb{C}^1$. Let $u(x,y) = \mathrm{Re}(f(x + iy))$ for points $(x,y) \in \mathbb{R}^2$ with $x^2 + y^2 < 1$.

(a) Show that $u(x,y)$ is represented by its power series

$$\sum_{j,k=0}^{\infty} c_{jk} x^j y^k$$

for $|x| + |y| < 1$.

(b) Show that in the example $f(z) = 1/(1 - z)$, the series for $u = (1 - x)/((1 - x)^2 + y^2)$ does not converge (absolutely) for $x = y = r > \frac{1}{2}$. [Hint: Show that here

$$\sum_k c_{kk} r^{2k}$$

diverges.]

4. The Lagrange–Green Identity

We recall the Gauss divergence theorem:

$$\int_\Omega D_k u(x)\,dx = \int_{\partial\Omega} u(x) \frac{dx_k}{dn}\,dS_x = \int_{\partial\Omega} u(x)\zeta_k\,dS_x, \tag{4.1}$$

where d/dn denotes differentiation in the direction of the exterior unit normal $\zeta = (\zeta_1, \ldots, \zeta_n)$ of $\partial\Omega$ and $dx = dx_1 \ldots dx_n$, dS_x = surface element with integration on x. We always assume the boundary $\partial\Omega$ of our region to be sufficiently regular so that the divergence theorem applies to all $u \in C^1(\bar{\Omega})$. The theorem can be generalized to $u \in C^1(\Omega) \cap C^0(\bar{\Omega})$ by approximating Ω from the interior. More generally, we have the formula for integration by parts,

$$\int_\Omega v^{\mathrm{T}} D_k u\,dx = \int_{\partial\Omega} v^{\mathrm{T}} u \zeta_k\,dS_x - \int_\Omega (D_k v^{\mathrm{T}}) u\,dx, \tag{4.2}$$

where u, v are column vectors belonging to $C^1(\Omega)$ with T denoting transposition.

Let now L be a linear differential operator

$$Lu = \sum_{|\alpha| \leq m} a_\alpha(x) D^\alpha u. \tag{4.3}$$

Let u, v be column vectors and a_α be square matrices in $C^m(\bar{\Omega})$. Then by repeated application of (4.2) it follows that

$$\int_\Omega v^{\mathrm{T}} \sum_{|\alpha| \leq m} a_\alpha(x) D^\alpha u\,dx$$

$$= \int_\Omega \sum_{|\alpha| \leq m} (-1)^{|\alpha|} D^\alpha(v^{\mathrm{T}} a_\alpha(x)) u\,dx + \int_{\partial\Omega} M(v, u, \zeta)\,dS_x. \tag{4.4}$$

Here M in the surface integral is linear in the ζ_k with coefficients which are bilinear in the derivatives of v and u, the total number of differentiations in each term being at most $m-1$. The expression M is not determined uniquely but depends on the order of performing the integration by parts. This is the *Lagrange–Green identity* for L which we also write in the form

$$\int_\Omega v^{\mathrm{T}} Lu\,dx = \int_\Omega (\tilde{L}v)^{\mathrm{T}} u\,dx + \int_{\partial\Omega} M(v,u,\zeta)\,dS_x, \tag{4.5}$$

where $\tilde{L}$ is the (formally) adjoint operator to L, defined by

$$\tilde{L}v = \sum_{|\alpha| \leq m} (-1)^{|\alpha|} D^\alpha\big(a_\alpha(x)^{\mathrm{T}} v\big). \tag{4.6}$$

The characteristic forms of L and $\tilde{L}$ differ at most in sign.

The simplest example corresponds to the Laplace operator $L=\Delta$ for scalars u and v. Then one integration by parts yields

$$\int_\Omega v\Delta u\,dx = \int_{\partial\Omega} \sum_i v u_{x_i} \zeta_i\,dS_x - \int_\Omega \sum_i v_{x_i} u_{x_i}\,dx. \tag{4.7}$$

We write this as

$$\int_\Omega v\Delta u\,dx = \int_{\partial\Omega} v \frac{du}{dn}\,dS_x - \int_\Omega \sum_i v_{x_i} u_{x_i}\,dx. \tag{4.8}$$

Integrating once more by parts we obtain

$$\int_\Omega v\Delta u\,dx = \int_\Omega u\Delta v\,dx + \int_{\partial\Omega} \left(v\frac{\partial u}{\partial n} - u\frac{\partial v}{\partial n}\right) dS_x. \tag{4.9}$$

5. The Uniqueness Theorem of Holmgren

It is clear from the arguments used in the proof of the Cauchy–Kowalevski theorem that an analytic Cauchy problem with data prescribed on an analytic noncharacteristic surface S has at most one *analytic* solution u, since the coefficients of the power series for u are determined uniquely. This does not exclude the possibility that other *nonanalytic* solutions of the same problem might exist. However, uniqueness can be proved for the Cauchy problem for a *linear* equation with analytic coefficients and for data (not necessarily analytic) prescribed on an analytic noncharacteristic surface S. The method of proof (due to Holmgren) makes use of the Cauchy–Kowalevski theorem and the Lagrange–Green identity. (Extension of the uniqueness theorem to *nonanalytic* equations is much more difficult).

The principle of Holmgren's uniqueness argument is simple. Let u be a solution of a first order linear system

$$Lu = \sum_{k=1}^{n} a^k(x) \frac{\partial u}{\partial x_k} + b(x)u = 0 \tag{5.1a}$$

in a "lens-shaped" region R bounded by two hypersufaces S and Z (see Fig. 3.1). Here $x \in \mathbb{R}^n$, $u \in \mathbb{R}^N$ and the a^k, b are $N \times N$ matrices. Assume that u has Cauchy data $u = 0$ on Z and that S is non-characteristic; that is, the matrix

$$A = \sum_{k=1}^{n} a^k(x)\zeta_k \tag{5.1b}$$

is non-degenerate for $x \in S$, and ζ = unit normal of S at x. Let v be a solution of the *adjoint* equation

$$\tilde{L}v = -\sum_{k=1}^{n} \frac{\partial}{\partial x_k}((a^k)^T v) + b^T v = 0 \quad \text{for } x \in \mathbb{R} \tag{5.1c}$$

(T for transposition) with Cauchy data

$$v = w(x) \quad \text{for } x \in S. \tag{5.1d}$$

Applying the Lagrange–Green identity (4.5) we find that

$$\int_S w^T A u \, dS = 0. \tag{5.1e}$$

Let now Γ be the set of functions w on S for which the Cauchy problem (5.1c,d) has a solution v. If Γ is dense in $C^0(S)$ (that is if every continuous function on S can be approximated uniformly by functions in Γ) we conclude that (5.1e) hold for every $w \in C^0(S)$. But then $Au = 0$ on S, and hence also, since A is non-degenerate, $u = 0$ on S. For if $Au \neq 0$ for some $z \in S$, then also $Au \neq 0$ for all x in a neighborhood ω of z on S. We can find a continuous non-negative scalar function $\phi(x)$ on S with support in ω and with $\phi(z) > 0$. Then

$$\int_S \phi (Au)^T (Au) \, dS > 0 \tag{5.1f}$$

for $w = \phi A u$ contrary to (5.1e). Now in the case where the matrices a^k and b are *real analytic*, and S and w are real analytic, the Cauchy–Kowalevski theorem guarantees the existence of a solution v of $\tilde{L}v = 0$ with $v = w$ on S

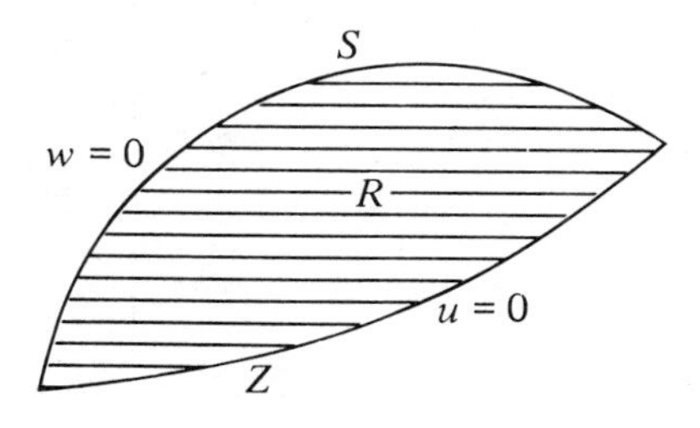

Figure 3.1

in a *sufficiently small* neighborhood of S, though we cannot be sure that that neighborhood includes all of R. To bridge the gap between S and Z and to conclude that $u = 0$ throughout R, we have to cover all of R by an analytic family of non-characteristic surfaces S_λ. Making these notions more precise we are led to the following definition of a family of hypersurfaces forming an *analytic field* of surfaces, and to the general uniqueness theorem below.

Definition. A family of hypersurfaces S_λ in $\mathbb{R}^n$ with parameter λ ranging over an open interval $\Lambda = (a,b)$ forms an *analytic field*, if the S_λ can be transformed bi-analytically into the cross sections of a cylinder whose base is the unit ball Ω in $\mathbb{R}^{n-1}$. This means that there shall exist a $1-1$ mapping $F: \Omega \times \Lambda \to \mathbb{R}^n$, where $x = F(y)$ is real analytic in $\Omega \times \Lambda$ and has a non-vanishing Jacobian; the S_λ for $\lambda \in \Lambda$ shall be the sets

$$S_\lambda = \{x \mid x = F(y); (y_1, \dots, y_{n-1}) \in \Omega; y_n = \lambda\}. \tag{5.2a}$$

(Our conditions imply that the set

$$\Sigma = \bigcup_{\lambda \in \Lambda} S_\lambda, \tag{5.2b}$$

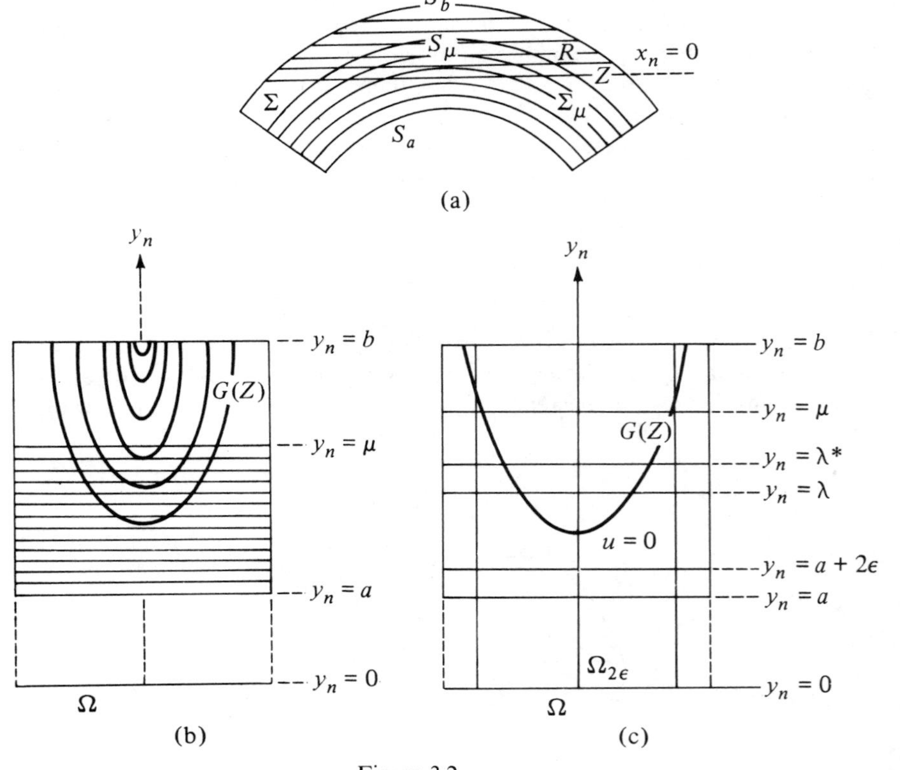

Figure 3.2

called the *support* of the field, is open, and that the transformation $x = F(y)$ has a real analytic inverse $y = G(x)$ mapping Σ onto $\Omega \times \Lambda$. In particular, $\lambda(x) = G_n(x)$ is real analytic in Σ.)

Uniqueness Theorem *(See Fig. 3.2a,b,c). Let the S_λ for $\lambda \in \Lambda = (a,b)$ form an analytic field in $\mathbb{R}^n$ with support Σ. Consider the m-th order linear system*

$$Lu = \sum_{|\alpha| \leqslant m} A_\alpha(x) D^\alpha u = 0 \tag{5.3a}$$

where $x \in \mathbb{R}^n$, $u \in \mathbb{R}^N$, and the coefficient matrices $A_\alpha(x)$ are real analytic in Σ. Introduce the sets

$$R = \{x \mid x \in \Sigma;\ x_n \geqslant 0\} \tag{5.3b}$$

$$Z = \{x \mid x \in \Sigma;\ x_n = 0\} \tag{5.3c}$$

and for $\mu \in \Lambda$

$$\Sigma_\mu = \{x \mid x \in S_\lambda \quad \text{for some } \lambda \text{ with } a < \lambda \leqslant \mu\}. \tag{5.3d}$$

We assume that Z and all S_λ are non-characteristic, with respect to L, and that $\Sigma_\mu \cap R$ for any $\mu \in \Lambda$ is a closed subset of the open set Σ. Let u be a solution of (5.3a) of class $C^m(R)$ and have vanishing Cauchy data on Z. Then $u = 0$ in R.*

Proof. Since Z is non-characteristic it follows from the vanishing of the Cauchy data of u on Z that also $D^\alpha u = 0$ on Z for $|\alpha| = m$. Define $u(x) = 0$ at all x of Σ not belonging to R. Then the extended function u is of class C^m and a solution of (5.3a) throughout Σ. Moreover, for any $\mu \in \Lambda$ the closure of the set of points $x \in \Sigma_\mu$ where $u(x) \neq 0$ is a closed subset of Σ. We only have to prove $u \equiv 0$ in Σ, ignoring Z. We apply the mapping $x = F(y)$ associated with the analytic field. Analyticity and non-characteristic behavior are preserved. Renaming the new independent variables x instead of y, and letting (5.3a) stand for the transformed differential equations, we now have to deal with the family of surfaces

$$S_\lambda = \{x \mid (x_1, \ldots, x_{n-1}) \in \Omega;\ x_n = \lambda\} \tag{5.4a}$$

whose support is the set $\Sigma = \Omega \times \Lambda$. The S_λ are non-characteristic, that is,

$$\det A_\alpha(x) \neq 0 \quad \text{for } \alpha = (0, \ldots, 0, m). \tag{5.4b}$$

$u(x)$ is a solution of class C^m of (5.3a) for $x \in \Sigma$. Any limit x of points where $u \neq 0$ either lies in Σ or has $x_n = b$.

We introduce an auxiliary solution v of an adjoint Cauchy problem. Let $w(x_1, \ldots, x_{n-1})$ denote a vector (fixed for the present) whose components are polynomials. Let $v = v(x,\lambda)$ denote the solution of the adjoint equation

$$\tilde{L}v = \sum_{|\alpha| \leqslant m} (-1)^{|\alpha|} D^\alpha (A_\alpha^T(x) v) = 0 \tag{5.4c}$$

* That means that u is of class C^m in all interior points of R, that all $D^\alpha u$ for $|\alpha| \leqslant m$ can be extended continuously to all of R, and that the extended values satisfy $D^\alpha u = 0$ on Z for $|\alpha| \leqslant m - 1$.

with Cauchy data prescribed on the plane $x_n = \lambda$:

$$D_n^k v(x_1, \ldots, x_{n-1}, \lambda, \lambda,) = 0 \quad \text{for } 0 \leqslant k < m - 1 \tag{5.4d}$$

$$D_n^{m-1} v(x_1, \ldots, x_{n-1}, \lambda, \lambda) = w(x_1, \ldots, x_{n-1}). \tag{5.4e}$$

In this problem the initial surface S_λ and hence the solution v depend on the parameter λ. We transform this Cauchy problem for v into one with data given on a fixed plane $x_n = 0$ by replacing x_n by $x_n + \lambda$ and considering λ just as an additional independent variable. Writing

$$V(x, \lambda) = v(x_1, \ldots, x_{n-1}, x_n + \lambda, \lambda) \tag{5.4f}$$

$$a_\alpha(x, \lambda) = A_\alpha(x_1, \ldots, x_{n-1}, x_n + \lambda) \tag{5.4g}$$

we have

$$\sum_{|\alpha| \leqslant m} (-1)^{|\alpha|} D^\alpha (a_\alpha^{\mathrm{T}}(x, \lambda) V) = 0 \tag{5.4h}$$

$$D_n^k V(x_1, \ldots, x_{n-1}, 0, \lambda) = 0 \quad \text{for } 0 \leqslant k < m - 1 \tag{5.4i}$$

$$D_n^{m-1} V(x_1, \ldots, x_{n-1}, 0, \lambda) = w(x_1, \ldots, x_{n-1}). \tag{5.4j}$$

Here the coefficient matrices $a_\alpha(x, \lambda)$ are real analytic in the set

$$(x_1, \ldots, x_{n-1}) \in \Omega; \qquad a < x_n + \lambda < b; \qquad a < \lambda < b \tag{5.4k}$$

and the initial plane $x_n = 0$ is non-characteristic with respect to (5.4h). Let Ω_ε for $0 < \varepsilon < \operatorname{Min}(1, (b - a)/2)$ denote the closed ball of radius $1 - \varepsilon$ and center at the origin in $\mathbb{R}^{n-1}$, and let Λ_ε denote the closed interval $[a + \varepsilon, b - \varepsilon]$. The set (5.4k) has the compact subset consisting of the (x, λ) with

$$(x_1, \ldots, x_{n-1}) \in \Omega_\varepsilon; \; x_n = 0; \; \lambda \in \Lambda_\varepsilon.$$

By the general Cauchy–Kowalevski theorem on p. 77 there exists a $\delta = \delta(\varepsilon) > 0$ and a solution $V(x, \lambda)$ of (5.4h,i,j) defined for all (x, λ) with

$$(x_1, \ldots, x_{n-1}) \in \Omega_\varepsilon; \qquad |x_n| < \delta; \qquad \lambda \in \Lambda_\varepsilon.$$

It follows from (5.4f) that the Cauchy problem (5.4c,d,e) has a real analytic solution $v(x, \lambda)$ for

$$(x_1, \ldots, x_{n-1}) \in \Omega_\varepsilon; \qquad |x_n - \lambda| < \delta; \qquad \lambda \in \Lambda_\varepsilon. \tag{5.41}$$

Take a value $\mu \in \Lambda$. Let ε be so small that

$$u(x) = 0 \quad \text{for } a < x_n < a + 2\varepsilon \tag{5.4m}$$

$$u(x) = 0 \quad \text{for } x \notin \Omega_{2\varepsilon}, \quad a < x_n < \mu. \tag{5.4n}$$

Let λ, λ^* denote values in the interval $(a + \varepsilon, \mu)$ with $|\lambda - \lambda^*| < \delta$. Apply the Lagrange–Green identity (4.5) to the slice of $\Omega \times \Lambda$ bounded by the planes

$x_n = \lambda$ and $x_n = \lambda^*$. We find for $v = v(x, \lambda)$ that

$$I(\lambda) = \int_{x_n=\lambda} w^T A_\alpha(x) u(x)\, dx_1 \cdots dx_{n-1}$$

$$= \int_{x_n=\lambda^*} M(v, u, \zeta)\, dx_1 \cdots dx_{n-1} \tag{5.4o}$$

where $\alpha = (0,\ldots,0,m)$ and the integrations are extended over Ω_ε. Now the last integral in (5.4o) depends on λ only through v, and hence* is a real analytic function of λ for $\lambda \in (a + \varepsilon, \mu)$, $|\lambda - \lambda^*| < \delta$. Hence $I(\lambda)$ is real analytic in λ for $\lambda \in (a + \varepsilon, \mu)$. By (5.4m) $I(\lambda) = 0$ for $\lambda \in (a + \varepsilon, a + 2\varepsilon)$. Hence $I(\lambda) = 0$ for $\lambda \in (a + \varepsilon, \mu)$. Because of the arbitrariness of μ and ε it follows that $I(\lambda) = 0$ for $\lambda \in (a,b)$. Since w is an arbitrary polynomial vector it follows that $A_\alpha(x)u(x) = 0$ for $x \in \Sigma$. But then also $u(x) = 0$ in Σ by (5.4b). □

From the general theorem just proved one can derive uniqueness theorems for "curved" initial surfaces Z by applying analytic deformations.†

Let L be a linear m-th order differential operator acting on functions $u(x_1,\ldots,x_n)$, and let Z be an $(n-1)$-dimensional manifold in $\mathbb{R}^n$. A closed set $R \subset \mathbb{R}^n$ is called a *domain of determinacy* for Z (with respect to L) if every solution u of class C^m of $Lu = 0$ in R vanishes if its Cauchy data on Z vanish.‡ The uniqueness theorem just proved permits to construct domains of determinacy with the help of suitable non-characteristic analytic fields, as will be shown by examples.

Let Z be a ball in the plane $x_n = 0$:

$$Z = \left\{x \mid \sum_{k=1}^{n-1} x_k^2 < r^2;\ x_n = 0\right\}. \tag{5.5}$$

Let L have real analytic coefficients in a neighborhood of Z in $\mathbb{R}^n$, and let Z be noncharacteristic with respect to L. Then the lens-shaped set

$$R = \left\{x \mid 0 \leqslant x_n \leqslant \varepsilon\left(r^2 - \sum_{k=1}^{n-1} x_k^2\right)\right\}$$

is a domain of determinacy for Z for all sufficiently small positive ε. To see this one only has to consider the analytic field formed by the portions of paraboloids

$$S_\lambda = \left\{x \mid x_n = \lambda + \varepsilon\left(r^2 - \sum_{k=1}^{n-1} x_k^2\right);\ \sum_{k=1}^{n-1} x_k^2 \leqslant r^2\right\}$$

* We can differentiate the last integral arbitrarily often with respect to λ and see that it belongs to some class $C_{M,r}$ uniformly in λ.

† Such theorems can also be derived directly for curved Z by the arguments applied above, imbedding Z into an analytic non-characteristic field. Here Z itself need not be analytic. Also the assumption that Z is non-characteristic can be replaced by the weaker requirement that the set of non-characteristic points is dense on Z.

‡ Z may be said to contain the *domain of dependence* of any point of R in the sense of pp. 5, 41, though a precise domain of dependence for the Cauchy problem cannot be defined in many instances, for example for elliptic equations.

for $\lambda \in \Lambda = (-\varepsilon r^2 - \varepsilon, 0)$, corresponding to $y_1 = x_1, \ldots, y_{n-1} = x_{n-1}, y_n = \lambda$. For ε sufficiently small the support of the field is contained in a given neighborhood of Z and the S_λ are non-characteristic, since their normals lie close to those of Z (Fig. 3.3).

Let next L be the wave operator:

$$Lu = u_{x_n x_n} - \sum_{k=1}^{n-1} u_{x_k x_k}. \tag{5.6a}$$

Set

$$\rho = \sqrt{\sum_{k=1}^{n-1} x_k^2}.$$

Let Z be again the $(n-1)$-dimensional ball of radius r

$$Z = \{x \mid \rho^2 < r^2; x_n = 0\}. \tag{5.6b}$$

Here we can see that the backward characteristic cone with base Z:

$$R = \{x \mid 0 \leqslant x_n \leqslant r - \rho\} \tag{5.6c}$$

is a domain of determinacy for Z. One only has to make use of the field formed by the homothetic hyperboloids of revolution

$$S_\lambda = \{x \mid (r - x_n)^2 - \rho^2 = -\lambda; \rho^2 < r^2; x_n < r\} \tag{5.6d}$$

with $\lambda \in \Lambda = (-r^2 - \varepsilon, 0)$, $\varepsilon > 0$, corresponding to $y_1 = x_1, \ldots, y_{n-1} = x_{n-1}, y_n = \lambda$. The S_λ are non-characteristic (Fig. 3.4).

Let the initial manifold Z contain two disjoint subsets Z_1 and Z_2. It can happen that there exists a domain of determinacy R for Z_1 that contains points of Z_2 (Fig. 3.5). In that case, it is impossible to solve the Cauchy problem with "arbitrary" data on Z, since prescribing the initial data on the portion Z_1 of Z leaves nothing to prescribe arbitrarily on Z_2 (at least for solutions u with domain containing R). The data can be continued uniquely from Z_1 to Z_2; the Cauchy problem is not well posed. This situation arises quite generally for all linear *elliptic* equations with analytic coefficients. Here

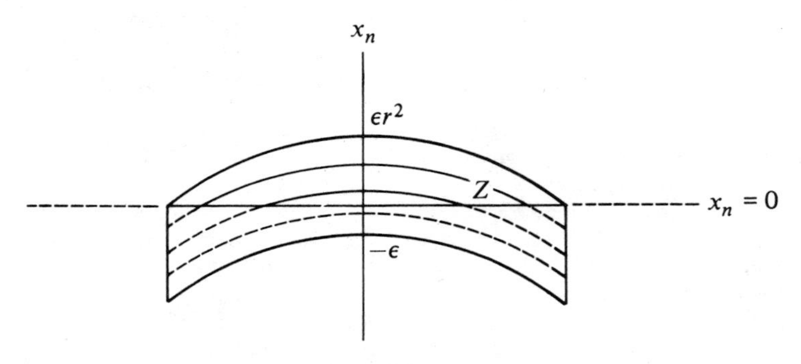

Figure 3.3

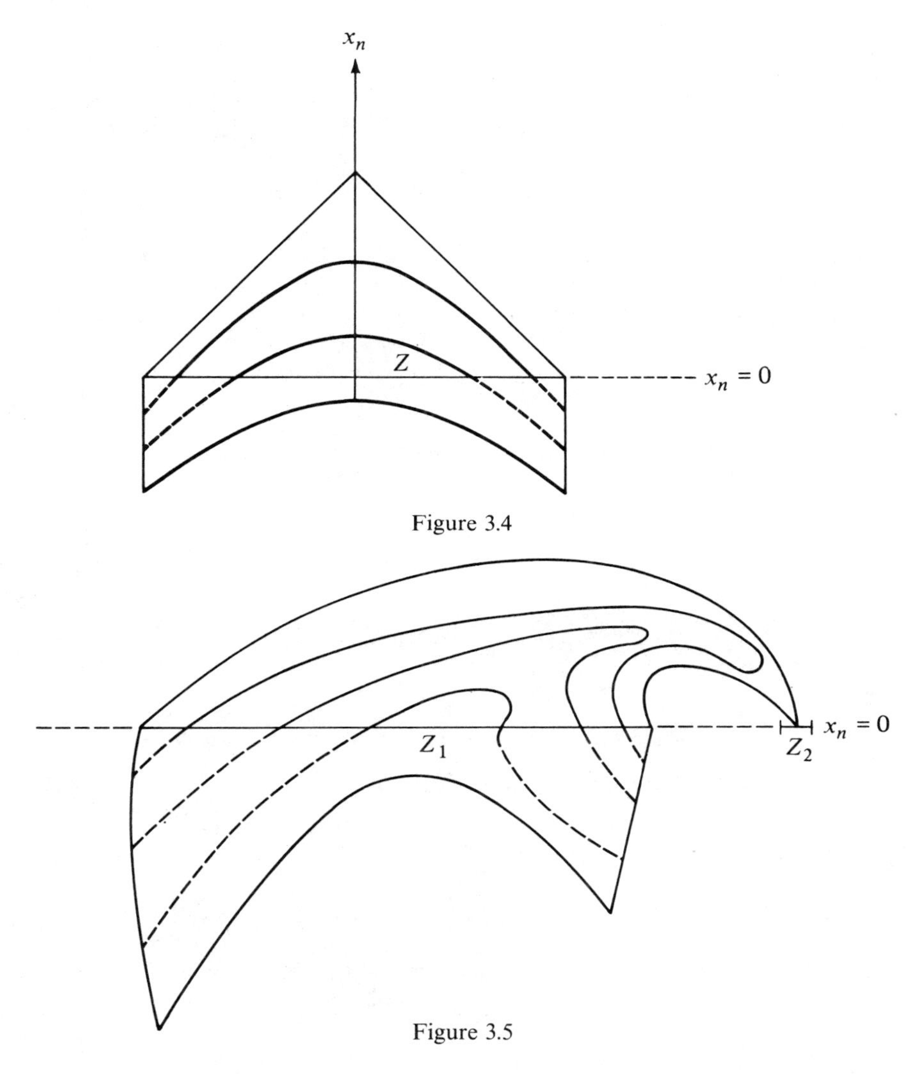

Figure 3.4

Figure 3.5

all surfaces are non-characteristic. We will only describe informally how an appropriate local analytic field can be obtained. Take the field (5.6d) for $\lambda < 0$, whose support is in the cone $x_n < r - \rho$. Deform the cone analytically into a conical region Σ that intersects the plane $x_n = 0$ in two disjoint sets Z_1 and Z_2 (see Fig. 3.5). Let R denote the portion of Σ with $x_n \geqslant 0$. Then for a solution u $C^m(R)$ the set R will be a domain of determinacy for Z_1 that contains Z_2. Notice that the Cauchy data propagate in a manner similar to analytic functions along the initial surface Z, even for solutions u that are only assumed to exist locally on one side of Z. Of course, a similar continuation process takes place when the initial surface is curved. The phenomenon is not restricted to elliptic equations, as shown by problem 3 below.

PROBLEMS

1. Let $Lu = 0$ be a linear equation or system of equations with constant coefficients, for which the plane $x_n = 0$ is non-characteristic. Let $u = u(x_1, \ldots, x_n)$ be a solution of class C^m for all x with $x_n \geqslant 0$. Show that u is determined uniquely by its Cauchy data on $x_n = 0$.

2. Let $Lu = 0$ be a linear system with constant coefficients for functions $u = u(x_1, \ldots, x_n)$. Let Z be a solid ellipsoid in the plane $x_n = 0$. Let y be a point of $\mathbb{R}^n$ with $y_n > 0$, and let R be the cone with vertex y and base Z (that is, R is the union of the closed line segments joining y to a point of Z). Let every *plane of support* of R at y (that is, every hyperplane passing through y and not containing points of Z) be non-characteristic. Then R is a domain of determinacy for Z. (Hint: Find an appropriate field of hyperboloids).

3. (Time-like Cauchy problem for the wave equation). Take the wave equation

$$Lu = u_{x_3x_3} - u_{x_1x_1} - u_{x_2x_2} = 0 \tag{5.7a}$$

for $u = u(x_1,x_2,x_3)$. Let Z be the disc

$$x_2^2 + x_3^2 < 1; x_1 = 0 \tag{5.7b}$$

in the "time-like' plane $x_1 = 0$. Show that every cone with base Z and vertex $y = (y_1,y_2,y_3)$ is a domain of determinacy for Z, if

$$2y_1^2 + y_2^2 < 2, \qquad y_3 = 0.$$

Let u be a solution of (5.7a) of class C^2 in the "semi-double" characteristic cone

$$x_3 + \sqrt{x_1^2 + x_2^2} \leqslant \sqrt{2}; x_1 \geqslant 0. \tag{5.7c}$$

Then the Cauchy data on Z uniquely determine those in the points

$$(0,x_2,0), \qquad |x_2| < \sqrt{2}.$$

(Hint: Use problem 2 and the convexity of the set (5.7c).)

4. For the heat equation

$$Lu = u_{x_2} - u_{x_1x_1} = 0 \tag{5.8a}$$

take for Z the time-like segment

$$x_1 = 0; \qquad a < x_2 < b.$$

Show that the semi-strip

$$x_1 \geqslant 0; \qquad a \leqslant x_2 \leqslant b \tag{5.8b}$$

is a domain of determinacy for Z.

5. Let $Lu = 0$ be the wave equation (5.6a) for $n = 3$ and Z be the ball (5.6b). Show that no domain of determinacy for Z contains points x with $x_n > r - \rho$. [Hint: Use "plane wave solutions" of the form

$$u = f(x_1 \cos\theta + x_2 \sin\theta + x_3)$$

where $f(s) = 0$ for $s < r, f(s) > 0$ for $s > r$.]

6. Distribution Solutions*

The Lagrange–Green identity (4.5) has other applications. The identity reduces to

$$\int_\Omega v^T Lu\,dx = \int_\Omega (\tilde{L}v)^T u\,dx \tag{6.1}$$

in the case when all boundary terms vanish, e.g., when either u or v have Cauchy data zero. In particular, let

$$Lu = w \quad \text{in } \Omega \quad \text{and} \quad D^\beta v = 0 \quad \text{on } \partial\Omega \text{ for } |\beta| < m. \tag{6.2}$$

Then the identity

$$\int_\Omega v^T w\,dx = \int_\Omega (\tilde{L}v)^T u\,dx \tag{6.3}$$

holds. We can use this identity to define *generalized solutions* u of $Lu = w$. For example, if u is continuous we can require this identity to hold for all $v \in C^m$, vanishing near $\partial\Omega$. This leads to the notion *of distribution solutions* in the sense of Laurent Schwartz.

The idea is to replace a function $f(x)$ which is defined on an open set Ω in $\mathbb{R}^n$ by the integrals formed with this function for different weights w. We associate with the scalar point function $f(x)$ the *functional*

$$f[\phi] = \int_\Omega \phi(x) f(x)\,dx, \tag{6.4}$$

where $\phi \in \mathfrak{D} = C_0^\infty(\Omega)$ is the space of "test functions," that is, functions having derivatives of all orders and *compact support*.† This *functional* f exists for any continuous or locally integrable *function* f. The integral for $f[\phi]$ defines a linear functional on $\mathfrak{D}$, with values in $\mathbb{R}$. The values of the functional $f[\phi]$ for varying ϕ determine the function $f(x)$ uniquely when $f(x)$ is continuous. Indeed, if the continuous function g is such that $f[\phi] = g[\phi]$ for all $\phi \in \mathfrak{D}$, then,

$$\int \phi(x)(f(x) - g(x))\,dx = 0 \tag{6.5}$$

for all $\phi \in \mathfrak{D}$. If here $f \not\equiv g$, say $f - g > 0$ at a point P, then also in a neighborhood of P; choosing a test function which is nonnegative and vanishes outside this neighborhood would contradict (6.5). Thus $f - g = 0$.

Consider next derivatives of f. If $f \in C^1(\Omega)$ we find by integration by parts

$$\int_\Omega \phi D_k f\,dx = -\int_\Omega (D_k\phi) f\,dx. \tag{6.6}$$

The left-hand side is obtained by applying the *functional* $D_k f$ to test

*([3], [7], [16])

†That is, $\phi = 0$ outside a closed and bounded subset of Ω. Generally, the support of a function $\phi(x)$ is the closure of the set of x for which $\phi(x) \neq 0$.

functions ϕ. Thus $D_k f[\phi] = -f[D_k\phi]$. We can use the right-hand side of (6.6) to define the left-hand side when f has no derivative or is not even continuous. As long as the functional associated with f is defined for all test functions ϕ, the functional associated with $D_k f$ makes sense. More generally, we are led to the notion of a distribution.

Definition. A *distribution* is a linear functional $f[\phi]$ defined for all $\phi \in \mathfrak{D} = C_0^\infty(\Omega)$ which is continuous on $\mathfrak{D}$ in the following sense: Let the ϕ_r be a sequence in $\mathfrak{D}$. Then

$$\lim_{k\to\infty} f[\phi_k] = 0$$

provided

(a) all ϕ_k vanish outside the same compact subset of Ω, and
(b)

$$\lim_{k\to\infty} D^\alpha \phi_k(x) = 0$$

uniformly in x for each α (not necessarily uniformly in α).

Each continuous (or even locally integrable) function $f(x)$ generates a distribution

$$f[\phi] = \int \phi(x) f(x)\, dx. \tag{6.7}$$

More generally, we write any distribution f *symbolically* as

$$f[\phi] = \int_\Omega \phi(x) f(x)\, dx. \tag{6.8}$$

A specially important distribution, not generated by an integrable point function, is the so-called *Dirac function* with singularity ξ, denoted by δ_ξ, which is defined by

$$\delta_\xi[\phi] = \phi(\xi) \quad \text{for } \xi \in \Omega. \tag{6.9}$$

It is given symbolically by

$$\int_\Omega \phi(x)\delta_\xi(x)\, dx = \phi(\xi). \tag{6.10}$$

Two distributions $f[\phi]$ and $g[\phi]$ naturally are called *equal*, if $f[\phi] = g[\phi]$ for all $\phi \in \mathfrak{D}$. More generally we say that two distributions f, g agree in an open subset ω of Ω if $f[\phi] = g[\phi]$ for all $\phi \in \mathfrak{D}$ that have their support in ω. This permits us in some cases to assign point values to a distribution in a subset of Ω. Thus for the Dirac function δ_ξ defined by (6.9) we have

$$\delta_\xi(x) = 0 \quad \text{for } x \in \Omega,\ x \neq \xi. \tag{6.11}$$

Indeed taking $f = \delta_\xi$, $g \equiv 0$, we have $f[\phi] = g[\phi] = 0$ for any ϕ with support in the set ω obtained by deleting ξ from Ω.

For a distribution f we define the "derivative" $D_k f$ as the distribution given by

$$D_k f[\phi] = -f[D_k \phi] \tag{6.12}$$

and more generally $D^\alpha f$ by

$$D^\alpha f[\phi] = (-1)^{|\alpha|} f[D^\alpha \phi]. \tag{6.13}$$

One easily verifies that formulas (6.12), (6.13) actually define distributions $D_k f, D^\alpha f$. As an example we have from (6.9) that $D_k \delta_\xi[\phi] = -\phi_{x_k}(\xi)$. In particular, (6.13) yields a definition for derivatives of a continuous function $f(x)$, not necessarily as a function with point values, but as a generalized function for which weighted integrals are defined.

Still more generally than (6.13) we can apply any scalar linear differential operator L with C^∞ coefficients to a distribution $u[\phi]$, we define the distribution $Lu[\phi]$ by

$$Lu[\phi] = u[\tilde{L}\phi] \tag{6.14}$$

in accordance with the Lagrange–Green integral formula

$$\int_\Omega \phi Lu\, dx = \int_\Omega (\tilde{L}\phi) u\, dx \tag{6.15}$$

valid by (4.5) for a test function ϕ and a scalar function $u \in C^m$. (Similar definitions can easily be given for the case where u is a vector, and L corresponds to a system.) We are interested particularly in the distribution solutions L of the equation

$$Lu = \delta_\xi \tag{6.16}$$

where δ_ξ is the Dirac operator defined by (6.9). They are the so-called *fundamental solutions* with pole ξ for the operator L. We notice that adding to a fundamental solution u any ordinary solution $v \in C^m(\Omega)$ of the homogeneous equation $Lv = 0$ again yields a fundamental solution.

In many cases the equations defining a particular distribution as a linear operator in $\mathcal{D}$ define the operator for a much wider variety of functions ϕ. For example, the operator $\delta_\xi[\phi]$ can be defined by (6.9) or (6.10) for all functions $\phi(x)$ that are just continuous in Ω. Similarly for an mth-order linear differential operator L with coefficients in $C^m(\Omega)$, we can define $Lu(\phi)$ by (6.14) for a locally integrable function $u(x)$ and for $\phi \in C^m(\Omega)$. More precisely we call $u(x)$ a *weak solution* of the equation $Lu(x) = w(x)$ in Ω, if

$$\int \phi(x) w(x)\, dx = \int (\tilde{L}\phi(x)) u(x)\, dx \tag{6.17}$$

for all $\phi \in C_0^m(\Omega)$ (that is for all ϕ of class C^m with compact support in Ω).* Other types of generalized solutions will be encountered in the sequel.

*It is sufficient to require (6.17) to hold for all $\phi \in \mathcal{D}$ since every $\phi \in C_0^m$ can be approximated uniformly with its derivatives of order $\leqslant m$ by $\phi \in \mathcal{D}$.

EXAMPLE. Let Ω be the x-axis and L the ordinary differential operator d^2/dx^2. Then $u(x)=\frac{1}{2}|x-\xi|$ is a fundamental solution for L. Indeed the distribution associated with u'' by (6.13) is

$$u''[\phi]=u[\phi'']=\int_{-\infty}^{+\infty}\tfrac{1}{2}|x-\xi|\phi''(x)\,dx=\phi(\xi)=\delta_\xi[\phi]$$

when $\phi\in C_0^\infty(\mathbb{R})$.

PROBLEMS

1. Show that for a continuous function f the expression $u=f(x-ct)$ is a weak solution of the partial differential equation

$$u_t+cu_x=0.$$

(Hint: Transform for $\phi\in C_0^1(\mathbb{R}^2)$ the integral

$$\int\int(\phi_t+c\phi_x)u\,dx\,dt$$

to the coordinates $y_1=x-ct, y_2=x$. Use $\phi=\psi(y_1)X(y_2)$.)

2. Show that the function $u(x_1,x_2)$ defined by

$$u(x_1,x_2)=\begin{cases}1 & \text{for } x_1>\xi_1,\ x_2>\xi_2\\ 0 & \text{for all other } x_1,x_2\end{cases}$$

is a fundamental solution with the pole (ξ_1,ξ_2) of the operator $L=\partial^2/\partial x_1\partial x_2$ in the x_1x_2-plane.

3. Show that the function

$$u(x_1,x_2)=\begin{cases}\frac{1}{2} & \text{for } |x_1-\xi_1|<\xi_2-x_2\\ 0 & \text{otherwise}\end{cases}$$

is a fundamental solution for $L=(\partial^2/\partial x_2^2)-(\partial^2/\partial x_1^2)$ with pole (ξ_1,ξ_2).

4. Verify the special cases of Green's identity

$$\int_\Omega(v(\Delta u+cu)-u(\Delta v+cv))\,dx=\int_{\partial\Omega}\left(v\frac{du}{dn}-u\frac{dv}{dn}\right)dS \tag{6.18}$$

$$\int_\Omega(v\Delta^2u-u\Delta^2v)\,dx=\int_{\partial\Omega}\left(v\frac{d\Delta u}{dn}-\Delta u\frac{dv}{dn}+\Delta v\frac{du}{dn}-u\frac{d\Delta v}{dn}\right)dS. \tag{6.19}$$

5. Show that $f[\phi]=\sum_{k=0}^{\infty}\phi^{(k)}(x+k)$ defines a distribution on $\mathbb{R}$ that is not representable as derivative of some order of a continuous function f.

6. (Convergence of distributions). Given a sequence of distributions f^k and a distribution f, we define $f^k\to f$ or $\lim_{k\to\infty}f^k=f$ to mean that $\lim_{k\to\infty}f^k[\phi]=f[\phi]$ uniformly in any set of $\phi\in\mathfrak{D}$ that have their support in the same compact subset of Ω and have common bounds M_α for all $|D^\alpha\phi(x)|$ in Ω.

   (a) Prove that $f^k\to f$ implies $D^\alpha f^k\to D^\alpha f$ for all α.

(b) Let $f^k(x)$ be a sequence of non-negative continuous functions defined in $\mathbb{R}^n$ for which

$$\int f^k(x)\,dx = 1; \qquad f^k(x) = 0 \quad \text{for } |x - \xi| > \frac{1}{k}.$$

Show that $f^k \to \delta_\xi$ in the sense of distributions.

(c) Let $f(x) \in C^0(\mathbb{R})$. Define

$$f_h(x) = \frac{f(x+h) - f(x)}{h}.$$

Show that

$$\lim_{h \to 0} f_h = Df$$

in the sense of distributions.

4 The Laplace Equation*

1. Green's Identity, Fundamental Solutions, and Poisson's Equation†

The *Laplace operator* acting on a function $u(x)=u(x_1,\ldots,x_n)$ of class C^2 in a region Ω is defined by

$$\Delta=\sum_{k=1}^{n} D_k^2 \tag{1.1}$$

For $u,v\in C^2(\bar{\Omega})$ we have‡ (see Chapter 3, (4.8), (4.9)) Green's identities.

$$\int_\Omega v\,\Delta u\,dx=-\int_\Omega \sum_i v_{x_i}u_{x_i}\,dx+\int_{\partial\Omega} v\frac{du}{dn}\,dS \tag{1.2a}$$

$$\int_\Omega v\,\Delta u\,dx=\int_\Omega u\,\Delta v\,dx+\int_{\partial\Omega}\left(v\frac{du}{dn}-u\frac{dv}{dn}\right)dS, \tag{1.2b}$$

where d/dn indicates differentiation in the direction of the *exterior* normal to $\partial\Omega$.

The special case $v=1$ yields the identity

$$\int_\Omega \Delta u\,dx=\int_{\partial\Omega}\frac{du}{dn}\,dS. \tag{1.3}$$

*([2], [6], [11], [13], [15], [27], [32])
†([19])

‡We assume here that Ω is an open bounded set, for which the divergence theorem, (4.1) of Chapter 3, is valid.

Another special case of interest is $v=u$. We find then from (1.2a) the *energy identity*

$$\int_\Omega \sum_i u_{x_i}^2\,dx+\int_\Omega u\,\Delta u\,dx=\int_{\partial\Omega} u\frac{du}{dn}\,dS. \tag{1.4}$$

If here $\Delta u=0$ in Ω and either $u=0$ or $du/dn=0$ on $\partial\Omega$, it follows that

$$\int_\Omega \sum_i u_{x_i}^2\,dx=0. \tag{1.5}$$

For $u\in C^2(\overline{\Omega})$ the integrand is nonnegative and continuous, and hence has to vanish. Thus $u=$const. in Ω. This observation leads to uniqueness theorems for two of the standard problems of potential theory:

The *Dirichlet problem*: Find u in Ω from prescribed values of Δu in Ω and of u on $\partial\Omega$.

The *Neumann problem*: Find u in Ω from prescribed values of Δu in Ω and of du/dn on $\partial\Omega$.

As always in discussing uniqueness of linear problems, we form the difference of two solutions, which is a solution of the same problem with data 0. We find that the difference is a constant, which, in the Dirichlet case, must have the value 0. Thus: *A solution $u\in C^2(\overline{\Omega})$ of the Dirichlet problem is determined uniquely. A solution $u\in C^2(\overline{\Omega})$ of the Neumann problem is determined uniquely within an additive constant.* (Notice also that the solution of the Neumann problem can only exist if the data satisfy condition (1.3)).

One of the principal features of the Laplace equation

$$\Delta u=0 \tag{1.6}$$

is its *spherical symmetry*. The equation is preserved under rotations about a point ξ, that is under orthogonal linear substitutions for $x-\xi$. This makes it plausible that there exist special solutions $v(x)$ of (1.6) that are invariant under rotations about ξ, that is have the same value at all points x at the same distance from ξ. Such solutions would be of the form

$$v=\psi(r), \tag{1.7}$$

where

$$r=|x-\xi|=\sqrt{\sum_i (x_i-\xi_i)^2} \tag{1.8}$$

represents the euclidean distance between x and ξ. By the chain rule of differentiation we find from (1.6) in n dimensions that ψ satisfies the ordinary differential equation

$$\Delta v=\psi''(r)+\frac{n-1}{r}\psi'(r)=0. \tag{1.9}$$

Solving we are led to

$$\psi'(r) = Cr^{1-n} \tag{1.10a}$$

$$\psi(r) = \begin{cases} \dfrac{Cr^{2-n}}{2-n} & \text{when } n>2 \quad (1.10b) \\ C\log r & \text{when } n=2 \quad (1.10c) \end{cases}$$

with $C = \text{const.}$, where we can still add a trivial constant solution to ψ.

The function $v(x) = \psi(r)$ satisfies (1.6) for $r>0$, that is for $x \neq \xi$, but becomes infinite for $x = \xi$. We shall see that v for a suitable choice of the constant C, is a *fundamental solution* for the operator Δ, satisfying the symbolic equation, (see Chapter 3, (6.16)),

$$\Delta v = \delta_\xi. \tag{1.11}$$

Let $u \in C^2(\bar{\Omega})$ and ξ be a point of Ω. We apply Green's identity (1.2b) with v given by (1.7), (1.10b,c). Since v is singular at $x = \xi$ we cut out from Ω a ball $B(\xi,\rho)$ contained in Ω with center ξ, radius ρ, and boundary $S(\xi,\rho)$. The remaining region $\Omega_\rho = \Omega - B(\xi,\rho)$ is bounded by $\partial\Omega$ and $S(\xi,\rho)$. Since $\Delta v = 0$ in Ω_ρ we have

$$\int_{\Omega_\rho} v\,\Delta u\,dx = \int_{\partial\Omega}\left(v\frac{du}{dn} - u\frac{dv}{dn}\right)dS + \int_{S(\xi,\rho)}\left(v\frac{du}{dn} - u\frac{dv}{dn}\right)dS. \tag{1.12}$$

Here on $S(\xi,\rho)$ the "exterior" normal to our region Ω_ρ points *towards* ξ. Consequently by (1.10a,b,c), (1.3)

$$v = \psi(\rho), \qquad \frac{dv}{dn} = -\psi'(\rho) \tag{1.13a}$$

$$\int_{S(\xi,\rho)} v\frac{du}{dn}\,dS = \psi(\rho)\int_{S(\xi,\rho)} \frac{du}{dn}\,dS = -\psi(\rho)\int_{B(\xi,\rho)} \Delta u\,dx \tag{1.13b}$$

$$\int_{S(\xi,\rho)} u\frac{dv}{dn}\,dS = -C\rho^{1-n}\int_{S(\xi,\rho)} u\,dS. \tag{1.13c}$$

Since both u and Δu are continuous at ξ the right-hand side of (1.13b) tends to 0 for $\rho \to 0$ by (1.10b,c), while that of (1.13c) tends to $-C\omega_n u(\xi)$, where ω_n denotes the surface "area" of the unit sphere in $\mathbb{R}^n$. [The values $\omega_2 = 2\pi$, $\omega_3 = 4\pi$ are familiar; generally $\omega_n = 2\sqrt{\pi}^{\,n}/\Gamma(\frac{1}{2}n)$]. Thus (1.12) becomes for $\rho \to 0$

$$\int_\Omega v\,\Delta u\,dx = \int_{\partial\Omega}\left(v\frac{du}{dn} - u\frac{dv}{dn}\right)dS + C\omega_n u(\xi) \tag{1.14}$$

(note that v, though ∞ at $x = \xi$, is integrable near ξ). We now choose

$C=1/\omega_n$ in (1.10b,c), so that

$$\psi(r)=\begin{cases} \dfrac{r^{2-n}}{(2-n)\omega_n} & \text{for} \quad n>2 \qquad (1.15\text{a}) \\ \dfrac{\log r}{2\pi} & \text{for} \quad n=2. \qquad (1.15\text{b}) \end{cases}$$

We write the corresponding v in its dependence on x and ξ as

$$v=K(x,\xi)=\psi(r)=\psi(|x-\xi|). \tag{1.16}$$

Then (1.14) becomes

$$u(\xi)=\int_{\Omega} K(x,\xi)\,\Delta u\,dx-\int_{\partial\Omega}\left(K(x,\xi)\frac{du(x)}{dn_x}-u(x)\frac{dK(x,\xi)}{dn_x}\right)dS_x \tag{1.17}$$

for $\xi\in\Omega$, where the subscript "x" in S_x and dn_x indicates the variable of integration respectively differentiation. (Notice that for $\xi\notin\bar{\Omega}$ the left-hand side of identity (1.17) has to be replaced by 0. This follows from (1.2b), since $v=K(x,\xi)\in C^2(\bar{\Omega})$ and $\Delta v=0$ for ξ outside $\bar{\Omega}$.)

Taking in particular for u in (1.17) a test function $\phi\in C_0^\infty(\Omega)$, we find that

$$\phi(\xi)=\int K(x,\xi)\,\Delta\phi(x)\,dx \tag{1.18}$$

Hence $v=K(x,\xi)$ defines a distribution for which

$$v[\Delta\phi]=\phi(\xi).$$

Since $L=\Delta$ is (formally) *selfadjoint* (that is $L=\tilde{L}$), we can interpret (1.18) as stating that the functional v applied to the Laplacian of a test function ϕ has the value $\phi(\xi)$ (see Chapter 3, (6.14)). or that v in the distribution sense satisfies (1.11), and is a fundamental solution with pole ξ.

Let $u\in C^2(\bar{\Omega})$ be "harmonic" in Ω, that is, be a solution of $\Delta u=0$; then by (1.17) for $\xi\in\Omega$

$$u(\xi)=-\int_{\partial\Omega}\left(K(x,\xi)\frac{du(x)}{dn_x}-u(x)\frac{dK(x,\xi)}{dn_x}\right)dS_x. \tag{1.19}$$

Formula (1.19) expressing u in Ω in terms of its Cauchy data u and du/dn_x on $\partial\Omega$ represents the solution of the Cauchy problem for Ω, provided such a solution exists. Actually, by the uniqueness theorem for the Dirichlet problem proved earlier a solution of $\Delta u=0$ is determined already by the values of u alone on $\partial\Omega$. Thus we cannot prescribe* both u and du/dn_x on $\partial\Omega$. *The Cauchy problem for the Laplace equation in Ω generally has no solution*. Formula (1.19), however, is useful in discussing *regularity* of harmonic functions. Since $K(x,\xi)=\psi(|x-\xi|)$ is in C^∞ in x and ξ for $x\neq\xi$,

*The Cauchy data cannot even be prescribed as arbitrary *analytic* functions on $\partial\Omega$. This does not contradict the Cauchy–Kowalevski existence theorem, since here we require u to exist in all of Ω, and not just in a sufficiently small neighborhood of $\partial\Omega$.

we can form derivatives of u with respect to ξ of all orders under the integral sign for $\xi \in \Omega$, and find that $u \in C^{\infty}(\Omega)$. More precisely, we can even conclude that $u(\xi)$ is real analytic in Ω. To see this we only have to continue u suitably into a complex neighborhood. We observe that $K = \psi(r)$, with r given by the algebraic expression

$$r = \sqrt{\sum_i (x_i - \xi_i)^2} ,$$

is defined and differentiable for complex x, ξ as long as $r \neq 0$. In particular $r \neq 0$ when x is confined to real points on $\partial\Omega$ and ξ to complex points in a sufficiently small complex neighborhood of a real point of Ω. Formula (1.19) then defines $u(\xi)$ as a differentiable function of ξ in that complex neighborhood. It follows that u is real analytic in the set Ω. We need not assume that $u \in C^2(\overline{\Omega})$. To belong to C^{∞} or to be analytic is a "local" property. A solution of the Laplace equation in any open set Ω is of class C^2 in any closed ball contained in Ω and hence real analytic in the open ball. Thus *harmonic functions are real analytic in the interior of their domain of definition*.

We can conclude from this that the Cauchy problem for the Laplace equation generally is unsolvable, even *locally*. Take for the initial surface a portion σ of the plane $x_n = 0$. Prescribe the Cauchy data

$$u = 0, \qquad u_{x_n} = g(x_1, \ldots, x_{n-1}) \tag{1.20}$$

on σ. Let $\xi \in \sigma$ and let B be a ball with center ξ whose intersection with $x_n = 0$ lies in σ. Denote by B^+ the hemispherical portion of B lying in $x_n \geqslant 0$. There cannot exist a solution u of the Laplace equation of class C^2 in $\overline{B}^+$ with Cauchy data (1.20) for $x_n = 0$, unless g is real analytic. To see this one continues u into the whole of B *by reflection*, defining $u(x) = u(x_1, \ldots, x_n)$ for $x \in B$ with $x_n < 0$ by

$$u(x_1, \ldots, x_{n-1}, x_n) = -u(x_1, \ldots, x_{n-1}, -x_n). \tag{1.21}$$

One easily verifies that the extended u belongs to $C^2(B)$; (the values of u and its first and second derivatives fit the reflected values along $x_n = 0$, because $u = \Delta u = 0$ there). Moreover, $\Delta u = 0$ in B, since the Laplace equation is unchanged when we replace u by $-u$ or x_n by $-x_n$. Consequently u is real analytic in the whole of B. In particular then $u_{x_n}(x_1, \ldots, x_{n-1}, 0)$ is a real analytic function of $x_1, \ldots, x_{n-1}$.

Let $w(x) = w(x_1, \ldots, x_n)$ be any solution of $\Delta w = 0$ of class $C^2(\overline{\Omega})$. Then

$$G(x, \xi) = K(x, \xi) + w(x) \tag{1.22}$$

again is a fundamental solution of the Laplace equation with pole ξ. (See Chapter 3, p. 91.) More precisely identity (1.17) stays valid when we replace K by G:

$$u(\xi)=\int_{\Omega} G(x,\xi)\,\Delta u\,dx-\int_{\partial\Omega}\left(G(x,\xi)\frac{du(x)}{dn_x}-u(x)\frac{dG(x,\xi)}{dn_x}\right)dS_x \quad (1.23)$$

as follows immediately from (1.17), (1.2b).

As an application we take the case where Ω is a ball $B(\xi,\rho)$ of center ξ and radius ρ, choosing for G the function

$$G(x,\xi)=K(x,\xi)-\psi(\rho)=\psi(|x-\xi|)-\psi(\rho). \quad (1.24)$$

Then on $\partial\Omega$

$$G=0, \qquad \frac{dG}{dn_x}=\psi'(\rho)=\frac{1}{\omega_n}\rho^{1-n},$$

and (1.23) becomes

$$u(\xi)=\int_{|x-\xi|<\rho}(\psi(|x-\xi|)-\psi(\rho))\,\Delta u(x)\,dx+\frac{1}{\omega_n\rho^{n-1}}\int_{|x-\xi|=\rho}u(x)\,dS_x. \quad (1.25)$$

For $\Delta u=0$ this is *Gauss's law of the arithmetic mean*:

$$u(\xi)=\frac{1}{\omega_n\rho^{n-1}}\int_{|x-\xi|=\rho}u(x)\,dS_x, \quad (1.26)$$

where $\omega_n\rho^{n-1}$ is the surface area of the sphere $|x-\xi|=\rho$. In words: *For a function u harmonic in a closed ball the value of u at the center equals the average of the values of u on the surface*. Since the $\psi(r)$ given by (1.15a,b) is a monotone increasing function of r for all dimensions n, we obtain more generally from (1.25):

If $\Delta u(x)\geqslant 0$ in the ball $|x-\xi|\leqslant\rho$, then

$$u(\xi)\leqslant\frac{1}{\omega_n\rho^{n-1}}\int_{|x-\xi|=\rho}u(s)\,dS_x. \quad (1.27)$$

A function u continuous in Ω is called *subharmonic*, if for each $\xi\in\Omega$ the inequality (1.27) holds for all sufficiently small ρ. Thus functions in $C^2(\Omega)$ with $\Delta u\geqslant 0$ are subharmonic.

Formula (1.18) expresses that $K(x,\xi)$ is a fundamental solution for Δ. Another aspect of this property is *Poisson's formula*

$$u(\xi)=\Delta_\xi\int_{\Omega}K(x,\xi)u(x)\,dx \quad (1.28)$$

valid for $u\in C^2(\overline{\Omega})$ and $\xi\in\Omega$. Here Δ_ξ denotes the Laplace operator taken with respect to the variables ξ. Thus for given $u\in C^2(\overline{\Omega})$ we have in

$$w(\xi)=\int_{\Omega}K(x,\xi)u(x)\,dx \quad (1.29)$$

a special solution w of the *inhomogeneous Laplace equation* ("Poisson's differential equation")

$$\Delta_\xi w(\xi) = u(\xi) \tag{1.30}$$

for $\xi \in \Omega$. *Formally* (1.28) follows directly from (1.16), (1.11):

$$\begin{aligned}\Delta_\xi \int K(x,\xi)u(x)\,dx &= \int (\Delta_\xi \psi(|x-\xi|))u(x)\,dx\\ &= \int (\Delta_x \psi(|x-\xi|))u(x)\,dx\\ &= \int \delta_\xi(x)u(x)\,dx = u(\xi).\end{aligned}$$

For a rigorous derivation we first assume that $u \in C_0^2(\Omega)$. Then by (1.17) for $\xi \in \Omega$, since $u=0$ near $\partial\Omega$,

$$u(\xi) = \int_\Omega K(x,\xi)\,\Delta_x u(x)\,dx = \int K(x,\xi)\,\Delta_x u(x)\,dx. \tag{1.31}$$

(As always the domain of integration is the whole space, when no other domain is indicated). Actually (1.31) holds for all ξ, since for given $u \in C_0^2(\Omega)$ nothing changes when Ω is replaced by any larger open set containing ξ. With $y = x - \xi$ as variable of integration we have by (1.16)

$$\begin{aligned}u(\xi) &= \int \psi(|x-\xi|)\,\Delta_x u(x)\,dx = \int \psi(|y|)\,\Delta_y u(y+\xi)\,dy\\ &= \int \psi(|y|)\,\Delta_\xi u(y+\xi)\,dy = \Delta_\xi \int \psi(|y|)u(y+\xi)\,dy\\ &= \Delta_\xi \int K(x,\xi)u(x)\,dx \end{aligned}\tag{1.31a}$$

confirming (1.28). Let next $u \in C^2(\bar{\Omega})$. Let b be any ball with $\bar{b} \subset \Omega$. We can find a concentric ball B such that $\bar{b} \subset B$, $\bar{B} \subset \Omega$, and a "cutoff function" $\zeta(x) \in C_0^2(\Omega)$, which has the value 1 everywhere in B. (See Figure 4.1.) Then

$$u = \zeta u + (1-\zeta)u,$$

where $\zeta u \in C_0^2(\Omega)$ and $(1 - \zeta)u$ vanishes in B. Thus for $\xi \in b$ by (1.31a)

$$\Delta_\xi \int_\Omega K(x,\xi)\zeta(x)u(x)\,dx = \zeta(\xi)u(\xi) = u(\xi) \tag{1.32a}$$

$$\Delta_\xi \int_\Omega K(x,\xi)(1-\zeta(x))u(x)\,dx = \Delta_\xi \int_{\Omega - B} K(x,\xi)(1-\zeta(x))u(x)\,dx = 0, \tag{1.32b}$$

since $K(x,\xi)$ is regular and harmonic in ξ for $x \in \Omega - B$, $\xi \in b$. Adding (1.32a,b) yields (1.28) for ξ in any b, and hence for all $\xi \in \Omega$.

For ξ outside $\bar{\Omega}$ we find by direct differentiation under the integral sign that $\Delta_\xi w(\xi) = 0$ for the w defined by (1.29). For ξ on $\partial\Omega$ the second

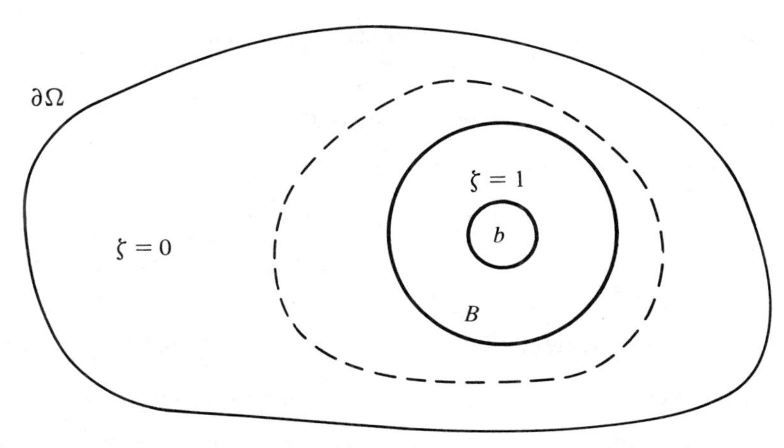

Figure 4.1

derivatives of $w(\xi)$ may cease to exist. One easily verifies however that $w(\xi) \in C^1(\mathbb{R}^n)$, since the first ξ-derivatives of $K(x,\xi)$ are still integrable with respect to x.

PROBLEMS

1. Let u belong to $C^2(\Omega)$ and be subharmonic. Show that $\Delta u \geqslant 0$. [Hint: Use (1.25).]

2. Let $L = \Delta + c$ in $n=3$ dimensions, where $c = \text{const.}$ (L = "reduced wave operator").

(a) Find all solutions of $Lu=0$ with spherical symmetry.

(b) Prove that

$$K(x,\xi) = -\frac{\cos(\sqrt{c}\, r)}{4\pi r}, \qquad r = |x-\xi| \tag{1.33}$$

is a fundamental solution for L with pole ξ. Show that for a solution u of $Lu=0$ formula (1.19) holds with K defined by (1.33). [Hint: Use formula (6.18) of Chapter 3.]

(c) Show that a solution u of $Lu=0$ in the ball $|x-\xi| \leqslant \rho$ for $\sin(\sqrt{c}\,\rho) \neq 0$ has the modified mean value property

$$u(\xi) = \frac{\sqrt{c}\,\rho}{\sin(\sqrt{c}\,\rho)} \frac{1}{4\pi\rho^2} \int_{|x-\xi|=\rho} u(x)\, dS_x. \tag{1.34}$$

(Here for $c = -\gamma < 0$, the factor $\sqrt{c}\,\rho/\sin(\sqrt{c}\,\rho)$ stands for $\sqrt{\gamma}\,\rho/\sinh(\sqrt{\gamma}\,\rho)$.) [Hint: Use the fundamental solution

$$G(x,\xi) = K(x,\xi) + k\frac{\sin(\sqrt{c}\, r)}{r} \tag{1.35}$$

with a suitable constant k.]

(d) Show that a solution u of $Lu=0$ of class $C^2(\bar{\Omega})$ vanishing on $\partial\Omega$ vanishes in Ω, provided $c<0$. Show that for $c>0$ there are solutions vanishing on a sphere but not in the interior.
(e) Show that solutions of $Lu=0$ in Ω are real analytic in Ω.

3. (a) Show that for $n=2$ the function

$$v=\frac{1}{8\pi}r^2\log r, \qquad r=|x-\xi| \tag{1.36}$$

is a fundamental solution for the operator Δ^2. [Hint: $\Delta v=(1+\log r)/2\pi$.]
(b) Show that for $u\in C^2(\bar{\Omega})$ and $\xi\in\Omega$

$$u(\xi)=\int_\Omega v\,\Delta^2 u\,dx-\int_{\partial\Omega}\left(v\frac{d\Delta u}{dn}-\Delta u\frac{dv}{dn}+\Delta v\frac{du}{dn}-u\frac{d\Delta v}{dn}\right)dS. \tag{1.37}$$

[Hint: Apply (1.2b) with u replaced by Δu, and use (1.17).]

4. (a) Find all solutions with spherical symmetry of the biharmonic equation $\Delta^2 u=0$ in n dimensions.
(b) Find a fundamental solution.

5. Prove that $\Delta u(x) = \Delta u(x_1,\dots,x_n) = 0$ also implies that

$$\Delta(|x|^{2-n}u(x/|x|^2))=0 \tag{1.38}$$

for $x/|x|^2$ in the domain of definition of u.

6. Let x,y denote coordinates in the plane. Let $u(x,y)$ be a solution of the two-dimensional Laplace equation $u_{xx}+u_{yy}=0$ in an open simply connected set Ω.
(a) Prove that there exists a *conjugate harmonic* function $v(x,y)$ such that the Cauchy–Riemann equations

$$u_x=v_y, \qquad u_y=-v_x \tag{1.39}$$

are satisfied. [Hint: for a fixed point (x_0,y_0) in Ω define v by

$$v(x,y)=\int_{(x_0,y_0)}^{(x,y)}(u_x\,dy-u_y\,dx), \tag{1.40}$$

where the integral is taken along any path joining (x_0,y_0) to (x,y).]
(b) Introduce the complex-valued function $f=u+iv$ of the complex argument $z=x+iy$. Prove Cauchy's theorem that for any closed curve C in Ω

$$\int_C f(z)\,dz=\int_C(u+iv)(dx+i\,dy)=0.$$

[Hint: There exist functions $\phi(x,y)$, $\psi(x,y)$ in Ω with

$$d\phi=u\,dx-v\,dy, \qquad d\psi=v\,dx+u\,dy.]$$

7. By Newton's law the gravitational attraction exerted on a unit mass located at $\xi=(\xi_1,\xi_2,\xi_3)$ by a solid Ω with density $\mu=\mu(x)$ is given by the vector

$$F(\xi)=\gamma\iiint_\Omega\frac{\mu(x)(x-\xi)}{|x-\xi|^3}\,dx$$

($\gamma=$universal gravitational constant).

(a) Prove that $F = \operatorname{grad} u$, where the "potential" u is given by

$$u(\xi) = \gamma \int\int\int \frac{\mu(x)}{|x-\xi|}\, dx.$$

(b) Prove that the attraction $F(\xi)$ exerted by Ω on a far away unit mass is approximately the same as if the total mass of Ω were concentrated at its center of gravity

$$x^0 = \int\int\int_\Omega \mu(x)x\, dx \Big/ \int\int\int_\Omega \mu(x)\, dx.$$

[Hint: Approximate $|\xi - x|^{-3}$ by $|\xi - x^0|^{-3}$ for large $|\xi|$.]

(c) Calculate the potential u and attraction F of a solid sphere Ω of radius a with center at the origin and of constant density μ. Use here that u must have spherical symmetry, must be harmonic outside Ω, satisfy Poisson's differential equation in Ω, be of class C^1 everywhere, and vanish at ∞. [Answer: $u(\xi) = 2\pi\gamma\mu a^2 - \frac{2}{3}\pi\gamma\mu|\xi|^2$ for $|\xi| < a$, $u(\xi) = \frac{4}{3}\pi\gamma\mu a^3|\xi|^{-1}$ for $|\xi| > a$.]

8. Let u, v be conjugate harmonic functions satisfying (1.39) in a simply connected region Ω. Show that on the boundary curve $\partial\Omega$

$$\frac{du}{dn} = \frac{dv}{ds}, \qquad \frac{dv}{dn} = -\frac{du}{ds},$$

where dn denotes differentiation in the direction of the exterior normal and ds differentiation in the counterclockwise tangential direction. Show how these relations can be used to reduce the Neumann problem for u to the Dirichlet problem for v and conversely.

9. Extend (1.28) to $u \in C^1(\bar{\Omega})$. [Hint: Write (1.28) in terms of first derivatives of K and u. Approximate u.]

2. The Maximum Principle*

One of the important tools in the theory of harmonic functions is the maximum principle. Similar principles hold for solutions of more general second-order elliptic equations, and for complex analytic functions. In this section we assume that Ω is a bounded, open, and connected set in $\mathbb{R}^n$. We first prove a *weak form of the principle*:

Let $u \in C^2(\Omega) \cap C^0(\bar{\Omega})$, and let $\Delta u \geqslant 0$ in Ω. Then

$$\max_{\bar{\Omega}} u = \max_{\partial\Omega} u. \tag{2.1}$$

(Notice that a continuous function u assumes its maximum somewhere in the closed and bounded set $\bar{\Omega}$. Formula (2.1) asserts that u assumes its maximum certainly on the boundary of Ω, possibly also in Ω.)

Indeed under the stronger assumption $\Delta u > 0$ in Ω, relation (2.1) follows from the fact that u cannot assume its maximum at any point $\xi \in \Omega$; for

*([28])

then $\partial^2 u/\partial x_k^2 \leqslant 0$ at ξ for all k, and hence $\Delta u \leqslant 0$. In the case $\Delta u \geqslant 0$ in Ω we make use of the auxiliary function $v = |x|^2$ for which $\Delta v > 0$ in Ω. Then for any constant $\varepsilon > 0$ the function $u + \varepsilon v$ belongs to $C^2(\Omega) \cap C^0(\overline{\Omega})$, and satisfies $\Delta(u + \varepsilon v) > 0$ in Ω; hence

$$\max_{\overline{\Omega}} (u + \varepsilon v) = \max_{\partial\Omega} (u + \varepsilon v).$$

Then also

$$\max_{\overline{\Omega}} u + \varepsilon \min_{\overline{\Omega}} v \leqslant \max_{\partial\Omega} u + \varepsilon \max_{\partial\Omega} v.$$

For $\varepsilon \to 0$ we obtain (2.1).

In the special case where u is harmonic in Ω, relation (2.1) also applies to $-u$. Since $\min u = -\max(-u)$ we obtain

$$\min_{\overline{\Omega}} u = \min_{\partial\Omega} u. \tag{2.2}$$

These relations imply a maximum principle for the absolute value (using that $|a| = \max(a, -a)$ for real a):

If $u \in C^2(\Omega) \cap C^0(\overline{\Omega})$ and $\Delta u = 0$ in Ω, then

$$\max_{\overline{\Omega}} |u| = \max_{\partial\Omega} |u|. \tag{2.3}$$

In particular $u = 0$ in Ω if $u = 0$ on $\partial\Omega$. This implies an improved uniqueness theorem for the Dirichlet problem (see p. 95) not requiring u to have derivatives on the boundary:

A function $u \in C^2(\Omega) \cap C^0(\overline{\Omega})$ is determined uniquely by the values of Δu in Ω and of u on $\partial\Omega$.

Stricter versions of the maximum principle *flow from the mean value theorem* (1.27):

Let $u \in C^2(\Omega)$ and $\Delta u \geqslant 0$ in Ω. Then either u is a constant or

$$u(\xi) < \sup_{\Omega} u \quad \textit{for all } \xi \in \Omega. \tag{2.4}$$

More generally (2.4) holds for any nonconstant u that is subharmonic in Ω.

For the proof we set $M = \sup u$, and decompose Ω into the two disjoint sets

$$\Omega_1 = \{\xi \in \Omega \,|\, u(\xi) = M\} \qquad \Omega_2 = \{\xi \in \Omega \,|\, u(\xi) < M\}.$$

The set Ω_2 is open because of the continuity of u. Using the fact that u is subharmonic we can show that Ω_1 is open as well. For if $\xi \in \Omega_1$, we have from (1.27) for all sufficiently small ρ

$$0 \leqslant \int_{|x-\xi|=\rho} u(x)\,dS_x - \omega_n \rho^{n-1} u(\xi)$$

$$= \int_{|x-\xi|=\rho} (u(x)-u(\xi))\,dS_x = \int_{|x-\xi|=\rho} (u(x)-M)\,dS_x.$$

Since $u(x)-M$ is continuous and $\leqslant 0$, it follows that $u(x)-M=0$ on every sufficiently small sphere with center ξ. Hence all x in a neighborhood of ξ belong to Ω_1, and Ω_1 is open. By definition a *connected* open set cannot be decomposed into two disjoint open nonempty sets. Thus either Ω_1 or Ω_2 is empty, proving the principle.

As an immediate consequence we have:

If $u \in C^2(\Omega) \cap C^0(\bar{\Omega})$ *and* $\Delta u \geqslant 0$ *in* Ω, *then either* $u = \text{const.}$ *or*

$$u(\xi) < \max_{\partial\Omega} u \quad \text{for all } \xi \in \Omega. \tag{2.5}$$

Again (2.5) holds more generally for all u that are continuous in $\bar{\Omega}$ and subharmonic in Ω. (We need only observe that max u cannot be assumed at a point of Ω.)

Problems

1. Let Ω denote the unbounded set $|x|>1$. Let $u \in C^2(\bar{\Omega})$, $\Delta u = 0$ in Ω, and $\lim_{x\to\infty} u(x) = 0$. Show that

$$\max_{\bar{\Omega}} |u| = \max_{\partial\Omega} |u|.$$

[Hint: Apply maximum principle to spherical shell.]

2. Let $u(x) \in C^2(\Omega) \cap C^0(\bar{\Omega})$ be a solution of

$$\Delta u + \sum_{k=1}^{n} a_k(x) u_{x_k} + c(x) u = 0, \tag{2.6}$$

where $c(x) < 0$ in Ω. Show that $u=0$ on $\partial\Omega$ implies $u=0$ in Ω. [Hint: Show that $\max u \leqslant 0$, $\min u \geqslant 0$.]

3. Prove the weak maximum principle (2.1) for solutions of the two-dimensional elliptic equation

$$Lu = au_{xx} + 2bu_{xy} + cu_{yy} + 2du_x + 2eu_y = 0, \tag{2.7}$$

where a,b,c,d,e are continuous functions of x,y in $\bar{\Omega}$ with $ac-b^2>0, a>0$. [Hint: Prove first the maximum principle for solutions of $Lu>0$, using that at a maximum point in Ω

$$u_{xx}u_{yy} - u_{xy}^2 \geqslant 0, \qquad u_{xx} \leqslant 0, \qquad u_{yy} \leqslant 0.$$

Apply this to $u+\varepsilon v$ where

$$v=\exp\left[M\left((x-x_0)^2+(y-y_0)^2\right)\right],$$

with (x_0,y_0) outside $\bar{\Omega}$ and M sufficiently large.]

4. Let $n=2$ and Ω be the half plane $x_2>0$. Prove that $\sup_\Omega u=\sup_{\partial\Omega} u$ for $u\in C^2(\Omega)\cap C^0(\bar{\Omega})$ which are harmonic in $\bar{\Omega}$, under the additional assumption that u is bounded from above in $\bar{\Omega}$. (The additional assumption is needed to exclude examples like $u=x_2$.) [Hint: Take for $\varepsilon>0$ the harmonic function

$$u(x_1,x_2)-\varepsilon\log\sqrt{x_1^2+(x_2+1)^2}\ .$$

Apply the maximum principle to a region $x_1^2+(x_2+1)^2<a^2, x_2\geqslant 0$ with large a. Let $\varepsilon\to 0$.]

3. The Dirichlet Problem, Green's Function, and Poisson's Formula

We derived the representation (1.23) for u in terms of Cauchy data, involving a fundamental solution $G(x,\xi)$. If here $G(x,\xi)=0$ for $x\in\partial\Omega$ and $\xi\in\Omega$, the term with du/dn_x drops out and we have solved the Dirichlet problem. We call a fundamental solution $G(x,\xi)$ with pole ξ a *Green's function* (for the Dirichlet problem for the Laplace equation in the domain Ω), if

$$G(x,\xi)=K(x,\xi)+v(x,\xi) \tag{3.1}$$

for $x\in\bar{\Omega}$, $\xi\in\Omega$, $x\neq\xi$, with K defined by (1.16), (1.15a,b), where $v(x,\xi)$ for $\xi\in\Omega$ is a solution of $\Delta_x v=0$, of class $C^2(\bar{\Omega})$, for which

$$G(x,\xi)=0 \quad \text{for } x\in\partial\Omega,\ \xi\in\Omega. \tag{3.2}$$

To construct G in general we have to find a harmonic v with $v=-K$ on $\partial\Omega$, which is again a Dirichlet problem. However in some cases G can be produced explicitly. This in particular is the case when Ω is a halfspace or a ball; then G can be obtained by reflection, leading to Poisson's integral formula. We note that for $n=2$ the Laplace equation is invariant under conformal mappings. Thus if we can solve the Dirichlet problem for a circular disk, we can solve it for any region which can be mapped conformally onto a disk.

To derive *Poisson's integral formula* for the solution of the Dirichlet problem for the ball of radius a and center 0,

$$\Omega=B(0,a)=\{x\,|\,|x|<a\}, \tag{3.3}$$

we make use of the fact that the sphere $\partial\Omega$ is the locus of points x for which the ratio of distances $r=|x-\xi|$ and $r^*=|x-\xi^*|$ from certain points ξ and ξ^* is constant. Here for ξ we can choose any point of Ω. Then ξ^* is the point obtained from ξ by "reflection" with respect to the sphere $\partial\Omega$, that is,

$$\xi^* = \frac{a^2}{|\xi|^2}\xi. \tag{3.4}$$

One easily verifies that

$$\frac{r^*}{r} = \frac{a}{|\xi|} = \text{const.} \quad \text{for } x \in \partial\Omega. \tag{3.5}$$

For $n > 2$ the fundamental solutions with poles ξ and ξ^*

$$K(x,\xi) = \frac{1}{(2-n)\omega_n} r^{2-n}, \qquad K(x,\xi^*) = \frac{1}{(2-n)\omega_n} r^{*2-n}$$

are then related for $x \in \partial\Omega$ by

$$K(x,\xi^*) = \left(\frac{a}{|\xi|}\right)^{2-n} K(x,\xi).$$

Thus the function

$$G(x,\xi) = K(x,\xi) - \left(\frac{|\xi|}{a}\right)^{2-n} K(x,\xi^*) \tag{3.6}$$

vanishes for $x \in \partial\Omega$. Moreover the second term is singular only when $x = \xi^*$. Since ξ^* lies outside $\overline{\Omega}$, the function $K(x,\xi^*)$ is harmonic in x throughout $\overline{\Omega}$. Thus G is a Green's function and formula (1.23) applies for $u \in C^2(\overline{\Omega})$. In the special case where $\Delta u = 0$ in the ball Ω, and $u \in C^2(\overline{\Omega})$ we find, after a simple computation, *Poisson's integral formula*

$$u(\xi) = \int_{|x|=a} H(x,\xi) u(x)\, dS_x \tag{3.7}$$

valid for $|\xi| < a$. Here H, given by

$$H(x,\xi) = \frac{1}{a\omega_n} \frac{a^2 - |\xi|^2}{|x-\xi|^n}, \tag{3.8}$$

is the *Poisson kernel*. We arrive at the same formula for $n = 2$.

For given boundary values $u = f$ on $\partial\Omega$ formula (3.7) solves the Dirichlet problem, provided that problem has a solution $u \in C^2(\overline{\Omega})$. We shall verify directly that for f continuous on $\partial\Omega$ the problem actually has a solution given by Poisson's formula:

Let f be continuous for $|x| = a$. Then the function $u(\xi)$ given by $f(\xi)$ for $|\xi| = a$ and by

$$u(\xi) = \int_{|x|=a} H(x,\xi) f(x)\, dS_x \tag{3.9}$$

for $|\xi| < a$, is continuous for $|\xi| \leqslant a$, and is in C^∞ and harmonic for $|\xi| < a$.

The proof follows easily from the following properties of H:

(a) $H(x,\xi)\in C^{\infty}$ for $|x|\leqslant a$, $|\xi|<a$, $x\neq\xi$.
(b) $\Delta_{\xi}H(x,\xi)=0$ for $|\xi|<a$, $|x|=a$.
(c) $\int_{|x|=a}H(x,\xi)\,dS_x=1$ for $|\xi|<a$.
(d) $H(x,\xi)>0$ for $|x|=a$, $|\xi|<a$.
(e) If $|\zeta|=a$, then

$$\lim_{\substack{\xi\to\zeta\\ |\xi|<a}} H(x,\xi)=0$$

uniformly in x for $|x-\zeta|>\delta>0$.

Here properties (a), (d), (e) are clear by inspection of (3.8). Property (b) follows either from (3.8) or by observing that $G(x,\xi)$ as defined by (3.6) is harmonic in x, that

$$G(x,\xi)=G(\xi,x), \tag{3.10}$$

and that $H(x,\xi)=dG(x,\xi)/dn_x$. To derive (c) directly would be tedious; but (c) follows simply by applying (3.7) to the function $u(x)=1$, which is harmonic and has boundary values 1.

By differentiation under the integral sign we find immediately from (a), (b) that the $u(\xi)$ defined by (3.9) belongs to C^{∞} and is harmonic for $|\xi|<a$. There remains to prove the continuity of u for $|\xi|\leqslant a$. Let $|\zeta|=a$, $|\xi|<a$. By (c)

$$\begin{aligned} u(\xi)-f(\zeta)&=\int_{|x|=a}H(x,\xi)(f(x)-f(\zeta))\,dS_x\\ &=I_1+I_2, \end{aligned} \tag{3.11}$$

where

$$I_1=\int_{\substack{|x-\zeta|<\delta\\ |x|=a}}\ldots,\qquad I_2=\int_{\substack{|x-\zeta|>\delta\\ |x|=a}}\ldots.$$

For a given $\varepsilon>0$ we can choose $\delta=\delta(\varepsilon)>0$ so small that

$$|f(x)-f(\zeta)|<\varepsilon\quad\text{for } |x-\zeta|<\delta,\ |x|=a,$$

since f is continuous. Then $|I_1|\leqslant\varepsilon$ by (c), (d). Let $\max|f(x)|=M$ for $|x|=a$. By (e) we can find a δ' such that

$$H(x,\xi)<\frac{\varepsilon}{2M\omega_n a^{n-1}}\quad\text{for } |\xi-\zeta|<\delta',\ |x-\zeta|>\delta, \tag{3.12}$$

where δ' depends on ε and $\delta=\delta(\varepsilon)$, and hence only on ε. (See Figure 4.2.) Then also $|I_2|<\varepsilon$. Hence

$$|u(\xi)-f(\zeta)|<2\varepsilon\quad\text{for } |\xi-\zeta|<\delta',\ |\xi|<a,$$

which shows that u is continuous at the boundary point ζ. This completes the proof of the theorem.

As an application we derive estimates for the derivatives of a harmonic function u in terms of estimates for u. We take at first the situation, where Ω is the ball $|x|<a$, where $u \in C^2(\bar{\Omega})$, and $\Delta u = 0$ in Ω. Using (3.7), (3.8) we find by a simple computation that

$$u_{\xi_i}(0) = \frac{n}{\omega_n a^{n+1}} \int_{|x|=a} x_i u(x)\, dS_x \tag{3.13a}$$

and hence that

$$|u_{\xi_i}(0)| \leqslant \frac{n}{a} \max_{|x|=a} |u(x)|. \tag{3.13b}$$

Let next $\Delta u = 0$ in any open set Ω. Let $\xi \in \Omega$ and let ξ have distance $d(\xi)$ from the boundary $\partial\Omega$. We apply (3.13b) to a ball of any radius $a < d(\xi)$ with center at ξ. For $a \to d(\xi)$ we obtain the inequality

$$|u_{\xi_i}|(\xi) \leqslant \frac{n}{d(\xi)} \sup_{\Omega} |u|. \tag{3.13c}$$

Similar inequalities obviously hold for higher derivatives also. Let ω denote a compact subset of Ω. Then there exists a positive lower bound for $d(\xi)$ in ω. All harmonic u with a common upper bound M for $|u|$ in Ω, will have a common upper bound in ω for the absolute values of their derivatives up to a fixed order.

The inequalities for derivatives of u in terms of u lead to completeness and compactness properties of the set of harmonic functions. Let there be

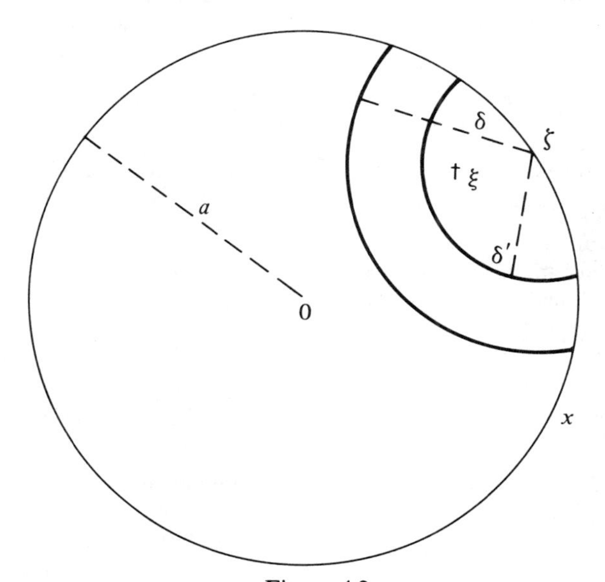

Figure 4.2

given a sequence $u_k \in C^2(\Omega) \cap C^0(\bar{\Omega})$ satisfying $\Delta u_k = 0$ in Ω. Assume first that the u_k converge uniformly on $\partial\Omega$ towards a function f. Then the u_k converge uniformly in $\bar{\Omega}$ to a continuous function u by the maximum principle. In any compact subset ω of Ω for a fixed multi-index α the derivatives $D^\alpha u_k$ converge uniformly to $D^\alpha u$. Hence $u \in C^\infty(\Omega)$ and $\Delta u = 0$. (We conclude in particular that uniform limits of harmonic functions are again harmonic.) Under the weaker assumption that $|u_k| \leqslant M$ on $\partial\Omega$ for all k, we have a common bound for the u_k and their first derivatives in the compact subset ω. By Arzela's theorem there exists then a *subsequence* of the u_k which converges uniformly in ω towards a function u. We conclude as before that u is harmonic in ω. (Compare the corresponding derivative estimates and compactness theorems for analytic functions, p. 71.)

Problems

1. Verify the symmetry property (3.10) for the Green's function (3.6).

2. Let $n=2$ and Ω be the halfplane $x_2 > 0$. For $\xi = (\xi_1, \xi_2)$ define by reflection $\xi^* = (\xi_1, -\xi_2)$.
   (a) Show that $G(x,\xi) = K(x,\xi) - K(x,\xi^*)$ is a Green's function for Ω. Derive formally the corresponding Poisson formula
$$u(\xi) = u(\xi_1,\xi_2) = \frac{1}{\pi} \int_{-\infty}^{\infty} \frac{\xi_2 f(x_1)}{(x_1-\xi_1)^2 + \xi_2^2}\, dx_1. \tag{3.14}$$
   (b) Show that (3.14) actually represents a bounded solution of the Dirichlet problem for Ω, if $f(x_1)$ is bounded and continuous.
   (c) Show that the maximum principle is satisfied by the solution (3.14) (see problem 4 of p. 106).

3. For $n=2$ find a Green's function for the quadrant $x_1 > 0$, $x_2 > 0$ by repeated reflection.

4. (Liouville's theorem.) Prove that a harmonic function defined and bounded in all of $\mathbb{R}^n$ is a constant. [Hint: Apply (3.13c) to balls of increasing radius.]

5. (a) Let B denote the ball $|x| < a$ in $\mathbb{R}^n$ and B_+ the semi-ball $\{x \mid |x| < a; x_n > 0\}$. Let u be continuous in $\bar{B}_+$, harmonic in B_+, and vanish for $x_n = 0$. Extend u as an odd function of x_n to all of $\bar{B}$. Show that the extended u is harmonic in B. [Hint: Let u^* be the harmonic function in B with boundary values u on ∂B. Then u^* is odd in x_n, and $u^* = u$ in B_+.]
   (b) Let $u(x_1, \ldots, x_n)$ be continuous and bounded in the half-space $x_n \geqslant 0$, and be harmonic for $x_n > 0$ and vanish for $x_n = 0$. Show that $u \equiv 0$. [Hint: Use problem 4.] (Compare problem 4, p. 106.)

6. Let $f(\theta)$ be a C^4-function of period 2π with Fourier series
$$f(\theta) = \sum_{n=0}^{\infty} (a_n \cos(n\theta) + b_n \sin(n\theta)).$$

(a) Prove that

$$u = \sum_{n=0}^{\infty} (a_n \cos(n\theta) + b_n \sin(n\theta)) r^n \tag{3.15}$$

represents in polar coordinates r, θ the solution of the Laplace equation $\Delta u = 0$ in the disk $x^2 + y^2 < 1$ with boundary values f.

(b) Derive Poisson's integral formula (3.7), (3.8) for $n=2$ from (3.15) by substituting for the a_n, b_n their Fourier expressions in terms of f and interchanging summation and integration.

7. (Harnack's inequality.) Let $u \in C^2$ for $|x| < a$; $u \in C^0$ for $|x| \leqslant a$; $u \geqslant 0$, $\Delta u = 0$ for $|x| < a$. Show that for $|\xi| < a$

$$\frac{a^{n-2}(a - |\xi|)}{(a + |\xi|)^{n-1}} u(0) \leqslant u(\xi) \leqslant \frac{a^{n-2}(a + |\xi|)}{(a - |\xi|)^{n-1}} u(0). \tag{3.16}$$

8. Show that the constant n in the inequality (3.13b) can be replaced by

$$\gamma_n = \frac{2n\omega_{n-1}}{(n-1)\omega_n} \tag{3.17}$$

and that this is the smallest possible constant in the inequality. [Hint: Use $u = \operatorname{sgn} x_n$ on $|x| = a$ in (3.13a).]

9. Let $\Delta u(x) = 0$ and $|u(x)| \leqslant M$ for $|x - \xi| < a$.

(a) Show that

$$|D^\alpha u(\xi)| \leqslant \left(\frac{m}{a}\gamma_n\right)^m M \quad \text{for } |\alpha| = m, \tag{3.18}$$

where γ_n is defined by (3.17) for dimension n. [Hint: Apply (3.13c) successively to the kth derivatives in the balls $|x - \xi| \leqslant a(m-k)/m$ for $k = 0, 1, \ldots, m-1$.]

(b) Show that $u(x)$ is represented by its series in terms of powers of $x - \xi$ for

$$|x_1 - \xi_1| + \cdots + |x_n - \xi_n| \leqslant \frac{a}{e\gamma_n} \tag{3.19}$$

(compare with problem 5., p. 79).

4. Proof of Existence of Solutions for the Dirichlet Problem Using Subharmonic Functions ("Perron's Method")

The proof is based on the fact that we can characterize the solution w of the Dirichlet problem by an *extremal property*; we only have to prove that a function with the extremal property exists. Assume we had a function $w \in C^2(\Omega) \cap C^0(\bar{\Omega})$, such that $\Delta w = 0$ in Ω and that w takes the prescribed values f on $\partial\Omega$. Let u be any function which is subharmonic in Ω and continuous in $\bar{\Omega}$, for which $u \leqslant f$ on $\partial\Omega$. Then $u - w \leqslant 0$ on $\partial\Omega$, and hence $u - w \leqslant 0$ in Ω by the maximum principle for subharmonic functions on p. 105. Thus the value of the solution w of the Dirichlet problem at any point ξ of Ω is the largest value taken at ξ by any subharmonic function with boundary values $\leqslant f$, and this will be our extremal property. The construction of w from its extremal property proceeds by a number of easily verified elementary lemmas. Essential use is made of the maximum principle and of the solvability of the Dirichlet problem for a ball, assured by Poisson's integral formula (3.7), (3.8). This somewhat limits the extension of the method to more general partial differential equations.

In what follows we denote again by $B(\xi,\rho)$ the open ball of center ξ and radius ρ in $\mathbb{R}^n$, by $\bar{B}(\xi,\rho)$ its closure, and by $S(\xi,\rho)$ its boundary. For a continuous $u = u(x)$ we denote by

$$M_u(\xi,\rho) = \frac{\rho^{1-n}}{\omega_n} \int_{S(\xi,\rho)} u(x)\, dS_x \tag{4.1}$$

the "arithmetic mean" of u on $S(\xi,\rho)$. We consider an open, bounded, and connected set Ω in $\mathbb{R}^n$. In accordance with our previous definition we call u subharmonic in Ω, if $u \in C^0(\Omega)$, and if for every ξ in Ω the inequality

$$u(\xi) \leqslant M_u(\xi,\rho) \tag{4.2}$$

holds for all sufficiently small ρ. We denote by $\sigma(\Omega)$ the set of functions subharmonic in Ω. The maximum principle proved on p. 105 constitutes

Lemma I. *For $u \in \sigma(\Omega) \cap C^0(\bar{\Omega})$ we have*

$$\max_{\bar{\Omega}} u \leqslant \max_{\partial\Omega} u. \tag{4.3}$$

Definition. For $u \in C^0(\Omega)$ and $\bar{B}(\xi,\rho) \subset \Omega$ we define $u_{\xi,\rho}$ as that function in $C^0(\Omega)$ for which

$$u_{\xi,\rho}(x) = u(x) \quad \text{for } x \in \Omega,\ x \notin B(\xi,\rho) \tag{4.4a}$$

$$\Delta_x u_{\xi,\rho}(x) = 0 \quad \text{for } x \in B(\xi,\rho). \tag{4.4b}$$

(That is, $u_{\xi,\rho}$ is obtained from u by replacing u in the ball $B(\xi,\rho)$ by the harmonic function that agrees with u on the boundary of the ball. The existence and uniqueness of u has been established earlier.)

Lemma II. *For $u \in \sigma(\Omega)$ and $\bar{B}(\xi,\rho) \subset \Omega$ we have*

$$u(x) \leqslant u_{\xi,\rho}(x) \quad \text{for all } x \in \Omega \tag{4.5a}$$

$$u_{\xi,\rho} \in \sigma(\Omega). \tag{4.5b}$$

PROOF. By definition (4.5a) holds for $x \notin B(\xi,\rho)$, with the equal sign. With

u also its restriction to $B(\xi,\rho)$ is subharmonic in B, and so is the harmonic function $-u_{\xi,\rho}$. Thus $u-u_{\xi,\rho}\in\sigma(B(\xi,\rho))$. Since $u-u_{\xi,\rho}$ is continuous in the ball $\bar{B}(\xi,\rho)$ and vanishes on its boundary, it follows from Lemma I that $u-u_{\xi,\rho}\leqslant 0$ in $B(\xi,\rho)$, which establishes (4.5a). In order to prove (4.5b) we have to show that for any $\xi\in\Omega$

$$u_{\xi,\rho}(\zeta)\leqslant M_{u_{\xi,\rho}}(\zeta,\tau) \tag{4.6}$$

for all sufficiently small τ. Now this is true for $\zeta\notin\bar{B}(\xi,\rho)$ by (4.4a), since $u\in\sigma(\Omega)$, and is true for $\zeta\in B(\xi,\rho)$ since then $u_{\xi,\rho}$ is harmonic near ζ. There remains the case where $\zeta\in S(\xi,\rho)$. But then by (4.2), (4.5a)

$$u_{\xi,\rho}(\zeta)=u(\zeta)\leqslant M_u(\zeta,\tau)\leqslant M_{u_{\xi,\rho}}(\zeta,\tau) \tag{4.7}$$

for all sufficiently small τ. □

Lemma III. *For $u\in\sigma(\Omega)$ inequality* (4.2) *holds whenever $\bar{B}(\xi,\rho)\subset\Omega$.*

PROOF. By (4.4a), (4.5a) and the mean value theorem for harmonic functions

$$u(\xi)\leqslant u_{\xi,\rho}(\xi)=M_{u_{\xi,\rho}}(\xi,\rho)=M_u(\xi,\rho).$$ □

Lemma IV. *A necessary and sufficient condition for u to be harmonic in Ω is that both u and $-u$ belong to $\sigma(\Omega)$.*

PROOF. The necessity is implied by the mean value theorem for harmonic functions. Let, on the other hand, $u\in\sigma(\Omega)$, $-u\in\sigma(\Omega)$. Then for $\bar{B}(\xi,\rho)\subset\Omega$ by Lemma II

$$u(x)\leqslant u_{\xi,\rho}(x), \qquad -u(x)\leqslant -u_{\xi,\rho}(x) \quad \text{for } x\in\Omega.$$

Thus $u=u_{\xi,\rho}$; hence u is harmonic in $B(\xi,\rho)$. □

Lemma IVa. *If $u\in C^0(\Omega)$, and if for each $\xi\in\Omega$*

$$u(\xi)=M_u(\xi,\rho)$$

for all sufficiently small ρ, then u is harmonic in Ω.

PROOF. The assumptions imply that $u\in\sigma(\Omega)$, $-u\in\sigma(\Omega)$. □

Definition. For $f\in C^0(\partial\Omega)$, define

$$\sigma_f(\bar{\Omega})=\{u\,|\,u\in C^0(\bar{\Omega})\cap\sigma(\Omega),\ u\leqslant f \text{ on } \partial\Omega\} \tag{4.8}$$

$$w_f(x)=\sup_{u\in\sigma_f(\bar{\Omega})} u(x) \quad \text{for } x\in\Omega. \tag{4.9}$$

Set

$$m=\inf f, \qquad \mu=\sup f. \tag{4.10}$$

We observe that the constant m belongs to $\sigma_f(\bar{\Omega})$, and hence that the set $\sigma_f(\bar{\Omega})$ is not empty. Also by Lemma I

$$u(x) \leqslant \mu \quad \text{for } u \in \sigma_f(\bar{\Omega}),\ x \in \bar{\Omega}. \tag{4.11}$$

Thus w_f is well defined.

Lemma V. *Let* $u_1, \ldots, u_k \in \sigma_f(\bar{\Omega})$ *and* $v = \max(u_1, \ldots, u_k)$. *Then* $v \in \sigma_f(\bar{\Omega})$.

Proof. One easily verifies that $v \in C^0(\bar{\Omega})$, since $u_1, \ldots, u_k \in C^0(\bar{\Omega})$. Then by (4.2) for any $\xi \in \Omega$ and all sufficiently small ρ.

$$\begin{aligned} v(\xi) &= \max(u_1(\xi), \ldots, u_k(\xi)) \\ &\leqslant \max\big(M_{u_1}(\xi,\rho), \ldots, M_{u_k}(\xi,\rho)\big) \leqslant M_v(\xi,\rho). \end{aligned}$$ □

Lemma VI. w_f *is harmonic in* Ω.

Proof. Let $\bar{B}(\xi,\rho) \subset \Omega$. Let $x^1, x^2, \ldots$ be a sequence of points in $B(\xi,\rho')$ where $\rho' < \rho$. By (4.9) we can find functions $u_k^j \in \sigma_f(\bar{\Omega})$ for $k,j = 1,2,3,\ldots$ such that

$$\lim_{j \to \infty} u_k^j(x^k) = w_f(x^k). \tag{4.12}$$

Relation (4.12) is preserved if any u_k^j is replaced by any $u \in \sigma_f(\bar{\Omega})$ for which $u \geqslant u_k^j$ in Ω, since then

$$u_k^j(x^k) \leqslant u(x^k) \leqslant w_f(x^k).$$

Define the sequence $u^j(x)$ for $x \in \bar{\Omega}$, $j = 1,2,\ldots$ by

$$u^j(x) = \max\big(u_1^j(x), \ldots, u_j^j(x)\big).$$

Then $u^j \in \sigma_f(\bar{\Omega})$ by Lemma V and $u^j(x) \geqslant u_k^j(x)$ for $x \in \Omega$, $j \geqslant k$. Thus

$$\lim_{j \to \infty} u^j(x^k) = w_f(x^k) \quad \text{for all } k. \tag{4.13}$$

Replacing, if necessary, u^j by $\max(u^j, m)$, we can bring about that

$$m \leqslant u^j(x) \leqslant \mu \quad \text{for } x \in \Omega \tag{4.14}$$

(see (4.11)). Finally replacing u^j by $u^j_{\xi,\rho}$, we can arrive at a sequence $u^j \in \sigma_f(\bar{\Omega})$, for which (4.13), (4.14) hold, and which are harmonic in $B(\xi,\rho)$. Since the u^j lie between fixed bounds, we conclude from the compactness property of harmonic functions (see p. 110) that there exists a subsequence of the $u^j(x)$ converging to a harmonic function $W(x)$ for x in the compact subset $\bar{B}(\xi,\rho')$ of $B(\xi,\rho)$. It follows from (4.13) that

$$w_f(x^k) = W(x^k) \quad \text{for all } k.$$

(Observe that here the harmonic function W could depend on the choice of the sequence x^k and on that of the subsequence of the js.) Taking first for the x^k a sequence converging to a point x in $B(\xi,\rho')$ with $x^1 = x$, we conclude from the continuity of the corresponding W that $\lim_{k\to\infty} w_f(x^k)$ always exists, hence that w_f is continuous in $B(\xi,\rho')$. Taking next for the x^k a sequence dense in $B(\xi,\rho')$ we find that w_f agrees with a harmonic function W in all x^k, and hence, by continuity, in all $x \in B(\xi,\rho')$. This means that w_f is harmonic in a neighborhood of ξ, and hence throughout Ω. □

The harmonic function w_f is our candidate for the solution of the Dirichlet problem. We have to show that it has the prescribed boundary values f on $\partial\Omega$. This requires some additional assumption on the nature of the boundary $\partial\Omega$ of Ω. The assumption is formulated conveniently as existence of certain *barrier functions*, that is of subharmonic functions that are zero at one boundary point and negative at all others.

Barrier Postulate. Let there exist for each $\eta \in \partial\Omega$ a function ("barrier function") $Q_\eta(x) \in C^0(\bar{\Omega}) \cap \sigma(\Omega)$, for which

$$Q_\eta(\eta)=0, \qquad Q_\eta(x)<0 \quad \text{for } x \in \partial\Omega,\ x \neq \eta. \tag{4.15}$$

Lemma VII. *For* $\eta \in \partial\Omega$

$$\liminf_{\substack{x\in\Omega\\ x\to\eta}} w_f(x) \geqslant f(\eta). \tag{4.16}$$

Proof. Let ε and K be positive constants. Then

$$u(x)=f(\eta)-\varepsilon+KQ_\eta(x) \tag{4.17}$$

belongs to $C^0(\bar{\Omega}) \cap \sigma(\Omega)$, and satisfies

$$u(x) \leqslant f(\eta)-\varepsilon \quad \text{for } x \in \partial\Omega, \qquad u(\eta)=f(\eta)-\varepsilon. \tag{4.17a}$$

Since f is continuous there exists a $\delta=\delta(\varepsilon)>0$ such that $f(x)>f(\eta)-\epsilon$ for $x \in \partial\Omega$, $|x-\eta|<\delta$. Then by (4.17a)

$$u(x) \leqslant f(x) \tag{4.18}$$

for $x \in \partial\Omega$, $|x-\eta|<\delta$. Since $Q_\eta(x)$ has a negative upper bound for $|x-\eta| \geqslant \delta$, we can find a $K=K(\varepsilon)$ so large that (4.18) holds as well for $x \in \partial\Omega$, $|x-\eta| \geqslant \delta$. Thus $u \in \sigma_f(\bar{\Omega})$ and consequently

$$u(x) \leqslant w_f(x) \quad \text{for } x \in \Omega.$$

Then also

$$f(\eta)-\varepsilon=\lim_{x\to\eta} u(x) \leqslant \liminf_{\substack{x\in\Omega\\ x\to\eta}} w_f(x).$$

For $\varepsilon\to 0$ we obtain (4.16). □

Lemma VIII. *For* $\eta \in \partial\Omega$

$$\lim_{\substack{x\in\Omega\\ x\to\eta}} w_f(x)=f(\eta). \tag{4.19}$$

PROOF. In view of Lemma VII it is sufficient to show that

$$\limsup_{\substack{x \in \Omega \\ x \to \eta}} w_f(x) \leqslant f(\eta). \tag{4.20}$$

We consider the function $-w_{-f}(x)$ which is defined in Ω by

$$-w_{-f}(x) = -\sup u(x) \quad \text{for } u \in \sigma_{-f}(\overline{\Omega}).$$

Writing $u = -U$ we have

$$-w_{-f}(x) = \inf U(x) \tag{4.21}$$

taken over all U for which

$$-U \in C^0(\overline{\Omega}) \cap \sigma(\Omega), \qquad -U \leqslant -f \quad \text{on } \partial\Omega. \tag{4.22}$$

Then for any $u \in \sigma_f(\overline{\Omega})$ and U satisfying (4.22)

$$u - U \leqslant 0$$

on $\partial\Omega$, and hence in Ω by Lemma I. Thus also by (4.9), (4.21)

$$w_f(x) \leqslant -w_{-f}(x) \quad \text{for } x \in \Omega$$

Applying Lemma VII to $w_{-f}(x)$ we have then

$$\limsup_{\substack{x \in \Omega \\ x \to \eta}} w_f(x) \leqslant \limsup_{\substack{x \in \Omega \\ x \to \eta}} (-w_{-f}(x)) = -\liminf_{\substack{x \in \Omega \\ x \to \eta}} w_{-f}(x) \leqslant f(\eta). \qquad \square$$

Lemma VIII implies that $w_f \in C^0(\overline{\Omega})$ when we define $w_f = f$ on $\partial\Omega$. We thus have proved:

Theorem. *If the domain Ω has the postulated barrier property there exists a solution $w \in C^2(\Omega) \cap C^0(\overline{\Omega})$ of the Dirichlet problem for arbitrary continuous boundary values f.*

The barrier postulate can be verified for a large class of domains Ω. Take, for example, the case where the open set is *strictly convex* in the sense that through each point η of $\partial\Omega$ there passes a hyperplane π_η having only η in common with $\overline{\Omega}$. We then can use for $Q_\eta(x)$ a suitable linear function which vanishes on π_η and is negative on $\overline{\Omega}$ except at the point η. See also the following problem.

PROBLEMS

1. Show that the open set Ω satisfies the barrier postulate if for each $\eta \in \partial\Omega$ there exists a ball $B(\xi,\rho)$ such that $\overline{B}(\xi,\rho)$ and $\overline{\Omega}$ have just the point η in common. [Hint: Use for $-Q_\eta(x)$ a fundamental solution with spherical symmetry about the point ξ.]

2. Solve Problem 5a, p. 110 with the help of Lemma IV.

5. Solution of the Dirichlet Problem by Hilbert-Space Methods*

A variety of methods have been invented to solve the Dirichlet problem for the Laplace equation. Of greatest significance are methods that make little use of special features of the Laplace equation, and can be extended to other problems and other equations. Among these is the method to be described now, reducing the Dirichlet problem to a standard problem in Hilbert space, that of finding the normal to a hyperplane. Typically the solution of the Dirichlet problem proceeds in two steps. In the first step a *modified* ("generalized") Dirichlet problem is solved in a deceptively simple manner. The second step consists in showing that under suitable regularity assumptions on region and data the solution of the modified problem actually is a solution of the original problem. The second step, which involves more technical difficulties, will not be carried out here in full generality. From the point of view of applications one might even take the attitude that the modified problem already adequately describes the physical situation.

In this book we can only summarize the relevant properties of a Hilbert space. We start with the broader concept of a vector space S (over the real number field). This is a set of objects $u, v, \ldots$ closed under the operations of addition and of multiplication with real numbers ("scalars") $\lambda, \mu, \ldots$. For $u, v \in S$ and scalars λ, μ we have $\lambda u + \mu v$ defined as an element of S and obeying the usual arithmetic laws. The simplest example is the set of vectors in the finite-dimensional space $\mathbb{R}^n$.

We next assume that in S there is defined for $u, v \in S$ a real-valued inner product (u, v), such that

$$(u,v)=(v,u), \qquad (\lambda u+\mu v, w)=\lambda(u,w)+\mu(v,w) \tag{5.1}$$

$$(u,u)>0 \quad \text{for } u\neq 0, \qquad (0,0)=0, \tag{5.2}$$

where "0" denotes both the 0-element of S and the real number. (These assumptions characterize S as an "inner product space".) Property (5.2) suggests defining the length ("norm") of a vector u by

$$\|u\|=\sqrt{(u,u)}\ . \tag{5.3}$$

One easily verifies† then the Cauchy and triangle inequalities

$$|(u,v)| \leqslant \|u\|\,\|v\|, \qquad \|u+v\| \leqslant \|u\|+\|v\|. \tag{5.4}$$

*([9], [15], [10], [11])

† The first inequality follows from the fact that $(\lambda u + \mu v, \lambda u + \mu v) = (u, u)\lambda^2 + 2(u, v)\lambda\mu + (v, v)\mu^2$ is a non-negative quadratic form in λ, μ; the second follows then from $\|u + v\|^2 = \|u\|^2 + 2(u, v) + \|v\|^2 \leqslant (\|u\| + \|v\|)^2$.

A sequence $u^1, u^2, \ldots \in S$ is said to *converge* to the (necessarily unique) limit $u \in S$ if

$$\lim_{k \to \infty} \|u - u^k\| = 0. \tag{5.5}$$

A sequence $u^1, u^2, \ldots \in S$ is called a *Cauchy sequence* if

$$\lim_{j,k \to \infty} \|u^k - u^j\| = 0. \tag{5.6}$$

The space S is *complete* and is called a *Hilbert space* if every Cauchy sequence in S converges, (to an element of S). Every inner product space S can be *completed*, that is imbedded into a Hilbert space H in which S is dense. This completion is achieved in analogy to the construction of real numbers from rational ones. We define H as the set of Cauchy sequences $\{u^1, u^2, \ldots\}$ in S, identifying Cauchy sequences $\{u^1, u^2, \ldots\}$ and $\{v^1, v^2, \ldots\}$ for which

$$\lim_{k \to \infty} \|u^k - v^k\| = 0 \tag{5.7}$$

and identifying an element u of S with the Cauchy sequence $\{u, u, u, \ldots\}$ in H. We define multiplication by a scalar, addition, and inner product of elements of H by performing these operations for each element of the corresponding Cauchy sequence. It is easily seen (as for real numbers) that the resulting H is indeed complete and that S is dense in H; (as a matter of fact an element of H given by the Cauchy sequence $u^1, u^2, \ldots \in S$ is just the limit of the u^k).*

A *linear functional* ϕ on an inner product space S is a real-valued function defined on S for which

$$\phi(\lambda u + \mu v) = \lambda \phi(u) + \mu \phi(v). \tag{5.8}$$

The functional is *bounded* if there exists an estimate

$$|\phi(u)| \leqslant M \|u\| \tag{5.9}$$

valid with the same M for all $u \in S$. For any $v \in S$ the inner product $(u, v) = \phi(u)$ defines a bounded linear functional by (5.1), (5.4). In a Hilbert space H the converse holds as well:

Representation theorem. *Every bounded linear functional ϕ on H can be represented uniquely in the form $\phi(u) = (u, v)$ with a suitable element v of H.*

* More generally a *norm* in a vector space S is a functional $\|x\|$ with the properties

$$\|x\| > 0 \quad \text{for } x \in S, \quad x \neq 0; \quad \|0\| = 0$$

$$\|\lambda x\| = |\lambda| \, \|x\| \quad \text{for } \lambda \in \mathbb{R}, \quad x \in S; \quad \|x + y\| \leqslant \|x\| + \|y\| \quad \text{for } x, y \in S.$$

A complete normed vector space is a *Banach space*.

Associating with a functional ϕ, (that does not vanish identically), the *hyperplane* π in H consisting of the points u with $\phi(u)=1$, we can think of v as giving the direction of the *normal* to π, since for any two points u^1, u^2 in π the scalar product (u^1-u^2, v) vanishes. Interpreting the direction of v as that of the normal to the plane π, suggests that v also has the direction of the *shortest line* from the origin to π. Indeed the point

$$w=\frac{v}{\phi(v)}=\frac{v}{(v,v)} \tag{5.10}$$

lies on π, and is the (unique) point of π closest to 0, since for any u on π

$$\|u\|^2=\|u-w+w\|^2=\|w\|^2+\|u-w\|^2+2(u-w,w)\geqslant\|w\|^2 \tag{5.11}$$

since

$$(u-w,w)=\frac{(u-w,v)}{(v,v)}=\frac{\phi(u)-\phi(w)}{(v,v)}=0.$$

This extremal property of w or v is made use of in the standard proof of the representation theorem. (See Problem 1.)

We now reformulate the Dirichlet problem for the Laplace equation as the problem of representing a certain bounded functional ϕ in a Hilbert space as an inner product (u,v). In the usual version of the problem one looks for a function U with domain Ω satisfying $\Delta U=0$ in Ω and $U=f$ on $\partial\Omega$. Assuming the prescribed f to be defined not only on $\partial\Omega$ but throughout $\bar{\Omega}$, we look instead for the function $v=U-f$ for which

$$v=0 \quad \text{on } \partial\Omega, \qquad \Delta v=-w \quad \text{in } \Omega \tag{5.12}$$

with $w=\Delta f$ prescribed. For simplicity we restrict ourselves to an open bounded connected set Ω, to which the divergence theorem applies. In the space of functions of class $C^1(\bar{\Omega})$ we define the bilinear form (u,v) by

$$(u,v)=\int_\Omega \sum_k u_{x_k} v_{x_k}\,dx. \tag{5.13}$$

With this definition the space $C^1(\bar{\Omega})$ is not an inner product space, since $(u,u)=0$ has the non-vanishing solutions $u=\text{const}$. Denote by $\tilde{C}_0^1(\bar{\Omega})$ the subspace of functions u in $C^1(\bar{\Omega})$ that vanish on $\partial\Omega$. Obviously (u,v) can be used as inner product on $\tilde{C}_0^1$ with the corresponding squared norm given by the *Dirichlet integral*

$$\|u\|^2=(u,u)=\int_\Omega \sum_k u_{x_k}^2\,dx. \tag{5.14}$$

Let $v\in C^2(\bar{\Omega})$ be a solution of (5.12), where the prescribed w belongs to $C^0(\bar{\Omega})$. Then for any $u\in\tilde{C}_0^1(\bar{\Omega})$ we have by the divergence theorem

$$(u,v)=-\int_\Omega u\,\Delta v\,dx=\int_\Omega uw\,dx. \tag{5.15}$$

This suggests that v can be found by simply representing the known linear functional in u

$$\phi(u)=\int uw\,dx \tag{5.16}$$

as an inner product (u,v). To make use of the representation theorem we have to complete $\tilde{C}_0^1(\Omega)$ into a Hilbert space $H_0^1(\Omega)$ with respect to the *Dirichlet norm* (5.14), and to prove that the $\phi(u)$ defined by (5.16) gives rise to a *bounded* linear functional in that space. Our *modified version of the Dirichlet problem* is then the following:

Find a $v\in H_0^1(\Omega)$ *such that*

$$(u,v)=\phi(u) \quad \textit{for all } u\in H_0^1(\Omega), \tag{5.17}$$

where (u,v) *and* ϕ *are defined by* (5.13), (5.16) *in* $\tilde{C}_0^1$ *and by extension in* H_0^1.

To show that the functional ϕ is bounded, we have to derive an inequality (5.9). Since by Cauchy

$$\left(\int_\Omega uw\,dx\right)^2\leqslant\int_\Omega u^2\,dx\int_\Omega w^2\,dx, \tag{5.18}$$

it is sufficient to show that there exists an N such that

$$\int_\Omega u^2\,dx\leqslant N\int_\Omega \sum_k u_{x_k}^2\,dx=N\|u\|^2. \tag{5.19}$$

We show this *Poincaré inequality* first for $u\in\tilde{C}_0^1(\bar{\Omega})$. For simplicity we assume that $\partial\Omega$ is intersected by every parallel to the x_1-axis in a *finite* number of points. Since Ω is bounded it can be enclosed in a cube

$$\Gamma: \qquad |x_i|\leqslant a \quad \text{for } i=1,2,\ldots,n.$$

We continue u as identically zero outside Ω. Then for any $x=(x_1,\ldots,x_n)\in\Gamma$

$$\begin{aligned}u^2(x)&=\left(\int_{-a}^{x_1}u_{x_1}(\xi_1,x_2,\ldots,x_n)\,d\xi_1\right)^2\\&\leqslant(x_1+a)\int_{-a}^{x_1}u_{x_1}^2\,d\xi_1\leqslant 2a\int_{-a}^{a}u_{x_1}^2\,d\xi_1.\end{aligned}$$

Thus

$$\int_{-a}^{a}u^2(x_1,\ldots,x_n)\,dx_1\leqslant 4a^2\int_{-a}^{a}u_{x_1}^2\,d\xi_1.$$

Integrating over $x_2,\ldots,x_n$ from $-a$ to a we find

$$\int_\Gamma u^2\,dx\leqslant 4a^2\int_\Gamma u_{x_1}^2\,dx, \tag{5.19a}$$

which implies (5.19) with $N=4a^2$. An element u of H_0^1 is represented by a Cauchy sequence $u^1,u^2,\ldots\in\tilde{C}_0^1(\bar{\Omega})$ for which

$$\lim_{j,k\to\infty}\|u^k-u^j\|=0.$$

By (5.19), this implies that also

$$\lim_{j,k\to\infty}\int_\Omega (u^k-u^j)^2\,dx=0. \tag{5.20}$$

By (5.18), (5.16) the numbers $\phi(u^k)$ then form a Cauchy sequence, and we can define ϕ for the element u of H_0^1 by

$$\phi(u)=\lim_{k\to\infty}\phi(u^k). \tag{5.21}$$

Since also

$$\|u\|=\lim_{k\to\infty}\|u^k\|$$

by definition, the inequality (5.9) for $u\in\tilde{C}_0^1(\bar{\Omega})$ implies the same inequality for $u\in H_0^1$. One finds then that ϕ can be extended to the Hilbert space H_0^1 as a bounded linear functional. The representation theorem guarantees the existence of a v in $H_0^1(\Omega)$ for which (5.17) holds, and thus solves our modified Dirichlet problem.*

As mentioned earlier there remains the task of identifying the solution $v\in H_0^1(\Omega)$ of the modified problem with a *function* v that satisfies (5.12) in the ordinary sense. Here the verification of the differential equation $\Delta v=-w$, say for $w\in C^1(\bar{\Omega})$, is not as difficult as to show that $v=0$ on $\partial\Omega$. One first observes that the solution v of the modified problem is representable by a Cauchy sequence $v^1,v^2,\ldots$ in $\tilde{C}_0^1(\bar{\Omega})$, with respect to the norm (5.14). This implies by (5.20) applied to the v^k that v can be identified with a function which is square integrable in the sense of Lebesgue in Ω, and for which

$$\lim_{j\to\infty}\int_\Omega (v^j-v)^2\,dx=0. \tag{5.21a}$$

Using for u a test function of class $C^\infty(\Omega)$ and of compact support we have

$$(v,u)=\lim_{j\to\infty}(v^j,u)=\lim_{j\to\infty}\int_\Omega\sum_k v^j_{x_k}u_{x_k}\,dx$$

$$=-\lim_{j\to\infty}\int_\Omega v^j\Delta u\,dx=-\int_\Omega v\Delta u\,dx.$$

Since test functions belong to H_0^1 it follows from (5.16), (5.17) that

$$\int v\Delta u\,dx=-\int wu\,dx. \tag{5.22}$$

* The modified problem can be solved more generally for any bounded open set Ω by defining $H_0^1(\Omega)$ as the completion with respect to the Dirichlet norm of the set $C_0^1(\Omega)$ of all $u\in C^1(\Omega)$ that have compact support in Ω. This does not involve any regularity assumptions on $\partial\Omega$.

Hence v is a solution of the P.D.E.

$$\Delta v = -w \tag{5.23}$$

in the sense of distributions (see p. 69). From this we can prove that $v \in C^2(\Omega)$ and $\Delta v = -w$. Take any $z \in \Omega$ and ρ so small that the ball $B(z, 3\rho)$ lies in Ω. Take any test function ϕ with support in $B(z,\rho)$ and a fixed test function $\zeta(x)$ which has the value 1 in $B(z, 2\rho)$. Set

$$u(x) = \int K(x,\xi)\phi(\xi)\,d\xi, \tag{5.24}$$

where K is the fundamental solution given by (1.16). By Poisson's formula (1.31a)

$$\phi(x) = \Delta_x u(x) = \phi_1(x) + \phi_2(x),$$

where

$$\phi_1(x) = \Delta_x(\zeta(x)u(x)), \qquad \phi_2(x) = \Delta_x((1-\zeta(x))u(x)).$$

Thus

$$\int v(x)\phi(x)\,dx = \int v(x)\phi_1(x)\,dx + \int v(x)\phi_2(x)\,dx. \tag{5.25a}$$

Here, since $\zeta(x)u(x)$ again is a test function, by (5.22)

$$\int v(x)\phi_1(x)\,dx = -\int w(x)\zeta(x)u(x)\,dx \tag{5.25b}$$

$$\phi_2(x) = \Delta_x \int (1-\zeta(x))K(x,\xi)\phi(\xi)\,d\xi$$

$$= \int_{B(z,\rho)} F(x,\xi)\phi(\xi)\,d\xi. \tag{5.25c}$$

Here

$$F(x,\xi) = \Delta_x((1-\zeta(x))K(x,\xi))$$

belongs to C^∞ in x,ξ for $\xi \in B(z,\rho)$ and all x, since $K(x,\xi)$ is singular only for $x=\xi$ while $1-\zeta(x)=0$ in $B(z,2\rho)$. Moreover

$$\Delta_\xi F(x,\xi) = 0, \tag{5.26}$$

since $\Delta_\xi K(x,\xi) = 0$ for $x \neq \xi$. We find from (5.24), (5.25a,b,c) that

$$\int v(\xi)\phi(\xi)\,d\xi = \int \phi(\xi)\,d\xi\left(\int (-w(x)\zeta(x)K(x,\xi) + F(x,\xi)v(x))\,dx.\right)$$

Since this identity holds for all test functions ϕ with support in $B(z,\rho)$, we conclude that

$$\int \phi(\xi)(v(\xi) - V(\xi))\,d\xi = 0 \tag{5.26a}$$

where

$$V(\xi) = -\int K(x,\xi)\zeta(x)w(x)\,dx + \int F(x,\xi)v(x)\,dx. \tag{5.26b}$$

Here (5.26a) holds for all $\phi \in C_0^\infty(B(z,\rho))$, and then more generally by approximation for all $\phi \in C_0^1(B(z,\rho))$. In particular (5.26a) holds, if we replace ϕ by $\phi(v^j - V)$. For $j \to \infty$ we find that

$$\int \phi(\xi)(v(\xi) - V(\xi))^2\,d\xi = 0 \quad \text{for } \phi \in C_0^\infty(B(z,\rho)).$$

It follows from (5.26) and Poisson's equation (1.28) that v coincides almost everywhere in $B(z, \rho)$ with a function V of class C^2 for which

$$\Delta_\xi V(\xi) = -\zeta(\xi)w(\xi) = -w(\xi).$$

We can identify v with V. Since z is arbitrary we have $\Delta v = -w$ throughout Ω.

The proof that v has boundary values 0 will be given here only for the dimension $n = 2$. (For higher dimensions see p. 202.) Prescribe a number $\varepsilon > 0$. To v there corresponds a Cauchy sequence $v^k \in \tilde{C}_0^1(\bar{\Omega})$. We can find a number j such that

$$\|v^j - v^k\| \leqslant \varepsilon$$

for all $k > j$. Denote by $d(\xi)$ for $\xi \in \Omega$ the distance of ξ to a closest point ξ^* of the boundary curve $\partial\Omega$. We assume that $\partial\Omega$ is sufficiently regular and $d(\xi) = |\xi^* - \xi|$ sufficiently small. Then each x in $B(\xi, d(\xi))$ can be joined to a point of $\partial\Omega$ by a segment parallel to $\xi\xi^*$ and of length $\leqslant 4d(\xi)$. The union U of these segments covers $B(\xi, d(\xi))$ (See Figure 4.3). Poincaré's inequality (5.19a) applies* to $u = v^j - v^k$ in U and yields that

$$\iint_U (v^j - v^k)^2\,dx_1\,dx_2 \leqslant 16d^2(\xi)\|v^j - v^k\|^2 \leqslant 16d^2(\xi)\varepsilon^2$$

for $k > j$. By (5.21a) then

$$\iint_{B(\xi, d(\xi))} (v^j - v)^2\,dx_1\,dx_2 \leqslant 16d^2(\xi)\varepsilon^2. \tag{5.27}$$

For the solution v of $\Delta v = -w$ we have by (1.25) for $\rho < d(\xi)$

$$v(\xi) = -\iint_{B(\xi,\rho)} (\psi(|x-\xi|) - \psi(\rho))w(x)\,dx_1\,dx_2 + \frac{1}{2\pi\rho}\int_{S(\xi,\rho)} v(x)\,ds, \tag{5.28}$$

*In the proof of (5.19a) we did not need that u vanishes everywhere on the boundary of Ω. It is sufficient to know that in Ω each coordinate has a range of length $\leqslant 2a$, and that each point of Ω can be joined in Ω by a parallel to the x_1-axis to a point where u vanishes.

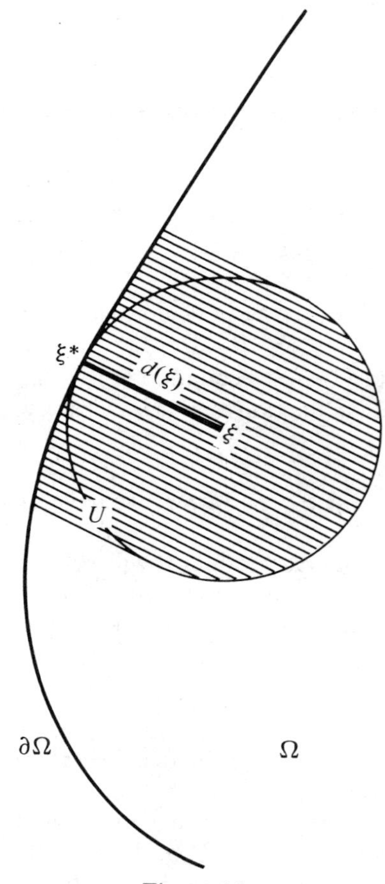

Figure 4.3

where $\psi(\rho) = (1/2\pi)\log \rho$. Let $M = \max_{\bar{\Omega}}|w|$. Multiplying (5.28) by $2\pi\rho$ and integrating with respect to ρ from 0 to $d(\xi)$ we find that

$$\begin{aligned}\pi d^2(\xi)|v(\xi)| &\leqslant \iint_{B(\xi, d(\xi))} |v|\,dx + \mathrm{O}\big(M d^4(\xi)\big)\\ &\leqslant \iint_{B(\xi, d(\xi))} |v - v^j|\,dx + \iint_{B(\xi, d(\xi))} |v^j|\,dx + \mathrm{O}\big(M d^4(\xi)\big).\end{aligned}$$

Since the continuous function v^j vanishes on $\partial\Omega$ we have $|v^j(x)| < \varepsilon$ for $x \in B(\xi, d(\xi))$ provided $d(\xi)$ is sufficiently small. In addition by (5.27)

$$\left(\iint_{B(\xi, d(\xi))}\left|v - v^j\right|\,dx\right)^2 \leqslant \pi d^2(\xi) \iint_{B(\xi, d(\xi))} (v - v^j)^2\,dx \leqslant 16\pi d^4(\xi)\varepsilon^2.$$

It follows then that for sufficiently small $d(\xi)$

$$v(\xi) = \mathrm{O}\big(\varepsilon + M d^2(\xi)\big).$$

This implies that $v(\xi)$ tends to 0 as ξ approaches the boundary $\partial\Omega$.

PROBLEMS

1. Prove the representation theorem in Hilbert space H by finding a point in a plane with the minimum distance from the origin. [Hint: Let $m=\inf\|u\|$ for $\phi(u)=1$. Take a minimizing sequence w^k with $\phi(w^k)=1$, $\|w^k\|\to m$. From

$$m^2 \leqslant \left\| \frac{w^k+w^j}{2} \right\|^2 = \frac{1}{2}\|w^k\|^2 + \frac{1}{2}\|w^j\|^2 - \frac{1}{4}\|w^k-w^j\|^2$$

prove that the w^k form a Cauchy sequence converging to an element w of H with $\phi(w)=1$, $\|w\|=m$. Show that $(w,u)=0$ for $\phi(u)=0$, and finally that $(v,u)=\phi(u)$ for $v=w/(w,w)$ and all u.]

2. Show that a solution $U\in C^2(\bar{\Omega})$ of

$$\Delta U=0 \quad \text{in } \Omega, \qquad U=f \quad \text{on } \partial\Omega$$

minimizes the Dirichlet integral (U,U) among all functions in $C^1(\bar{\Omega})$ with boundary values f ("Dirichlet's principle").

3. Consider the Dirichlet problem for $n=2$ with Ω being the unit disk referred to polar coordinates r,θ. Let $f(\theta)$ denote the boundary values assumed to be continuous, and $u(r\cos\theta, r\sin\theta)$ the solution of $\Delta u=0$ with boundary values f, known to be of class $C^2(\Omega)\cap C^0(\bar{\Omega})$.

(a) Show that for $r<1$

$$u=\sum_{k=0}^{\infty}(a_k\cos k\theta+b_k\sin k\theta)r^k, \tag{5.29}$$

where a_k,b_k are the Fourier coefficients of f. [Hint: prove first that u has a Fourier expansion of the form (5.29) for $r<1$ with certain a_n,b_n. Show that by continuity the a_n,b_n are the coefficients of f.]

(b) Show that the Dirichlet integral of u is given by

$$\|u\|^2=\pi\sum_{k=1}^{\infty}k\left(a_k^2+b_k^2\right). \tag{5.30}$$

(c) Show that there are continuous f for which the Dirichlet integral for the corresponding u is infinite. [Hint: Find sequences a_k,b_k for which the series in (5.30) diverges, while the series

$$\sum_{k=1}^{\infty}(|a_k|+|b_k|)$$

converges.] (This shows that the solution of Dirichlet problem cannot be determined from Dirichlet's principle for all continuous boundary values.)

4. (Best constant $1/\lambda$ in Poincaré's inequality). Show that if there exists a function $u\in C^2(\bar{\Omega})$ vanishing on $\partial\Omega$ for which the quotient

$$\frac{(u,u)}{\int_{\Omega}u^2\,dx}$$

reaches its smallest value λ, then u is an eigenfunction to the eigenvalue λ, so that $\Delta u+\lambda u=0$ in Ω. In fact λ must be the smallest eigenvalue belonging to an eigenfunction in $C^2(\bar{\Omega})$.

5 Hyperbolic Equations in Higher Dimensions*

1. The Wave Equation in n-Dimensional Space†

(a) *The method of spherical means*

The wave equation for a function $u(x_1,\ldots,x_n,t)=u(x,t)$ of n space variables $x_1,\ldots,x_n$ and the time t is given by

$$\Box u = u_{tt} - c^2\Delta u = 0 \tag{1.1}$$

with a positive constant c. The operator "$\Box$" defined by (1.1) is known as the *D'Alembertian*. For $n=3$ the equation can represent waves in acoustics or optics, for $n=2$ waves on the surface of water, for $n=1$ sound waves in pipes or vibrations of strings. In the *initial-value problem* we ask for a solution of (1.1) defined in the $(n+1)$-dimensional half space $t>0$ for which

$$u=f(x),\quad u_t=g(x)\quad \text{for } t=0. \tag{1.2}$$

The initial-value problem (1.1), (1.2) can be solved by the method of spherical means due to Poisson. We associate generally with a continuous function $h(x)=h(x_1,\ldots,x_n)$ in $\mathbb{R}^n$ its average $M_h(x,r)$ on a sphere with center x and radius r:

$$M_h(x,r)=\frac{1}{\omega_n r^{n-1}}\int_{|y-x|=r} h(y)\,dS_y. \tag{1.3}$$

Setting $y=x+r\xi$ with $|\xi|=1$, we get

$$M_h(x,r)=\frac{1}{\omega_n}\int_{|\xi|=1} h(x+r\xi)\,dS_\xi. \tag{1.4}$$

*([16], [21])

†([2], [6], [19])

Originally $M_h(x,r)$ is defined by (1.3) only for $r>0$. We can extend its definition to all real r using (1.4). The resulting $M_h(x,r)$ then is an *even* function of r, since replacing r by $-r$ in (1.4) can be compensated for by replacing the variable of integration ξ by $-\xi$. It is also clear from (1.4) that $M_h \in C^s(\mathbb{R}^{n+1})$ for $h \in C^s(\mathbb{R}^n)$ since we can differentiate under the integral sign. For $h \in C^2(\mathbb{R}^n)$ we find from (1.4), using the divergence theorem (4.1) of Chapter 3, p. 79, that

$$\begin{aligned}
\frac{\partial}{\partial r} M_h(x,r) &= \frac{1}{\omega_n} \int_{|\xi|=1} \sum_{i=1}^{n} h_{x_i}(x+r\xi)\xi_i \, dS_\xi \\
&= \frac{r}{\omega_n} \int_{|\xi|<1} \Delta_x h(x+r\xi)\, d\xi \\
&= \frac{r^{1-n}}{\omega_n} \Delta_x \int_{|y-x|<r} h(y)\, dy \\
&= \frac{r^{1-n}}{\omega_n} \Delta_x \int_0^r d\rho \int_{|y-x|=\rho} h(y)\, dS_y \\
&= r^{1-n} \Delta_x \int_0^r \rho^{n-1} M_h(x,\rho)\, d\rho.
\end{aligned}$$

Multiplying by r^{n-1} and differentiating with respect to r yields

$$\frac{\partial}{\partial r}\left(r^{n-1} \frac{\partial}{\partial r} M_h(x,r) \right) = \Delta_x r^{n-1} M_h(x,r). \tag{1.5}$$

Thus the spherical means $M_h(x,r)$ of any function $h \in C^2(\mathbb{R}^n)$ satisfy the partial differential equation

$$\left(\frac{\partial^2}{\partial r^2} + \frac{n-1}{r} \frac{\partial}{\partial r} \right) M_h(x,r) = \Delta_x M_h(x,r) \tag{1.6}$$

known as *Darboux's* equation. Using that the solution $M_h(x,r)$ of (1.6) is even in r, we find for its initial values

$$M_h(x,0) = h(x), \qquad \left(\frac{\partial}{\partial r} M_h(x,r) \right)_{r=0} = 0. \tag{1.7}$$

Forming spherical means we can transform the initial-value problem for the wave equation into one for a hyperbolic equation in two independent variables. Let $u(x,t)$ be a solution of (1.1), (1.2) of class C^2 in the half space $x \in \mathbb{R}^n$, $t \geqslant 0$. We form the spherical means of u as a function of x:

$$M_u(x,r,t) = \frac{1}{\omega_n} \int_{|\xi|=1} u(x+r\xi,t)\, dS_\xi. \tag{1.8}$$

Obviously u can be recovered from M_u, since

$$M_u(x,0,t) = u(x,t). \tag{1.9}$$

By (1.6)

$$\Delta_x M_u = \left(\frac{\partial^2}{\partial r^2} + \frac{n-1}{r}\frac{\partial}{\partial r} \right) M_u.$$

On the other hand by (1.1), (1.8)

$$\Delta_x M_u = \frac{1}{\omega_n} \int_{|\xi|=1} \Delta_x u(x+r\xi,t)\, dS_\xi$$
$$= \frac{1}{c^2}\frac{\partial^2}{\partial t^2}\frac{1}{\omega_n}\int_{|\xi|=1} u(x+r\xi,\ t\ )\, dS_\xi = \frac{1}{c^2}\frac{\partial^2}{\partial t^2} M_u.$$

Hence $M_u(x,r,t)$ as a function of the two scalar variables r,t for fixed x is a solution of the P.D.E.*

$$\frac{\partial^2}{\partial t^2} M_u = c^2 \left(\frac{\partial^2}{\partial r^2} + \frac{n-1}{r}\frac{\partial}{\partial r} \right) M_u \tag{1.10}$$

The P.D.E. (1.10) depending on the parameter n (here equal to the dimension of x-space) is known as the *Euler–Poisson–Darboux* equation. Our M_u as function of r,t by (1.2), (1.8) is a solution of (1.10) with the known initial values

$$M_u = M_f(x,r), \qquad \frac{\partial}{\partial t} M_u = M_g(x,r) \quad \text{for } t=0. \tag{1.11}$$

The initial-value problem (1.10), (1.11) can be solved most easily when the number of space dimensions is $n=3$†

Indeed by (1.10)

$$\frac{\partial^2}{\partial t^2}(rM_u) = c^2\left(r\frac{\partial^2}{\partial r^2} M_u + 2\frac{\partial}{\partial r} M_u \right) = c^2 \frac{\partial^2}{\partial r^2}(rM_u). \tag{1.12}$$

Thus $rM_u(x,r,t)$ as a function of r,t is a solution of the one-dimensional wave equation with initial values

$$rM_u = rM_f(x,r), \qquad \frac{\partial}{\partial t} rM_u = rM_g(x,r) \quad \text{for } t=0. \tag{1.13}$$

* The expression

$$\frac{\partial^2}{\partial r^2} + \frac{n-1}{r}\frac{\partial}{\partial r}$$

has been encountered already (Chapter 4, (1.9)) as the Laplacian of a function in $\mathbb{R}^n$ depending only on the distance r from a fixed point. Equation (1.10) thus asserts that the spherical means of a solution of (1.1) on spheres with center x again form a solution of (1.1). This is to be expected since the spherical means could plausibly be obtained by rotating the solution u in all possible ways about x and averaging over all rotations. Since the wave equation is invariant under rotations this procedure should lead again to a solution.

† This is more difficult for other values of n. See problem 2, p. 132.

Thus by our general formula ((4.13) of Chapter 2)

$$rM_u(x,r,t)=\frac{1}{2}\big[(r+ct)M_f(x,r+ct)+(r-ct)M_f(x,r-ct)\big]$$
$$+\frac{1}{2c}\int_{r-ct}^{ct+r}\xi M_g(x,\xi)\,d\xi.$$

Using that $M_f(x,r)$ and $M_g(x,r)$ are even in r we are led to

$$M_u(x,r,t)=\frac{(ct+r)M_f(x,ct+r)-(ct-r)M_f(x,ct-r)}{2r}$$
$$+\frac{1}{2rc}\int_{ct-r}^{ct+r}\xi M_g(x,\xi)\,d\xi.$$

Letting r tend to 0 and replacing differentiation with respect to r by differentiation with respect to ct, we find by (1.9) that

$$u(x,t)=tM_g(x,ct)+\frac{\partial}{\partial t}\big(tM_f(x,ct)\big)$$
$$=\frac{1}{4\pi c^2 t}\int_{|y-x|=ct}g(y)\,dS_y+\frac{\partial}{\partial t}\left(\frac{1}{4\pi c^2 t}\int_{|y-x|=ct}f(y)\,dS_y\right). \quad (1.14)$$

Any solution u of the initial-value problem (1.1), (1.2) of class C^2 for $t\geqslant 0$ in $n=3$ space dimensions is given by formula (1.14), hence is *unique*. Conversely for any $f\in C^3(\mathbb{R}^3)$ and $g\in C^2(\mathbb{R}^3)$ the $u(x,t)$ defined by (1.14) is of class C^2 and satisfies (1.1), (1.2). Indeed (1.2) follows by inspection, using (1.7). Moreover by (1.6) for $n=3$ and $r=ct$

$$\frac{\partial^2}{\partial t^2}\big(tM_g(x,ct)\big)=c\frac{\partial^2}{\partial r^2}\big(rM_g(x,r)\big)=cr\Delta_x M_g(x,r)=c^2\Delta_x\big(tM_g(x,ct)\big).$$

Thus $tM_g(x,ct)$, and similarly $(\partial/\partial t)tM_f(x,ct)$, satisfy the wave equation (1.1).*

Formula (1.14) displays the relevant features of the solution u of the initial-value problem for the wave equation in the case $n=3$. First of all, writing our spherical means in the form (1.4) we can carry out the t-differentiations under the integral sign, arriving at the expression

$$u(x,t)=\frac{1}{4\pi c^2 t^2}\int_{|y-x|=ct}\Big(tg(y)+f(y)+\sum_i f_{y_i}(y)(y_i-x_i)\Big)\,dS_y. \quad (1.15)$$

(1.15) indicates that u can be less regular than the initial data. There is a possible loss of one order of differentiability: $u\in C^s$, $u_t\in C^{s-1}$ initially, guarantee only that $u\in C^{s-1}$, $u_t\in C^{s-2}$ at a later time. This is the *focussing effect*, present when $n>1$.† For example the second derivatives of u could

* Incidentally the first expression for u in (1.14) defines u as a solution of (1.1) for *all* x,t, since $M_f(x,ct)$, $M_g(x,ct)$ are defined for all t as even functions in t.

† Irregularities in the initial data are "focussed" from different localities into a smaller set, ("caustic") leading to stronger irregularities. This phenomenon does not occur for $n=1$ where u is no worse than its data, as shown by formula (4.13) of Chapter 2.

become infinite at some point for $t>0$, though they are bounded for $t=0$. In contrast to the *pointwise* behavior of u, we shall find that in the L_2-sense u does not deteriorate. This follows from the fact that the *energy norm* of u

$$E(t)=\frac{1}{2}\iiint\Big(u_t^2(x,t)+c^2\sum_i u_{x_i}^2(x,t)\Big)\,dx \tag{1.16}$$

does not change at all with t. Indeed

$$\frac{dE}{dt}=\int\Big(u_t u_{tt}+c^2\sum_i u_{x_i}u_{x_i t}\Big)\,dx=\int\Big(u_t\Box u+c^2\sum_i (u_t u_{x_i})_{x_i}\Big)\,dx$$
$$=0 \tag{1.16a}$$

if $u(x,t)=0$ for all sufficiently large $|x|$.

According to (1.15) the value $u(x,t)$ depends on the values of g and of f and its first derivatives on the sphere $S(x,ct)$ of center x and radius ct. Thus the *domain of* dependence for $u(x,t)$ is the surface $S(x,ct)$. (See Figure 5.1.) Conversely the initial data f,g near a point y in the plane $t=0$ only *influence* u at the time t in points (x,t) near the cone $|x-y|=ct$. (See Figure 5.2). Let f,g have their support contained in a set $\Omega\in\mathbb{R}^3$. In order that $u(x,t)\neq 0$ the point x has to lie on a sphere of radius ct with its center y in Ω. The union of all spheres $S(y,ct)$ for $y\in\Omega$ contains the support of u at the time t. This gives rise again to Huygens's construction for a disturbance confined originally to Ω. (See p. 31.) The support of u spreads with velocity c. It is contained in the region bounded by the envelope of the spheres of radius ct with centers on $\partial\Omega$. Actually the support of $u(x,t)$ can be smaller. Take, for example, for the region Ω containing the support of f,g the ball $B(0,\rho)$ of radius ρ and center 0. Then $S(x,ct)$ for $ct>\rho$ will have a point in common with Ω only when x lies in the spherical shell bounded by the spheres $S(0,ct+\rho)$ and $S(0,ct-\rho)$. For any fixed x and all sufficiently large t (namely $t>(|x|+\rho)/c$) we have $u(x,t)=0$. A disturbance originating in $B(0,\rho)$ is confined at the time t to a shell of

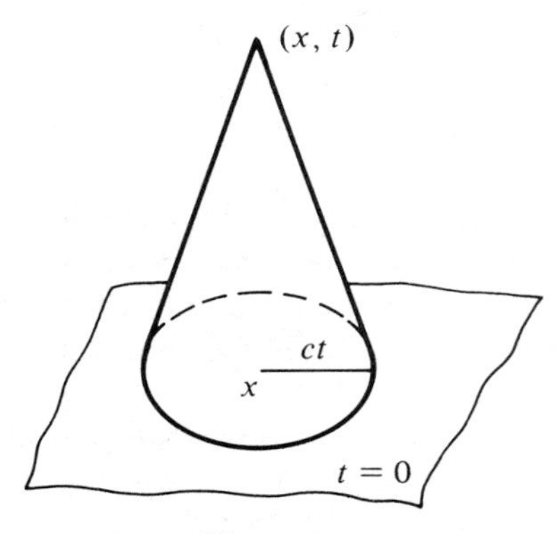

Figure 5.1

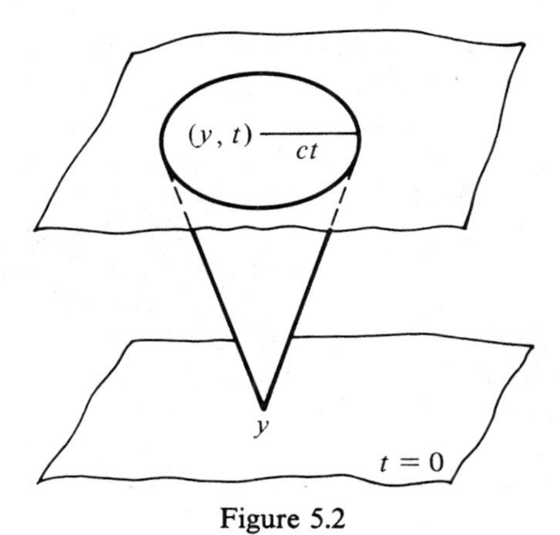

Figure 5.2

thickness 2ρ expanding with velocity c. (See Figure 5.3.) This accounts for the possibility of "sharp" signals being transmitted in accordance with equation (1.1) in three dimensions. This phenomenon is due to the fact that the domain of dependence for $u(x,t)$ is a *surface* in x-space rather than a solid region ("Huygens's principle in the strong form"). For most hyperbolic equations (even for the wave equation in an even number of dimensions) the principle does not hold. Disturbances propagate with finite

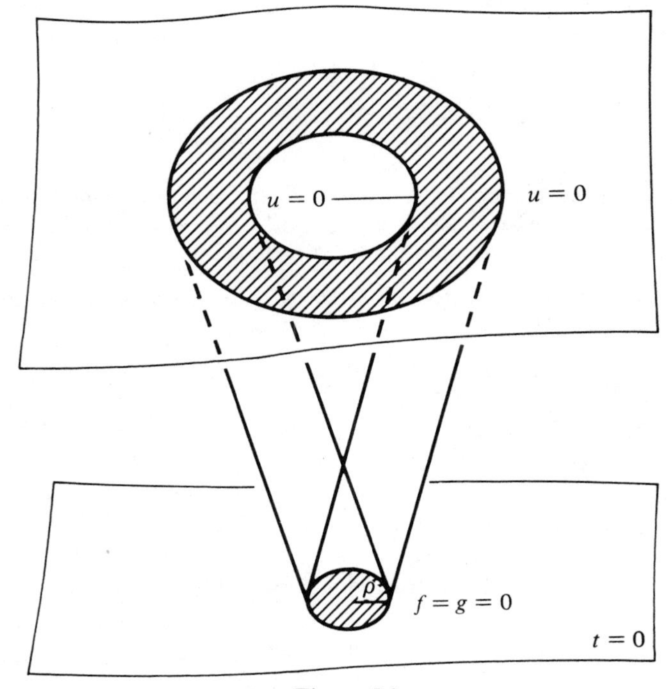

Figure 5.3

speed but after having reached a point never die out completely in a finite time at that point, like the surface waves arising from a stone dropped into water.

While the support of the solution with initial data of compact support expands, the solution *decays* in time.* Assume that f,g and the first derivatives of f are bounded, and vanish outside $B(0,\rho)$. Contributions to the integral in (1.15) arise only from that portion of the sphere $S(x,ct)$ that lies inside the ball $B(x,\rho)$. Elementary geometry shows that the area of intersection of any sphere in 3-space with a ball of radius ρ is at most $4\pi\rho^2$. Thus the integral is at most equal to the maximum of the absolute value of the integrand multiplied by $4\pi\rho^2$. It follows that u for large t is at most of the order of $1/t$.

PROBLEMS

1. (a) Show that for $n=3$ the general solution of (1.1) with spherical symmetry about the origin has the form

$$u=\frac{F(r+ct)+G(r-ct)}{r}, \qquad r=|x| \tag{1.17}$$

with suitable F,G.

(b) Show that the solution with initial data of the form

$$u=0, \qquad u_t=g(r) \tag{1.18}$$

($g=$even function of r) is given by

$$u=\frac{1}{2cr}\int_{r-ct}^{r+ct}\rho g(\rho)\,d\rho. \tag{1.19}$$

(c) For

$$g(r)=\begin{cases}1 & \text{for } 0<r<a\\ 0 & \text{for } r>a\end{cases} \tag{1.20}$$

find u explicitly from (1.19) in the different regions bounded by the cones $r=a\pm ct$ in xt-space. Show that u is discontinuous at $(0,a/c)$; (due to focussing of the discontinuity of u_t at $t=0$, $|x|=a$).

2. Consider the initial-value problem (1.1), (1.2) for the wave equation in $n=5$ dimensions. With $M_u(x,r,t)$ defined by (1.8), set

$$N(x,r,t)=r^2\frac{\partial}{\partial r}M_u(x,r,t)+3rM_u(x,r,t). \tag{1.21}$$

(a) Show that $N(x,r,t)$ is a solution of

$$\frac{\partial^2}{\partial t^2}N=c^2\frac{\partial^2}{\partial r^2}N$$

and find N from its initial data in terms of M_f and M_g.

*For the analogous situation of surface waves in water ($n=2$) compare Shakespeare (*Henry VI, part 1*):

> Glory is like a circle in the water
> Which never ceaseth to enlarge itself
> Till by broad spreading it disperse to nought.

(b) Show that

$$u(x,t)=\lim_{r\to 0}\frac{N(x,r,t)}{3r}$$
$$=\left(\frac{1}{3}t^2\frac{\partial}{\partial t}+t\right)M_g(x,ct)+\frac{\partial}{\partial t}\left(\frac{1}{3}t^2\frac{\partial}{\partial t}+t\right)M_f(x,ct). \tag{1.22}$$

3. For $x=(x_1,x_2,x_3)$ consider the equation of elastic waves (see (2.8) of Chapter 1)

$$Lu=\left(\frac{\partial^2}{\partial t^2}-c_1^2\Delta\right)\left(\frac{\partial^2}{\partial t^2}-c_2^2\Delta\right)u(x,t)=0 \tag{1.23}$$

with positive distinct constants c_1, c_2.

(a) Show that $M_u(x,r,t)$ defined by (1.8) satisfies

$$\Lambda rM_u=\left(\frac{\partial^2}{\partial t^2}-c_1^2\frac{\partial^2}{\partial r^2}\right)\left(\frac{\partial^2}{\partial t^2}-c_2^2\frac{\partial^2}{\partial r^2}\right)rM_u=0.$$

(b) Show that the general solution $v(r,t)$ of $\Lambda v=0$ is of the form

$$v=F_1(r+c_1t)+F_2(r-c_1t)+G_1(r+c_2t)+G_2(r-c_2t).$$

(c) Solve the general initial-value problem for (1.23) using (a) and (b).

4. For $n=3$ let $u(x,t)\in C^2$ and $\Box u=0$ for $x\in\mathbb{R}^3$, $t\geqslant 0$. Assume moreover that

$$U(t)=\sum_{|\alpha|\leqslant 2}\int|D^\alpha u(x,t)|\,dx<\infty\quad\text{for } t=0.$$

(a) Show that there exists a constant K independent of u such that

$$|u(x,t)|\leqslant\frac{K}{t}U(0)\quad\text{for } t>0.$$

[Hint: Write the integrand in (1.15) as

$$\sum_i[(ct)^{-1}(tg(y)+f(y))(y_i-x_i)+ctf_{y_i}(y)]\xi_i$$

where the $\xi_i=(y_i-x_i)/ct$ are the direction cosines of the exterior surface normal. Convert the integral to one extended over the solid sphere $|y-x|<ct$.]

(b) Show that

$$\lim_{t\to\infty}\frac{U(t)}{t}=0$$

implies that u vanishes identically. [Hint: Apply (a) to the function $v(x,t,T)=u(x,T-t)$ for large T.]

(b) *Hadamard's method of descent*

In this method solutions of a partial differential equation are obtained by considering them as special solutions of another equation which involves more independent variables, and can be solved. For example a solution $u(x_1,x_2,t)$ of (1.1), (1.2) with $n=2$, can be looked at as a solution of the same problem with $n=3$ which happens not to depend on x_3. Then $u(x_1,x_2,t)$ is given by formula (1.14) for $x_3=0$ with

$$g(y)=g(y_1,y_2), \qquad f(y)=f(y_1,y_2),$$

the surface integrals being extended over the sphere

$$|y-x|=\sqrt{(y_1-x_1)^2+(y_2-x_2)^2+y_3^2}=ct.$$

Observing that on that sphere

$$dS_y=\sqrt{1+\left(\frac{\partial y_3}{\partial y_1}\right)^2+\left(\frac{\partial y_3}{\partial y_2}\right)^2}\,dy_1\,dy_2=\frac{ct}{|y_3|}\,dy_1\,dy_2,$$

and that the points (y_1,y_2,y_3) and $(y_1,y_2,-y_3)$ make the same contribution to the integrals, we find that

$$u(x_1,x_2,t)=\frac{1}{2\pi c}\int\int_{r<ct}\frac{g(y_1,y_2)}{\sqrt{c^2t^2-r^2}}\,dy_1\,dy_2$$
$$+\frac{\partial}{\partial t}\frac{1}{2\pi c}\int\int_{r<ct}\frac{f(y_1,y_2)}{\sqrt{c^2t^2-r^2}}\,dy_1\,dy_2, \tag{1.24a}$$

where

$$r=\sqrt{(x_1-y_1)^2+(x_2-y_2)^2}\,. \tag{1.24b}$$

We observe that here the domain of dependence of the point (x_1,x_2,t) on the initial data consists of the *solid* disk $r\leqslant ct$ in the y_1y_2-plane. Thus Huygens's principle in the strong form does not hold for the wave equation in two dimensions. Disturbances will continue indefinitely, as exhibited by water waves.

The same method can be applied to other lower-dimensional equations as well. Consider, for example, a solution u of (1.1), (1.2) with $n=3$ of the special form

$$u(x_1,x_2,x_3,t)=e^{i\lambda x_3}v(x_1,x_2,t).$$

Then v is a solution of the 2-dimensional equation

$$v_{tt}=c^2(v_{x_1x_1}+v_{x_2x_2})-\lambda^2c^2v. \tag{1.25}$$

The solution v of (1.25) with initial values

$$v=\phi(x_1,x_2), \qquad v_t=\psi(x_1,x_2) \quad \text{for } t=0 \tag{1.26}$$

is obtained from formula (1.14) for $x_3=0$, taking

$$f(x_1,x_2,x_3)=e^{i\lambda x_3}\phi(x_1,x_2), \qquad g(x_1,x_2,x_3)=e^{i\lambda x_3}\psi(x_1,x_2).$$

Problems

1. Write out the solution of the initial-value problem (1.25), (1.26).

2. Show that the solution $w(x_1,t)$ of the initial-value problem for the *telegraph equation*

$$w_{tt}=c^2 w_{x_1x_1}-c^2\lambda^2 w \tag{1.27}$$

$$w(x_1,0)=0, \qquad w_t(x_1,0)=\psi(x_1) \tag{1.28}$$

is given by

$$w(x_1,t)=\frac{1}{2c}\int_{x_1-ct}^{x_1+ct} J_0(\lambda s)\psi(y_1)\,dy_1. \tag{1.29}$$

Here

$$s=c^2t^2-(x_1-y_1)^2 \tag{1.30}$$

while J_0 denotes the Bessel function defined by

$$J_0(z)=\frac{2}{\pi}\int_0^{\pi/2}\cos(z\sin\theta)\,d\theta. \tag{1.31}$$

[Hint: "Descend" to (1.27) from the two-dimensional wave equation satisfied by

$$u(x_1,x_2,t)=\cos(\lambda x_2)w(x_1,t).$$

Use formulas (1.24a, b).]

3. Solve the initial-value problem (1.1), (1.2) for $n=4$ by descent from the solution (1.22) for $n=5$.

(c) *Duhamel's principle and the general Cauchy problem*

Consider the *inhomogeneous* wave equation

$$\square u(x,t)=w(x,t) \tag{1.32}$$

for a function $u(x,t)$ with initial values

$$u(x,0)=f(x), \qquad u_t(x,0)=g(x). \tag{1.33}$$

Duhamel's principle permits reduction of the problem (1.32), (1.33) to a succession of problems of the type (1.1), (1.2) for the *homogeneous* wave equation. (The method, the analogue of the method of "variation of parameters" for ordinary differential equations, applies to more general linear partial equations.) It is sufficient to consider the problem (1.32), (1.33) for the special case where the initial data are

$$u(x,0)=u_t(x,0)=0. \tag{1.34}$$

We only have to subtract from u the solution of the problem (1.1), (1.2) which we assume to be known. We claim that the solution of (1.32), (1.34) is given by

$$u(x,t)=\int_0^t U(x,t,s)\,ds, \tag{1.35}$$

where $U(x,t,s)$ for each fixed $s\geqslant 0$ is the solution of

$$\square U(x,t,s)=0 \quad \text{for } t\geqslant s \tag{1.36}$$

with initial data prescribed on the plane $t=s$:

$$U(x,s,s)=0, \qquad U_t(x,s,s)=w(x,s). \tag{1.37}$$

Let indeed $U(x,t,s)$ be a solution of (1.36), (1.37) of class C^2 in its arguments for $x\in\mathrm{R}^n$, $0\leqslant s\leqslant t$. Then for the u given by (1.35)

$$u_t = U(x,t,t)+\int_0^t U_t(x,t,s)\,ds = \int_0^t U_t(x,t,s)\,ds \tag{1.38}$$

$$\begin{aligned} u_{tt} &= U_t(x,t,t)+\int_0^t U_{tt}(x,t,s)\,ds \\ &= w(x,t)+\int_0^t c^2\Delta_x U(x,t,s)\,ds = w(x,t)+c^2\Delta_x u(x,t), \end{aligned}$$

confirming (1.32). That u satisfies the initial conditions (1.34) is clear from (1.35), (1.38). Since the wave operator $\Box$ is invariant under translations, we have in

$$V(x,t,s)=U(x,t+s,s) \tag{1.39a}$$

a solution of the wave equation

$$\Box V(x,t,s)=0 \quad \text{for } t\geqslant 0 \tag{1.39b}$$

with initial values prescribed for $t=0$:

$$V(x,0,s)=0, \qquad V_t(x,0,s)=w(x,s). \tag{1.39c}$$

For $n=3$ formula (1.14) yields in

$$V(x,t,s)=\frac{1}{4\pi c^2 t}\int_{|y-x|=ct} w(y,s)\,dS_y$$

a C^2-solution of (1.39b, c) provided $w(x,t)\in C^2$ for $t\geqslant 0$ and all x. Substituting into (1.35) leads to the expression

$$\begin{aligned} u(x,t) &= \int_0^t V(x,t-s,s)\,ds \\ &= \frac{1}{4\pi c^2}\int_0^t \frac{ds}{t-s}\int_{|y-x|=c(t-s)} w(y,s)\,dS_y. \end{aligned} \tag{1.40}$$

Thus the value of the solution u of (1.32), (1.34) at the point (x,t) depends only on the values of w in points (y,s) with

$$|y-x|=c(t-s), \qquad 0<s<t, \tag{1.41}$$

that is, on the values of w in points of the upper half space lying on the backward characteristic cone with vertex (x,t). This truncated cone represents the domain of dependence of $u(x,t)$ on w.

The Cauchy problem for (1.32) (and similarly for other equations) with an arbitrary initial surface S in xt-space can be reduced to that for $t=0$. Let the hypersurface S be given by (see Fig 5.4)

$$t=\phi(x)=\phi(x_1,x_2,x_3). \tag{1.42}$$

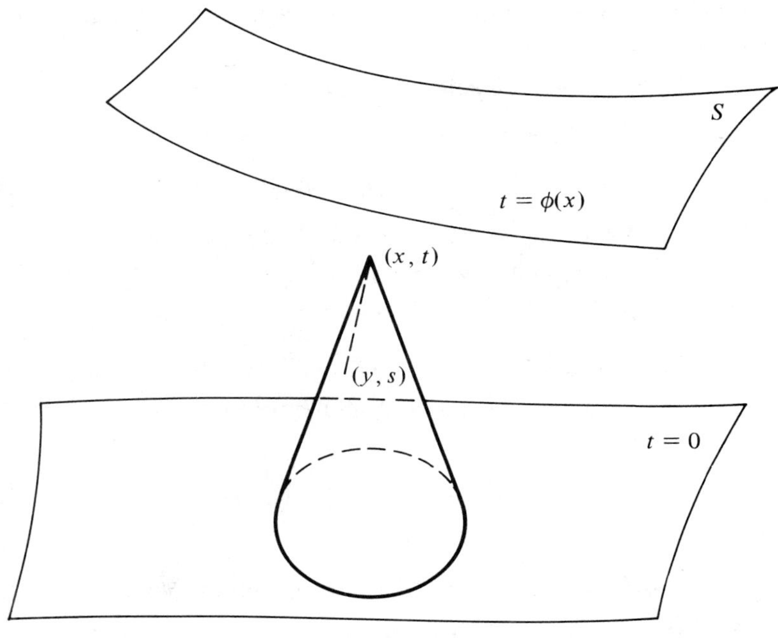

Figure 5.4

We prescribe Cauchy data

$$u = f(x), \qquad u_t = g(x) \quad \text{for } (x,t) \text{ on } S. \tag{1.43}$$

Of course, S will have to be noncharacteristic, that is,

$$\psi(x) = 1 - c^2 \sum_i \phi_{x_i}^2 \neq 0 \tag{1.44}$$

(see (2.28a, b) of Chapter 3). To simplify the expression we make the (unessential) assumption that $\phi, f, g \in C^\infty(\mathbb{R}^3)$, $w \in C^\infty(\mathbb{R}^4)$, and that $\phi > 0$. To find a solution u of

$$\square u(x,t) = w(x,t) \quad \text{for } t > \phi(x) \tag{1.45}$$

with Cauchy data (1.43), we first construct an *approximate solution* v *of order* 2, i.e., a function $v(x,t)$ which just on S satisfies

$$v = f, \qquad v_t = g, \qquad \square v - w = \frac{\partial}{\partial t}(\square v - w) = \frac{\partial^2}{\partial t^2}(\square v - w) = 0. \tag{1.46}$$

We can always find a special v satisfying (1.46) in the form of a 4th-degree polynomial in t:

$$v(x,t) = \sum_{i=0}^{4} a_i(x)(t - \phi(x))^i \tag{1.47}$$

(this is just the beginning of the formal Cauchy–Kowalevski expansion for

u in terms of powers of $t-\phi$). The coefficients $a_i(x)$ have to be found recursively from

$$a_0=f, \qquad a_1=g \tag{1.48a}$$

$$2\psi a_2+c^2(\Delta(a_1\phi-a_0)-\phi\Delta a_1)=w(x,\phi) \tag{1.48b}$$

$$6\psi a_3+c^2(\Delta(2a_2\phi-a_1)-2\phi\Delta a_2)=w_t(x,\phi) \tag{1.48c}$$

$$24\psi a_4+c^2(\Delta(6a_3\phi-2a_2)-6\phi\Delta a_3)=w_{tt}(x,\phi). \tag{1.48d}$$

The a_i exist because of (1.44). We change our problem (1.45), (1.43) to one for the function $U=u-v$. We then want

$$\Box U(x,t)=W(x,t) \quad \text{for } t>\phi(x) \tag{1.49a}$$

$$U=U_t=0 \quad \text{for } t=\phi(x). \tag{1.49b}$$

Here

$$W(x,t)=w(x,t)-\Box v(x,t) \tag{1.50}$$

belongs to $C^\infty(\mathbb{R}^4)$ and, by (1.46), satisfies

$$W(x,t)=W_t(x,t)=W_{tt}(x,t)=0 \quad \text{for } t=\phi(x). \tag{1.51}$$

Then all first and second derivatives of W vanish for $t=\phi(x)$ and the function W^* defined by

$$W^*(x,t)=\begin{cases} W(x,t) & \text{for } t\geqslant\phi(x)\\ 0 & \text{for } t\leqslant\phi(x)\end{cases} \tag{1.52}$$

belongs to $C^2(\mathbb{R}^4)$. We replace the Cauchy problem (1.49a, b) by the initial-value problem

$$\Box U(x,t)=W^*(x,t) \quad \text{for } t>0 \tag{1.53a}$$

$$U(x,0)=U_t(x,0)=0 \tag{1.53b}$$

for which we can find a solution of class C^2 for $t\geqslant 0$ with the help of formula (1.40). We claim that the U found from (1.53a, b) satisfies (1.49a, b) *provided* S is *spacelike* in the sense that

$$\psi(x)=1-c^2\sum_i \phi_{x_i}^2>0. \tag{1.54}$$

Clearly (1.53a) implies (1.49a), since $\phi>0$. Moreover, (1.49b) follows if we show that $U(x,t)=0$ for $t<\phi(x)$. Now by (1.40) $U(x,t)=0$ if $W^*(y,s)=0$ in all points of the backward characteristic cone (see Figure 5.4)

$$|x-y|<c(t-s), \qquad 0<s<t \tag{1.55}$$

with vertex (x,t). By (1.52) this will be the case if we can show that $s<\phi(y)$ whenever (1.55) holds and $t<\phi(x)$. Now $s\geqslant\phi(y)$ would imply that

$$|x-y|<c(t-s)<c(\phi(x)-\phi(y))=c\sum_i \phi_{x_i}(\zeta)(x_i-y_i)$$
$$\leqslant c\sqrt{\sum_i \phi_{x_i}^2(\zeta)}\ |x-y|$$

with an intermediate point ζ. Using (1.54) this leads to a contradiction, and (1.49b) is proved.

In this way the existence of a solution of the Cauchy problem with data on a spacelike surface S is established. The argument will not work if S is not spacelike, since then the cone (1.55) might contain points where $W^*\neq 0$. Data on "timelike" surfaces with $\psi(x)<0$ or surfaces which are partly spacelike and partly timelike cannot be prescribed arbitrarily, since then some points of S lie in the domain of influence of others.

Uniqueness for the Cauchy problem (1.45), (1.43) is trivial. Let u be a solution with homogeneous data $f=g=w=0$ and of class C^2 for $t\geqslant\phi(x)$. We can then continue u as identically 0 to all (x,t) with $0\leqslant t<\phi(x)$. Then u becomes a C^2-solution of $\square u=0$ with vanishing initial data for $t=0$, and hence vanishes identically.

Problem

Let S denote a spacelike hyperplane with equation $t=\gamma x_1$ in xt-space. Show that the Cauchy problem for $\square u=0$ with data on S can be reduced to the initial-value problem for the same equation by introducing new independent variables x',t' by the *Lorentz transformation*

$$x_1'=\frac{x_1-\gamma c^2 t}{\sqrt{1-\gamma^2c^2}},\qquad x_2'=x_2,\qquad x_3'=x_3,\qquad t'=\frac{t-\gamma x_1}{\sqrt{1-\gamma^2c^2}}.$$

(d) *Initial-boundary-value problems ("Mixed" problems)*

So far we have considered the "pure" initial-value problem for the wave equation, with x ranging over the whole space $\mathbb{R}^n$ for $t>0$. We next consider solutions $u(x,t)$ of

$$\square u=w(x,t)\quad \text{for } x\in\Omega,\ t>0, \tag{1.56}$$

where Ω is an open set in x-space. For simplicity we take $n=3$. We associate with the operator $\square$ the *energy integral*

$$E(t)=\int_\Omega \frac{1}{2}\left(u_t^2+c^2\sum_i u_{x_i}^2\right)dx \tag{1.57}$$

(see (1.16)*). Then, using (1.56)

$$\frac{dE}{dt} = \int_\Omega \left(u_t u_{tt} + c^2 \sum_i u_{x_i} u_{x_i t} \right) dx$$

$$= \int_\Omega \left(c^2 \sum_i (u_t u_{x_i})_{x_i} + u_t w \right) dx$$

$$= \int_\Omega u_t w \, dx + \int_{\partial\Omega} c^2 u_t \frac{du}{dn} \, dS_x, \tag{1.58}$$

where d/dn denotes differentiation in the direction of the exterior normal to $\partial\Omega$. If here

$$u = u_t = 0 \quad \text{for } x \in \Omega, \ t = 0 \tag{1.59a}$$

$$u = 0 \quad \text{or} \quad du/dn = 0 \quad \text{for } x \in \partial\Omega, \ t > 0 \tag{1.59b}$$

$$\square u = 0 \quad \text{for } x \in \Omega, \ t > 0 \tag{1.59c}$$

it follows that $E(t) = 0$ for $t > 0$, since $E(0) = 0$, $dE/dt = 0$. Since the integrand in (1.57) is a definite quadratic form in the first derivatives of u, we conclude that $u_t = u_{x_i} = 0$ for $x \in \Omega$, $t > 0$, hence that u is constant, and thus $u = 0$ because $u = 0$ initially. Consequently (subject to appropriate regularity assumptions) a solution u of (1.56) is determined uniquely given (a) the values of w for $x \in \Omega$, $t > 0$, (b) initial values

$$u = f(x), \qquad u_t = g(x) \quad \text{for } x \in \Omega, \ t = 0, \tag{1.60}$$

and (c) boundary values ("Dirichlet data")

$$u = h \quad \text{for } x \in \partial\Omega, \ t > 0 \tag{1.61}$$

(where instead of u, possibly du/dn can be prescribed on $\partial\Omega$).

In the special case where $w = h = 0$, the solution can be found by *expansion into eigenfunctions* for the Laplace operator for the region Ω, in analogy to (4.21) of Chapter 2. An eigenfunction $v(x)$ corresponding to the eigenvalue λ is a solution of

$$\Delta v(x) + \lambda v(x) = 0 \quad \text{for } x \in \Omega \tag{1.62a}$$

$$v(x) = 0 \quad \text{for } x \in \partial\Omega, \tag{1.62b}$$

where v does not vanish identically. Under appropriate regularity assumptions on Ω there exists a sequence of eigenvalues λ_k and a corresponding sequence of eigenfunctions $v_k(x)$ which form a complete orthonormal set on Ω. (See Courant–Hilbert [6].) This leads to an expansion

$$u(x,t) = \sum_k a_k(t) v_k(x) \tag{1.63}$$

for the solution $u(x,t)$ of our initial-value problem. Substituting into the

* In many applications E represents the sum of kinetic and potential energy.

equation $u=0$ and comparing coefficients, one finds that the $a_k(t)$ are solutions of the ordinary differential equation

$$a_k''(t)+c^2\lambda_k a_k(t)=0. \tag{1.64}$$

Using the initial conditions (1.60) and Fourier's formula

$$a_k=\int_\Omega uv_k\,dx \tag{1.65}$$

for the coefficients of an expansion in terms of orthonormal functions, one arrives at initial conditions

$$a_k(0)=\int_\Omega f(x)v_k(x)\,dx, \qquad a_k'(0)=\int_\Omega g(x)v_k(x)\,dx. \tag{1.66}$$

Equations (1.64), (1.66) easily permit us to determine the $a_k(t)$ as trigonometric functions of t:

$$a_k(t)=\int_\Omega\left[f(x)\cos(c\lambda_k^{1/2}t)+\frac{g(x)\sin(c\lambda_k^{1/2}t)}{c\lambda_k^{1/2}}\right]v_k(x)\,dx. \tag{1.67}$$

Some mixed problems can be solved "in closed form" when the space region Ω is the halfspace $x_n>0$. The principal tool here is the extension to the whole space by *reflection*. Let u be the solution of

$$\Box u=w(x,t) \quad \text{for } x_3\geqslant 0,\ t\geqslant 0 \tag{1.68a}$$

$$u=f(x), \qquad u_t=g(x) \quad \text{for } x_3\geqslant 0,\ t=0 \tag{1.68b}$$

$$u=h(x_1,x_2,t) \quad \text{for } x_3=0,\ t\geqslant 0. \tag{1.68c}$$

Relations (1.68a, b, c) imply certain *consistency conditions* between the data f,g,h,w, at least if a solution u of class C^3 is to exist in the closed quadrant $x_3\geqslant 0$, $t\geqslant 0$:

$$f=h, \qquad g=h_t, \qquad c^2\Delta f+w=h_{tt}, \qquad c^2\Delta g+w_t=h_{ttt} \quad \text{for } x_3=t=0. \tag{1.69}$$

We shall construct a solution u of class C^2 for $x_3\geqslant 0$, $t\geqslant 0$, provided the data f,g,h,w, have sufficiently many derivatives in their closed domains of definition and satisfy (1.69). Moreover the solution is unique.

For that purpose we first simplify the problem by making use of an *approximate solution* $v(x,t)$ *of order* 2 of (1.68a, c) (see p. 137), that is, a solution of

$$v=h, \qquad \frac{\partial^i}{\partial x_3^i}(\Box v-w)=0 \quad \text{for } i=0,2 \text{ when } x_3=0,\ t\geqslant 0. \tag{1.70}$$

We can always find such a v that is a polynomial in x_3

$$v=h(x_1,x_2,t)+a_2(x_1,x_2,t)x_3^2+a_4(x_1,x_2,t)x_3^4 \tag{1.71}$$

with suitable coefficients a_2, a_4. It remains to find a solution $U(x, t) = u(x,t) - v(x,t)$ of

$$\square U = W = w - \square v \quad \text{for } x_3 \geqslant 0,\ t \geqslant 0 \tag{1.72a}$$

$$U = F(x) = f(x) - v(x,0), \qquad U_t = G(x) = g(x) - v_t(x,0)$$
$$\text{for } x_3 \geqslant 0,\ t = 0 \tag{1.72b}$$

$$U = H = 0 \quad \text{for } x_3 = 0,\ t \geqslant 0. \tag{1.72c}$$

Observe that here by (1.70), (1.69)

$$W = W_{x_3x_3} = 0 \quad \text{for } x_3 = 0,\ t \geqslant 0 \tag{1.73a}$$

$$F = F_{x_3x_3} = G = G_{x_3x_3} = 0 \quad \text{for } x_3 = 0. \tag{1.73b}$$

If we extend F, G, W by "reflection" to all values of x_3, so as to be odd in x_3, then the resulting functions F^*, G^*, W^* will be defined for all x. Moreover by (1.73a, b) $F^* \in C^3$, $G^* \in C^2$ for all x, and $W^* \in C^2$ for all x and $t \geqslant 0$. We can then solve the *pure* initial-value problem

$$\square U = W^* \quad \text{for } t > 0$$
$$U = F^*, \qquad U_t = G^* \quad \text{for } t = 0$$

using formula (1.40) to obtain a C^2-solution. Obviously the restriction of U to values $x_3 \geqslant 0$, $t \geqslant 0$ satisfies (1.72a, b). Moreover U is odd in x_3, since

$$U(x_1, x_2, x_3, t) + U(x_1, x_2, -x_3, t)$$

will be a solution of the wave equation with initial values 0, thus vanish identically. This also implies that (1.72c) holds. Finally a C^2-solution u of (1.68a, b, c) is unique, since for $f = g = h = w = 0$ it could be extended by reflection to a solution of the wave equation with initial values that are 0 everywhere, and hence would vanish.

We observe that the consistency conditions (1.69) are satisfied automatically, when $h = 0$ and in addition f, g, w vanish for all sufficiently small x_3. When $w = h = 0$ and (1.69) holds, we can reduce the mixed problem to a pure initial value problem by directly continuing u as an odd function of x_3.

Problems

1. Let Ω denote an open bounded set in n-dimensional x-space described by an inequality $\phi(x) > 0$, so that $\phi(x) = 0$ on $\partial\Omega$. Let S_λ for $\lambda \geqslant 0$ denote the hypersurface in xt-space given by $t = \lambda\phi(x)$ for $x \in \Omega$. On S_λ define for a function $u(x,t)$ the *energy integral*

$$E(\lambda) = \int_{S_\lambda} Q_\lambda \, dx, \tag{1.74a}$$

where

$$Q_\lambda = \frac{1}{2}\left(u_t^2 + c^2 \sum_i u_{x_i}^2\right) + \lambda c^2 u_t \sum_i u_{x_i}\phi_{x_i}. \tag{1.74b}$$

(a) Prove $E(\lambda)=\text{const.}$ when $\Box u=0$. [Hint: Integrate $u_t\Box u$ over the lens-shaped region $0<t<\lambda\phi(x)$.]
(b) Show that Q_λ as a quadratic form in $u_t, u_{x_1}, \ldots, u_{x_n}$ is positive definite, when S_λ is spacelike.
(c) Show that the initial data on S_0 of a solution of $\Box u=0$ uniquely determine u on all S_λ with sufficiently small λ. (Compare Holmgren's theorem, p. 85.)

2. Let u be a solution of (1.68a, b, c) where $f=h=w=0$. Find the domain of dependence of $u(x,t)$ on g.

3. Consider the mixed problem for $u(x,t)=u(x_1,x_2,x_3,t)$

$$\Box u=0 \quad \text{for } x_3>0,\ t>0 \tag{1.75a}$$

$$u=f(x), \qquad u_t=g(x) \quad \text{for } x_3>0,\ t=0 \tag{1.75b}$$

$$Mu=0 \quad \text{for } x_3=0,\ t>0, \tag{1.75c}$$

where M denotes a first-order operator of the form

$$M=\frac{\partial}{\partial t}+\sum_i \alpha_i\frac{\partial}{\partial x_i} \tag{1.75d}$$

with constant coefficients α_i, and f,g vanish for all sufficiently small $x_3>0$. Prove there exists a solution u provided $\alpha_3\leqslant 0$. (Compare problem 2, p. 45.) [Hint: First determine $v=Mu$ for $x_3>0$, $t>0$ from its initial and boundary conditions as a solution of $\Box v=0$. Next find u for $x_3>0$, $t>0$ as a solution of $Mu=v$ with initial condition $u=f$ by the methods of Chapter 1. Verify that the u obtained satisfies (1.75a, b, c).]

2. Higher-Order Hyperbolic Equations with Constant Coefficients

(a) *Standard form of the initial-value problem*

For functions $u(x,t)=u(x_1,\ldots,x_n,t)$ we define the differentiation operators

$$D=(D_1,\ldots,D_n)=\left(\frac{\partial}{\partial x_1},\ldots,\frac{\partial}{\partial x_n}\right), \qquad \tau=\frac{\partial}{\partial t} \tag{2.1}$$

where D is the gradient vector with respect to the space variables. Using the Schwartz notation of Chapter 3 we can write the most general mth-order linear partial differential equation with constant coefficients in the form

$$P(D,\tau)u=w(x,t), \tag{2.2}$$

where $P(D,\tau)=P(D_1,\ldots,D_n,\tau)$ is a polynomial of degree m in its $n+1$ arguments. We associate with equation (2.2) in the half space $t>0$ the *initial conditions*

$$\tau^k u=f_k(x) \quad \text{for } k=0,\ldots,m-1 \text{ and } t=0. \tag{2.3}$$

We shall assume that the plane $t=0$ is noncharacteristic. This means that the coefficient $P(0,1)$ of τ^m in the polynomial P does not vanish. Dividing by a suitable constant we can bring about that

$$P(0,1)=1. \tag{2.4}$$

Problem (2.2), (2.3) for general data w, f_k can be reduced to the *standard problem* where the data have the special form

$$w=f_0=f_1=\cdots=f_{m-2}=0, \qquad f_{m-1}=g(x). \tag{2.5}$$

The solution of the standard problem (unique by Holmgren's theorem) will be denoted by $u_g(x,t)$. To achieve this reduction we first find a solution u of (2.2) with zero initial data. Such a solution is furnished according to *Duhamel's principle* by the formula

$$u(x,t)=\int_0^t U(x,t,s)\,ds, \tag{2.6}$$

where $U(x,t,s)$ for each parameter value $s>0$ is the solution of the initial-value problem

$$P(D,\tau)U(x,t,s)=0 \quad \text{for } t \geqslant s \tag{2.7a}$$

$$\tau^k U(x,t,s)=0 \quad \text{for } k=0,\ldots,m-2 \text{ and } t=s \tag{2.7b}$$

$$\tau^{m-1}U(x,t,s)=w(x,s) \quad \text{for } t=s. \tag{2.7c}$$

That it solves (2.2) is easily verified, using (2.4). Here for each $s \geqslant 0$ the function $U(x,t,s)$ is found by solving a standard problem; in fact

$$U(x,t,s)=u_g(x,t-s) \quad \text{where} \quad g(x)=w(x,s). \tag{2.8}$$

It remains to reduce the solution of the homogeneous equation

$$P(D,\tau)u=0 \tag{2.9}$$

with general initial conditions (2.3) to standard problems. For that purpose we arrange the polynomial $P(D,\tau)$ according to powers of τ:

$$P(D,\tau)=\tau^m+P_1(D)\tau^{m-1}+\cdots+P_m(D), \tag{2.10}$$

where $P_k(D)$ is a polynomial of degree $\leqslant k$ in $D_1,\ldots,D_n$. Using the differential equation (2.9) one easily verifies that the solution u with initial data (2.3) is representable in terms of the standard problems associated with each individual f_k by the formula

$$\begin{aligned} u=u_{f_{m-1}}&+(\tau+P_1(D))u_{f_{m-2}}+(\tau^2+P_1(D)\tau+P_2(D))u_{f_{m-3}}\\ &+\cdots\\ &+(\tau^{m-1}+P_1(D)\tau^{m-2}+P_2(D)\tau^{m-3}+\cdots+P_{m-1}(D))u_{f_0}. \end{aligned} \tag{2.11}$$

As an example we have for the solution of the wave equation

$$(\tau^2-c^2\Delta)u=0$$

with initial values

$$u=f, \qquad u_t=g \quad \text{for } t=0$$

the formula

$$u=u_g+\tau u_f$$

in agreement with (1.14).

A *system* of N linear partial differential equations of order m for N functions $u_1,\ldots,u_N$ can also be written in the form (2.2), where now u stands for the column vector with components $u_1,\ldots,u_N$, and $P(D,\tau)$ is a square $N\times N$ matrix whose elements are polynomials of degree $\leqslant m$ in $D_1,\ldots,D_n,\tau$. The data w,f_k in (2.2), (2.3) are column vectors. The solution u_g of a standard problem corresponds to the data (2.5). For a noncharacteristic initial plane $t=0$ the matrix $P(0,1)$ is nondegenerate, and we can assume that

$$P(0,1)=I \tag{2.12}$$

is the unit matrix. The solution of (2.2) with zero initial data still is described by Duhamel's formulas (2.6), (2.7a, b, c), and thus reduced to standard problems as in (2.8). The reduction of general initial data to standard ones is achieved by a modification of (2.11) which reads

$$u=u_{f_{m-1}}+\left(\tau u_{f_{m-2}}+u_{P_1 f_{m-2}}\right)+\cdots+\left(\tau^{m-1}u_{f_0}+\tau^{m-2}u_{P_1 f_0}+\cdots+u_{P_{m-1}f_0}\right). \tag{2.13}$$

In what follows we shall only have to deal with the standard problem

$$P(D,\tau)u=0 \quad \text{for } t\geqslant 0 \tag{2.14a}$$

$$\tau^k u=0 \quad \text{for } k=0,\ldots,m-2 \text{ and } t=0 \tag{2.14b}$$

$$\tau^{m-1}u=g(x) \quad \text{for } t=0. \tag{2.14c}$$

We call the differential equation or system of equations (2.14a) *hyperbolic* (with respect to the plane $t=0$), if the initial-value problem (2.14a, b, c) has a solution $u(x,t)$ of class C^m, for all $g(x)\in C_0^s(\mathbb{R}^n)$, where s is sufficiently large.* We also say that the plane $t=0$ is *spacelike*.

PROBLEM

Verify that formulas (2.11), respectively (2.13), give the solution of the initial-value problem (2.3), (2.9).

(b) *Solution by Fourier transformation*†

Following Cauchy a formal solution of the standard problem (2.14a, b, c) can be obtained by Fourier transformation with respect to the space

* Using the finiteness of the domain of dependence of u on g (implied, e.g., by Holmgren's theorem), one can show that in the hyperbolic case the problem (2.14a, b, c) has a solution for $g\in C^s(\mathbb{R}^n)$, even without the assumption of compact support.

†([8]).

variables. It will be an actual solution if the integrals involved converge adequately. We associate with a function $g(x)\in C_0^s(\mathbb{R}^n)$ its *Fourier transform* $\hat{g}$, defined by

$$\hat{g}(\xi)=(2\pi)^{-n/2}\int e^{-ix\cdot\xi}g(x)\,dx \tag{2.15}$$

$(x\cdot\xi=x_1\xi_1+\cdots+x_n\xi_n)$.* For $g\in C_0^s$ with sufficiently large s the reciprocal formula

$$g(x)=(2\pi)^{-n/2}\int e^{ix\cdot\xi}\hat{g}(\xi)\,d\xi \tag{2.16}$$

holds. We find from (2.15) by integration by parts for any $k=1,2,\ldots,n$ that

$$\begin{aligned}-i\xi_k\hat{g}(\xi)&=(2\pi)^{-n/2}\int D_k(e^{-ix\cdot\xi})\,g(x)\,dx\\&=-(2\pi)^{-n/2}\int e^{-ix\cdot\xi}D_kg(x)\,dx.\end{aligned}$$

We write this fundamental identity as

$$i\xi_k\hat{g}=\widehat{D_kg}. \tag{2.17}$$

By repeated application we find more generally for $g\in C_0^s$ and any multi-index $\alpha=(\alpha_1,\ldots,\alpha_n)$ with $|\alpha|\leqslant s$ that

$$(i\xi)^\alpha\hat{g}=\widehat{D^\alpha g}. \tag{2.18}$$

Thus differentiation for g is transformed into multiplication for $\hat{g}$.

Formula (2.18) permits us to show that $\hat{g}(\xi)$ decreases rapidly for $\xi\to\infty$ when s is large. Let $\xi=(\xi_1,\ldots,\xi_n)$, where, say,

$$|\xi_j|=\max_k|\xi_k|. \tag{2.19a}$$

Then

$$|\xi|=\sqrt{\sum_k\xi_k^2}\leqslant\sqrt{n}\,|\xi_j| \tag{2.19b}$$

$$\begin{aligned}(1+|\xi|)^s&=\sum_{k=0}^{s}\binom{s}{k}|\xi|^k\leqslant 2^s\sum_{k=0}^{s}n^{k/2}|\xi_j|^k\\&\leqslant 2^sn^{s/2}\sum_{|\alpha|\leqslant s}|\xi^\alpha|.\end{aligned} \tag{2.19c}$$

*Here, of course, $i=\sqrt{-1}$. Observe that generally $\hat{g}$ is complex valued, even when the variables x,ξ,g are restricted to real values. In what follows the independent variables x,ξ will be assumed to be real, unless the contrary is stated, but $g,\hat{g}$, and the coefficients of the polynomial P will be allowed to be complex valued.

Consequently by (2.18), (2.15),

$$(1+|\xi|)^s|\hat{g}(\xi)| \leqslant 2^s n^{s/2} \sum_{|\alpha|\leqslant s} |(i\xi)^\alpha \hat{g}(\xi)|$$
$$\leqslant 2^s n^{s/2} \sum_{|\alpha|\leqslant s} \int |D^\alpha g(x)|\,dx \leqslant M_s < \infty, \tag{2.20}$$

where M_s depends on n, s, the size of the support of g, and the maxima of the absolute values of the derivatives of g of orders $\leqslant s$. It follows in particular that

$$|\hat{g}(\xi)| \leqslant \frac{M_{n+1}}{(1+|\xi|)^{n+1}} \quad \text{for } g \in C^{n+1}(\mathbb{R}^n) \tag{2.21}$$

and hence that the integral in (2.16) converges absolutely. Formula (2.16) is valid for $s > n$.

Let now $u(x,t)$ be a solution of (2.14a, b, c). To begin with we work with a *single* partial differential equation, so that u is a scalar. We write tentatively

$$u(x,t) = (2\pi)^{-n/2} \int e^{ix\cdot\xi} \hat{u}(\xi,t)\,d\xi, \tag{2.22}$$

where $\hat{u}(\xi,t)$ is the Fourier transform of u with respect to x. Purely formally we obtain by differentiation

$$0 = P(D,\tau)u(x,t) = (2\pi)^{-n/2} \int e^{ix\cdot\xi} P(i\xi,\tau)\hat{u}(\xi,t)\,d\xi.$$

In addition for $t=0$

$$(2\pi)^{-n/2} \int e^{ix\cdot\xi} \tau^k \hat{u}(\xi,t)\,d\xi = \tau^k u(x,t)$$
$$= \begin{cases} 0 & \text{for } k=0,\ldots,m-2 \\ g(x) = (2\pi)^{-n/2} \int e^{ix\cdot\xi} \hat{g}(\xi)\,d\xi & \text{for } k=m-1. \end{cases}$$

These relations are satisfied formally, when $\hat{u}(\xi,t)$ for each $\xi \in \mathbb{R}^n$ is a solution of the ordinary differential equation

$$P(i\xi,\tau)\hat{u}(\xi,t) = 0 \qquad \left(\tau = \frac{d}{dt}\right) \tag{2.23a}$$

with initial values for $t=0$

$$\tau^k \hat{u}(\xi,t) = \begin{cases} 0 & \text{for } k=0,\ldots,m-2 \\ \hat{g}(\xi) & \text{for } k=m-1. \end{cases} \tag{2.23b}$$

This leads to the *formal* solution* of (2.14a, b, c)

$$u(x,t)=(2\pi)^{-n/2}\int e^{ix\cdot\xi}Z(\xi,t)\,\hat{g}(\xi)\,d\xi, \tag{2.24a}$$

where Z as a function of t denotes the solution of the ordinary differential equation problem

$$P(i\xi,\tau)Z(\xi,t)=0 \tag{2.24b}$$

with initial values for $t=0$

$$\tau^k Z(\xi,t)=\begin{cases}0 & \text{for } k=0,\ldots,m-2\\ 1 & \text{for } k=m-1.\end{cases} \tag{2.24c}$$

There is no problem with the existence of Z. Moreover we can verify directly that the u given by (2.24a) is of class C^m in x,t for $x\in\mathbb{R}^n$, $t\geqslant 0$, and actually satisfies (2.14a, b, c), if $g\in C_0^s$ with $s>n$, and g and Z are such that all differentiations with respect to x or t of orders $\leqslant m$ can be carried out under the integral sign in (2.24a). This is certainly the çase when the resulting integrals converge absolutely. For that it is sufficient that the expressions

$$(1+|\xi|)^{n+1}|\tau^k\xi^\alpha Z(\xi,t)\,\hat{g}(\xi)| \quad \text{for } |\alpha|+k\leqslant m \tag{2.25}$$

are bounded uniformly in ξ,t for all $\xi\in\mathbb{R}^n$ and for t restricted to any finite interval $0\leqslant t\leqslant T$.

Of course, the expressions (2.25) will be bounded in any bounded set in ξt-space. What matters is only the behavior for large $|\xi|$. Here, to a certain extent, $\hat{g}(\xi)$ can be controlled by assuming that s is large enough, as is shown by the estimate (2.20). It is just a question of the growth of $Z(\xi,t)$ and its t-derivatives. If we can show that there exists a constant N, such that

$$|\tau^k Z(\xi,t)|\leqslant(1+|\xi|)^k N \quad \text{for all } \xi\in\mathbb{R}^n, 0\leqslant t\leqslant T, 0\leqslant k\leqslant m, \tag{2.26}$$

we find for the expressions (2.25), using (2.20), the upper bound

$$(1+|\xi|)^{n+1+m-s}NM_s.$$

For the boundedness of the expressions (2.25) it is here sufficient to assume that

$$s\geqslant n+1+m. \tag{2.27}$$

Formula (2.24a) will then represent a solution of our standard Cauchy problem (2.14a, b, c).

The proper condition on the partial differential equation (2.14a), i.e., on

* More precisely our arguments show that if there exists a solution $u(x,t)$ of (2.14a, b, c) of compact support in x and sufficiently often differentiable, then u must be given by the expression (2.24a).

the polynomial P, under which an estimate of the form (2.26) holds, and hence the initial-value problem can be solved, is:

Gårding's hyperbolicity condition. Equation (2.14a) is hyperbolic if there exists a real number c such that

$$P(i\xi, i\lambda) \neq 0 \text{ for all } \xi \in \mathbb{R}^n \text{ and all complex } \lambda \text{ with } Im\lambda \leqslant -c. \tag{2.28}$$

Condition (2.28) is equivalent to the statement that all of the m roots λ of

$$P(i\xi, i\lambda) = 0 \tag{2.29}$$

lie in one and the same half plane

$$\mathrm{Im}\lambda > -c \tag{2.30}$$

of the complex number plane for all real vectors ξ.*

To establish the *sufficiency* of Gårding's condition we represent the solution Z of (2.24b, c) by a Cauchy integral:

$$Z(\xi, t) = \frac{1}{2\pi} \int_\Gamma \frac{e^{i\lambda t}}{P(i\xi, i\lambda)} d\lambda, \tag{2.31}$$

where the closed path of integration Γ runs around each root λ of (2.29) once in the counterclockwise direction. Indeed differentiation of Z as defined by (2.31) with respect to t results in multiplying the integrand by $i\lambda$ so that

$$P(i\xi, \tau)Z = \frac{1}{2\pi} \int_\Gamma P(i\xi, i\lambda) \frac{e^{i\lambda t}}{P(i\xi, i\lambda)} d\lambda$$

$$= \frac{1}{2\pi} \int_\Gamma e^{i\lambda t} d\lambda = 0$$

by Cauchy's theorem, while for $t=0$ by (2.10)

$$\tau^k Z = \frac{1}{2\pi} \int_\Gamma \frac{i^k \lambda^k}{i^m \lambda^m + P_1(i\xi) i^{m-1} \lambda^{m-1} + \ldots + P_m(i\xi)} d\lambda.$$

Expanding Γ to infinity we see that this expression has the value 0 for $k = 0, \ldots, m-2$, and the value 1 for $k = m-1$.

We first derive an upper bound for the roots λ of (2.29). Using the expansion (2.10) we have

$$P(i\xi, i\lambda) = (i\lambda)^m + P_1(i\xi)(i\lambda)^{m-1} + \cdots + P_m(i\xi) = 0 \tag{2.32}$$

* Gårding showed that his condition (2.28) is necessary as well as sufficient. An even stronger statement holds in the case where the polynomial $P(D, \tau)$ is irreducible (i.e., not representable as product of lower degree polynomials): If the equation (2.14a) is not hyperbolic, the initial-value problem (2.14a, b, c) for $g \in C_0^s(\mathbb{R}^n)$ *never* has a solution, unless g vanishes identically.

For each kth-degree polynomial P_k we have a trivial estimate

$$|P_k(i\xi)| \leqslant M(1+|\xi|)^k \quad \text{for all } \xi \in \mathbb{R}^n \tag{2.33}$$

with a suitable constant M. Then for a root λ of (2.29)

$$|\lambda|^m \leqslant M \sum_{k=1}^{m} (1+|\xi|)^k |\lambda|^{m-k}.$$

Setting $\theta = |\lambda|/(1+|\xi|)$, we have then

$$\theta^m \leqslant M(1+\theta+\theta^2+\cdots+\theta^{m-1}).$$

This implies that either $\theta < 1$ or $\theta^m \leqslant Mm\theta^{m-1}$ and hence $|\theta| < Mm$. Thus for the roots λ of (2.29)

$$\theta = \frac{|\lambda|}{1+|\xi|} < 1 + Mm. \tag{2.34}$$

Denote by $\lambda_k(\xi)$ for $k=1,\ldots,m$ the m (not necessarily distinct) roots λ of (2.29) taken in any order. Then

$$P(i\xi, i\lambda) = i^m \prod_{k=1}^{m} (\lambda - \lambda_k(\xi)). \tag{2.35}$$

Take for each $k=1,\ldots,m$ the open disk of center λ_k and radius 1 in the complex λ-plane. Let U denote the union of these m (possibly overlapping) disks. Take for the path of integration Γ in (2.31) the boundary of U, which possibly consists of several closed curves and is composed of pieces of the boundaries of the individual unit disks. Then Γ runs once around each of the λ_k and has total length $\leqslant 2m\pi$. Moreover each of the points λ of Γ has distance $\geqslant 1$ from each of the λ_k, so that by (2.35)

$$|P(i\xi, i\lambda)| \geqslant 1 \quad \text{for } \lambda \in \Gamma.$$

Since each point of Γ has distance 1 from some λ_k we have from (2.34) and the Gårding condition (2.30)

$$\operatorname{Im}\lambda \geqslant -c-1, \qquad |\lambda| \leqslant 1 + (1+Mm)(1+|\xi|) \leqslant (2+Mm)(1+|\xi|) \quad \text{on } \Gamma.$$

Thus

$$|e^{i\lambda t}| \leqslant e^{(1+c)t} \quad \text{for } t \geqslant 0, \lambda \in \Gamma.$$

It follows from (2.31) that

$$|\tau^k Z(\xi,t)| = \left| \frac{1}{2\pi} \int \frac{(i\lambda)^k e^{i\lambda t}}{P(i\xi, i\lambda)} d\lambda \right|$$

$$\leqslant m(2+Mm)^k (1+|\xi|)^k e^{(1+c)T} \tag{2.35a}$$

for $0 \leqslant t \leqslant T$, $\xi \in \mathbb{R}^n$, $0 \leqslant k \leqslant m$. This is an estimate of the type (2.26). It follows that for $g \in C_0^{n+m+1}(\mathbb{R}^n)$ the initial-value problem has a solution of class C^m for $t \geqslant 0$, provided the Cårding condition (2.28) is satisfied.

The integral (2.31) for Z is easily evaluated by the calculus of residues, in the case where all the roots λ_k are distinct. One finds that then

$$Z(\xi,t)=\sum_{k=1}^{m}\frac{e^{i\lambda_k t}}{P_\tau(i\xi,i\lambda_k)}. \tag{2.36}$$

As an example consider the n-dimensional wave equation corresponding to the operator

$$P(D,\tau)=\square=\tau^2-c^2\sum_{k=1}^{n}D_k^2. \tag{2.37a}$$

Here

$$P(i\xi,i\lambda)=-(\lambda^2-c^2|\xi|^2) \tag{2.37b}$$

has the real roots

$$\lambda_1=c|\xi|, \qquad \lambda_2=-c|\xi| \tag{2.37c}$$

satisfying the Gårding condition. Then by (2.36)

$$Z(\xi,t)=\frac{\sin(c|\xi|t)}{c|\xi|}. \tag{2.37d}$$

Thus the standard problem for the wave equation has the solution

$$u(x,t)=(2\pi)^{-n/2}\int e^{ix\cdot\xi}\frac{\sin(c|\xi|t)}{c|\xi|}\,\hat{g}(\xi)\,d\xi \tag{2.37e}$$

for $g\in C_0^{n+3}(\mathbb{R}^n)$.

If the polynomial $P(D,\tau)$ is *homogeneous* of degree m in D and τ (as in equation (2.37a)), we have for every solution (ξ,λ) of (2.29) and every real s

$$P(s\xi,s\lambda)=0, \qquad \operatorname{Im}(s\lambda)=s\operatorname{Im}\lambda.$$

Here $s\operatorname{Im}\lambda$ can be bounded from below for all s only, if $\operatorname{Im}\lambda=0$. Thus the Gårding condition for homogeneous P is that all roots λ of the equation

$$P(i\xi,i\lambda)=i^m P(\xi,\lambda)=0 \tag{2.38}$$

are real for all real ξ.

In many cases hyperbolicity can be inferred from properties of the *principal part* of P alone. We arrange the terms in the polynomial according to their degree, writing

$$P(D,\tau)=p_m(D,\tau)+p_{m-1}(D,\tau)+\cdots+p_0(D,\tau), \tag{2.39}$$

where $p_k(D,\tau)$ is a form of degree k in D and τ. Here $p_m(D,\tau)$ is identical with the principal part of $P(D,\tau)$ as defined in Chapter 3. We shall prove:

For the Gårding condition for P to be satisfied it is necessary that all roots λ of

$$p_m(\xi,\lambda)=0 \tag{2.40}$$

are real for all real ξ (i.e., that p_m satisfies the condition); *a sufficient requirement which implies the Gårding condition for P is that all roots* λ *of* (2.40) *are real and distinct for all real* $\xi \neq 0$.

To prove this statement we apply the substitution

$$\xi = \rho\eta, \qquad \lambda = \rho\mu, \tag{2.41}$$

where $\rho = |\xi|$ and η is a unit vector. Then $P(i\xi, i\lambda) = 0$ goes over into the equation

$$p_m(\eta,\mu) + \frac{1}{i\rho} p_{m-1}(\eta,\mu) + \cdots + \frac{1}{(i\rho)^m} p_0(\eta,\mu) = 0 \tag{2.42}$$

for μ, depending on the parameters ρ, η. By (2.4) the coefficient of μ^m in (2.42) has the value 1. The coefficients of the powers of μ not contributed by the principal part tend to 0 for $\rho \to \infty$, since η is bounded. Using the fact that the roots of a polynomial with highest coefficient 1 depend continuously on the coefficients, we see that for $\rho \to \infty$ the roots μ of (2.42) will tend* to the roots of

$$p_m(\eta,\mu) = 0. \tag{2.43}$$

Let there exist for a certain η a root μ_0 of (2.43) with $\operatorname{Im}\mu_0 \neq 0$. Assume $\operatorname{Im}\mu_0 = -\gamma < 0$ (otherwise replace η by $-\eta$ and μ_0 by $-\mu_0$). Then there exist roots μ of (2.42) for all sufficiently large ρ for which $\operatorname{Im}\mu < -\gamma/2$ and hence roots λ of (2.29) for which $\operatorname{Im}\lambda < -\rho\gamma/2$. This contradicts (2.30) for large ρ. Thus necessary for (2.30) is that the roots μ of (2.43) are real for all real η with $|\eta| = 1$, and then also for all real η. Assume next that the roots μ are real and distinct for real $\eta \neq 0$, in particular for $|\eta| = 1$. We now use the fact that roots of a polynomial equation with highest coefficient one are differentiable (even analytic) functions of the coefficients in any region not containing multiple roots. They will be uniformly Lipschitz continuous in any compact subregion. For large ρ and $|\eta| = 1$ the coefficients of equation (2.42) for μ differ from those of equation (2.43) by terms of order $1/\rho$. Hence the difference of the roots μ of (2.42) from appropriate roots of (2.43) is of order $1/\rho$ uniformly for $|\eta| = 1$. Since (2.43) has real roots, it follows that the imaginary parts of the roots μ of (2.42) are of order $1/\rho$, and hence the imaginary parts of the roots λ of (2.29) are bounded uniformly for all sufficiently large $\rho = |\xi|$. By (2.34) λ and $\operatorname{Im}\lambda$ also are bounded for bounded $|\xi|$. Thus (2.30) follows.

We call P *strictly hyperbolic* when its principal part $p_m(\xi, \lambda) = 0$ has real distinct roots for $\xi \neq 0$ hyperbolic. We see that strict hyperbolicity implies hyperbolicity. Thus, for example, any equation of the form

$$u_{tt} = c^2 \Delta u + ku \tag{2.44}$$

is hyperbolic.

*More precisely in a given neighborhood of a root μ of (2.43) of multiplicity γ there lie precisely γ roots of (2.42) if ρ is sufficiently large.

Formula (2.24a) for the solution $u(x,t)$ of the standard initial-value problem makes use of the values of $\hat{g}(\xi)$, which by (2.15) depend on the values of the given function g at all points. Actually by Holmgren's theorem the domain of dependence of $u(x,t)$ on the values of g is known to be finite; equivalently initial data g of compact support lead to solutions $u(x,t)$ of compact support in x. This is not obvious from the expression (2.24a), but can be deduced for strictly hyperbolic P from a version of the *Paley–Wiener theorem.* This involves a shift in the integrations in (2.24a) to complex ξ. For this we require estimates for the functions Z and $\hat{g}$ for *complex* arguments $\xi+i\zeta$ and real $t\geqslant 0$, where ξ and ζ are real.

Assume that the function $g(x)$ belongs to $C_0^s(\mathbb{R}^n)$ where $s\geqslant n+m+1$, and that the support of $g(x)$ lies in a ball $|x|<a$. For the complex vector $\xi+i\zeta$ we define $|\xi+i\zeta|$ by

$$|\xi+i\zeta|^2=\sum_{k=1}^{n}|\xi_k+i\zeta_k|^2=|\xi|^2+|\zeta|^2. \tag{2.45}$$

Then as in (2.19c) and by the same arguments

$$(1+|\xi+i\zeta|)^s\leqslant 2^s n^{s/2}\sum_{|\alpha|\leqslant s}|(\xi+i\zeta)^\alpha|. \tag{2.46}$$

We conclude from (2.15), (2.18) in analogy to (2.20) that

$$\begin{aligned}(1+|\xi+i\zeta|)^s|\hat{g}(\xi+i\zeta)|&<2^s n^{s/2}\sum_{|\alpha|\leqslant s}\int_{|x|<a}|e^{-ix\cdot(\xi+i\zeta)}D^\alpha g(x)|\,dx\\ &\leqslant 2^s n^{s/2}e^{a|\zeta|}\sum_{|\alpha|\leqslant s}\int|D^\alpha g(x)|\,dx\leqslant e^{a|\zeta|}M_s,\end{aligned} \tag{2.47}$$

since for real x with $|x|<a$

$$|e^{-ix\cdot(\xi+i\zeta)}|=e^{x\cdot\zeta}\leqslant e^{|x||\zeta|}\leqslant e^{a|\zeta|}. \tag{2.48}$$

We proceed to estimate

$$Z(\xi+i\zeta,t)=\frac{1}{2\pi}\int_\Gamma\frac{e^{i\lambda t}}{P(i(\xi+i\zeta),i\lambda)}\,d\lambda$$

using as path of integration Γ again the boundary of the union of the unit disks with centers at the roots λ_k of

$$P(i(\xi+i\zeta),i\lambda)=0. \tag{2.49}$$

It follows, as in (2.35a) for $k=0$ that

$$|Z(\xi+i\zeta,t)|\leqslant me^{(1+c)t}, \tag{2.50a}$$

where

$$c=-\min_k(\operatorname{Im}\lambda_k)\leqslant\max_k|\operatorname{Im}\lambda_k|. \tag{2.50b}$$

To estimate c we apply the substitution

$$\xi+i\zeta=\rho\eta, \qquad \lambda=\rho\mu, \tag{2.51}$$

where $\rho=|\xi+i\eta|$ and η is a complex vector with $|\eta|=1$. For a root λ of (2.49) we obtain again equation (2.42) for μ. The coefficients in equation (2.42) differ from those in the equation

$$p_m(\eta,\mu)=0 \tag{2.52}$$

by terms of order $1/\rho$. Since $\eta=(\xi+i\zeta)/\rho$, and $|\xi/\rho|,|\zeta/\rho|<1$ the coefficients in the equation (2.52) differ from those in the equation

$$p_m(\xi/\rho,\mu)=0 \tag{2.53}$$

by terms of order $|\zeta|/\rho$. Since the roots μ of (2.53) are real and distinct for $\xi/\rho \neq 0$, it follows for the roots μ of (2.42) that

$$\operatorname{Im}\mu=O\left(\frac{1+|\zeta|}{\rho}\right)$$

as long as ξ/ρ is bounded away from zero, and hence as long as $|\zeta|/\rho < \frac{1}{2}$. Thus the roots λ of (2.49) satisfy

$$\operatorname{Im}\lambda=O(1+|\zeta|). \tag{2.54}$$

Since also as in (2.34).

$$|\operatorname{Im}\lambda| \leqslant |\lambda| = O(1+|\xi+i\zeta|) = O(1+\rho) = O(1+|\zeta|)$$

for $|\zeta|/\rho > \frac{1}{2}$, we see that (2.54) is valid for all $\xi + i\zeta$. Thus there exists a constant M such that for the roots $\lambda=\lambda_k$ of (2.49)

$$|\operatorname{Im}\lambda_k| \leqslant M(1+|\zeta|),$$

and hence, using (2.50a,b), (2.47)

$$|Z(\xi+i\zeta),t)| \leqslant me^{(1+M+M|\zeta|)t}$$

$$|e^{ix\cdot(\xi+i\zeta)}Z(\xi+i\zeta,t)\,\hat{g}(\xi+i\zeta)| \leqslant \frac{mM_s e^{-x\cdot\zeta+t+Mt+(a+Mt)|\zeta|}}{(1+|\xi|)^s}. \tag{2.55}$$

By Cauchy's theorem we can in (2.24a) shift the domain of integration from that of real ξ to $\xi+i\zeta$ with fixed ζ without changing the value of the integral, due to the decay of the integrand for large $|\xi|$. Choose now ζ to be of the form $\sigma x/|x|$, where $\sigma>0$. It follows from (2.24a), (2.55) that

$$|u(x,t)| \leqslant (2\pi)^{-n/2} mM_s e^{t+Mt-\sigma(|x|-a-Mt)} \int (1+|\xi|)^{-s}\,d\xi.$$

If here

$$|x|>a+Mt$$

it follows for $\sigma\to\infty$ that $u(x,t)=0$. Hence u for each $t>0$ has bounded support lying in the ball $|x|\leqslant a+Mt$. The constant M here represents an upper bound for the *speed of propagation* of disturbances.

So far we have dealt with the standard problem for a single scalar equation $P(D,\tau)u=0$. The case of a system of equations with constant coefficients requires only minor adjustments. If P is an $N\times N$ matrix satisfying (2.12) a formal solution of (2.14a, b, c) is again furnished by (2.24a), where now, however, $Z(\xi,t)$ is an $N\times N$ matrix given by

$$Z(\xi,t)=\frac{1}{2\pi}\int_\Gamma e^{i\lambda t}(P(i\xi,i\lambda))^{-1}d\lambda. \tag{2.56}$$

Here Γ has to be a path in the λ-plane enclosing all singularities of the matrix P^{-1}, that is, all of the mN roots λ_k of the equation

$$Q(i\xi,i\lambda)=\det P(i\xi,i\lambda)=0. \tag{2.57}$$

Gårding's hyperbolicity condition for systems states that there exists a constant c such that

$$\operatorname{Im}\lambda>-c$$

for all roots λ of (2.57) for all real vectors ξ.

J. Hadamard introduced the important distinction between *well-posed* (also called *correctly-set*) problems and those that are *ill posed* (*improperly posed, incorrectly set*). The distinction applies specially to problems where a "solution" u is to be found from "data" g. Well-posed problems are those for which

(a) u exists for "arbitrary" g.
(b) u is determined "uniquely" by g.
(c) u depends "continuously" on g.

Here the words in quotation marks are somewhat vague and require that the spaces of admitted functions u and of functions g are specified. Typically well-posed problems are the Dirichlet problem for the Laplace equation and the initial-value problem for a hyperbolic equation with constant coefficients* (see [33]).

The initial-value problem for the scalar equation $P(D,\tau)u=0$ certainly is ill-posed when the principal part p_m of P does not satisfy Gårding's condition, that is, when there exists a real vector η and a nonreal scalar μ_0 such that

$$p_m(\eta,\mu_0)=0.$$

We can assume here that

$$\operatorname{Im}\mu_0=-\gamma<0,\qquad |\eta|=1.$$

*In the latter problem one can specify that $u\in C^m$ for $x\in\mathbb{R}^n$, $t\geqslant 0$ and that g has uniformly bounded derivatives of orders $\leqslant s$ with s chosen sufficiently large. Then u depends continuously on g in the sense that the maximum of $|u|$ can be estimated in terms of the maxima of the $|D^\alpha g|$ for $|\alpha|\leqslant s$. More generally the Cauchy problem with data on a *space like* surface is well posed.

Consider for any s exponential solutions of the form

$$(1+|\xi|)^{-s-m}e^{i(x\cdot\xi+\lambda t)}, \tag{2.58}$$

where $P(i\xi,i\lambda)=0$ and s is an arbitrary integer. Take here

$$\xi=\rho\eta, \qquad \lambda=\rho\mu.$$

For sufficiently large ρ we can find a λ for which $|\mu-\mu_0|<\gamma/2$. Then

$$|\lambda|<\left(|\mu_0|+\tfrac{1}{2}\gamma\right)\rho, \qquad \operatorname{Im}\lambda<-\tfrac{1}{2}\gamma\rho,$$

so that for $t=0$ and $|\alpha|\leqslant s$, $0\leqslant k\leqslant m$

$$\begin{aligned}|D^{\alpha}\tau^{k}u|&=(1+\rho)^{-s-m}|\lambda|^{k}|\xi^{\alpha}|\\&\leqslant(1+\rho)^{-s-m}\left(|\mu_0|+\tfrac{1}{2}\gamma\right)^{k}\rho^{k+|\alpha|}\\&\leqslant\left(1+|\mu_0|+\tfrac{1}{2}\gamma\right)^{m},\end{aligned}$$

while

$$u(0,t)=(1+\rho)^{-s-m}|e^{i\lambda t}|\geqslant(1+\rho)^{-s-m}e^{\gamma\rho t/2}.$$

Thus the initial data and their derivatives of orders $\leqslant s$ are bounded uniformly for all x, while $u(0,t)\to\infty$ for $\rho\to\infty$ and any fixed $t>0$. Here u does not depend "continuously" on its initial data.

PROBLEMS

1. For $n=3$ identify the solution of the standard initial-value problem for the wave equation given by (2.37e) with the solution $u=tM_g(x,ct)$ obtained from (1.14). [Hint: Compute $M_g(x,ct)$ in terms of $\hat{g}$ from (2.16).]

2. Solve the standard initial-value problem for the system of equations of elastic waves

$$\rho\frac{\partial^2 u_i}{\partial t^2}=\mu\Delta u_i+(\lambda+\mu)\frac{\partial}{\partial x_i}\left(\sum_i\frac{\partial u_k}{\partial x_k}\right) \tag{2.59}$$

(with positive constants ρ,λ,μ) in the form (2.24a), computing the matrix $Z(\xi,t)$ explicitly from (2.56). [Answer: $Z(\xi,t)$ is the matrix with elements

$$\frac{c_1\left(\delta_{ik}|\xi|^2-\xi_i\xi_k\right)\sin(c_2|\xi|t)+c_2\xi_i\xi_k\sin(c_1|\xi|t)}{c_1c_2|\xi|^3}, \tag{2.60}$$

where $c_1^2=(\lambda+2\mu)/\rho$, $c_2^2=\mu/\rho$.]

3. Show that for $n=1$, $m=1$ and any N the system of equations

$$u_t+Bu_x-Cu=0$$

is strictly hyperbolic, when the matrix B has real and distinct eigenvalues. (Compare (5.12) of Chapter 2.)

4. Prove that, when Gårding's condition is satisfied, the solution u of (2.14a, b, c) can be written

$$u(x,t)=(1-\Delta_x)^s\int K(x-y,t)\,g(y)\,dy, \tag{2.61a}$$

where

$$K(x-y,t)=(2\pi)^{-n/2}\int e^{i(x-y)\cdot\xi}(1+|\xi|^2)^{-s}Z(\xi,t)\,d\xi \tag{2.61b}$$

and s is any integer exceeding $n/2$. [Hint: Introducing $h(x)=(1-\Delta_x)^s g(x)$, we can substitute for $\hat{g}$ in (2.24a) the expression $(1+|\xi|^2)^{-s}\hat{h}(\xi)$. Interchanging the integrations yields

$$u(x,t)=\int K(x-y)h(y)\,dy$$

from which (2.61a) can be derived.]

(c) *Solution of a mixed problem by Fourier transformation*

In many cases mixed initial-boundary-value problems can be solved by Fourier transformation, when the domain of the solution is a half space. The method will be illustrated (following R. Hersh*) by a problem for the wave equation for $n=3$, which could also be solved by reflection. (Compare problem 3, p. 143.) We seek to find a $u(x,t)=u(x_1,x_2,x_3,t)$ for which

$$\Box u=0 \quad \text{for } x_3\geqslant 0,\ t>0 \tag{2.62a}$$

$$u=u_t=0 \quad \text{for } x_3>0,\ t=0 \tag{2.62b}$$

$$Mu=u_t+\alpha_1u_{x_1}+\alpha_2u_{x_2}+\alpha_3u_{x_3}=h(x_1,x_2,t) \quad \text{for } x_3=0,\ t>0. \tag{2.62c}$$

We assume here that the α_k are constant, that $\alpha_3<0$, and (to avoid inconsistencies for $t=x_3=0$) that there exists a positive ε such that

$$h(x_1,x_2,t)=0 \quad \text{for } t<\varepsilon. \tag{2.62d}$$

Let moreover $h\in C_0^s(\mathbb{R}^3)$ with a sufficiently large s.

The building blocks are again exponential solutions

$$u=e^{i(\lambda t+\xi_1x_1+\xi_2x_2+\xi_3x_3)} \tag{2.63a}$$

of (2.62a), for which the relation

$$\lambda^2-c^2(\xi_1^2+\xi_2^2+\xi_3^2)=0 \tag{2.63b}$$

will have to hold. For these u we have

$$Mu=i(\lambda+\alpha_1\xi_1+\alpha_2\xi_2+\alpha_3\xi_3)e^{i(\lambda t+\xi_1x_1+\xi_2x_2)} \quad \text{for } x_3=0. \tag{2.63c}$$

This leads to a formal solution of (2.62a,c) given by

$$u(x,t)=(2\pi)^{-3/2}\int\frac{e^{i(\lambda t+\xi_1x_1+\xi_2x_2+\xi_3x_3)}}{i(\lambda+\alpha_1\xi_1+\alpha_2\xi_2+\alpha_3\xi_3)}\hat{h}(\xi_1,\xi_2,\lambda)\,d\xi_1\,d\xi_2\,d\lambda. \tag{2.64}$$

* Mixed problems in several variables, *J. Math. Mech.* **12** (1963), 317–334.

Here $\hat{h}$ is the Fourier transform of h:

$$\hat{h}(\xi_1,\xi_2,\lambda)=(2\pi)^{-3/2}\int e^{-i(\lambda t+\xi_1 x_1+\xi_2 x_2)}h(x_1,x_2,t)\,dx_1\,dx_2\,dt \tag{2.65}$$

and ξ_3 in (2.64) is a function of (ξ_1,ξ_2,λ) satisfying (2.63b).

For convergence of the integral in (2.64) it is essential that the exponential in the integrand for each fixed t is bounded for x in the half space $x_3>0$. This is the case, when ξ_1,ξ_2 are real and

$$\operatorname{Im}\xi_3\geqslant 0. \tag{2.66}$$

This condition does not fix the solution ξ_3 of (2.63b) uniquely for all real ξ_1,ξ_2,λ. We also have to worry about possible vanishing of the denominator of the integrand. It is best to shift the integration with respect to λ in (2.64) in the complex plane, letting λ run through values with $\operatorname{Im}\lambda=-\delta$ with a fixed real number $\delta>0$. Under these circumstances the solution ξ_3 of (2.63b) cannot be real, and there is a unique solution ξ_3, for which (2.66) holds. Moreover the denominator in (2.64) cannot vanish since

$$\operatorname{Im}(\lambda+\alpha_1\xi_1+\alpha_2\xi_2+\alpha_3\xi_3)=-\delta+\alpha_3\operatorname{Im}\xi_3<-\delta, \tag{2.67}$$

using (2.66) and the important assumption $\alpha_3<0$. We obtain the estimate

$$\frac{e^{\delta t}}{\delta}|\hat{h}(\xi_1,\xi_2,\lambda)| \tag{2.68}$$

for the absolute value of the integrand in the expression (2.64) for u. We now make use of assumption (2.62d), which implies that

$$|e^{-i\lambda t}|=e^{-\delta t}\leqslant e^{-\delta\varepsilon}$$

in the integral (2.65) giving $\hat{h}$. We conclude, similarly as in (2.47), that for the complex λ and real ξ_1,ξ_2 in question

$$(1+\xi_1^2+\xi_2^2+|\lambda|^2)^s|\hat{h}(\xi_1,\xi_2,\lambda)|<Me^{-\varepsilon\delta} \tag{2.69}$$

with a suitable constant M. Thus for sufficiently regular h of compact support formula (2.64) furnishes an actual solution of (2.62a,c). It remains to show that it satisfies (2.62b) as well. Now by (2.68), (2.69)

$$|u(x,t)|\leqslant(2\pi)^{-3/2}M\frac{e^{\delta(t-\varepsilon)}}{\delta}\int(1+\xi_1^2+\xi_2^2+|\lambda|^2)^{-s}d\xi_1\,d\xi_2\,d\lambda.$$

Letting $\delta\to\infty$ it follows that

$$u(x,t)=0 \quad \text{for } x_3>0,\ 0\leqslant t<\varepsilon.$$

A similar result holds for u_t. This implies that (2.62b) holds.

(d) *The method of plane waves*

In what preceded the standard initial-value problem was solved by decomposing the initial function $g(x)$ into exponentials $\exp(ix\cdot\xi)$, according to

Fourier's formula. For those the initial-value problem is easily solved; by superposition we then obtained the solution for general g. A disadvantage of this method is that the resulting solution u is expressed in terms of the Fourier transform $\hat{g}$ instead of directly in terms of g. For homogeneous partial differential equation with constant coefficients a different type of decomposition of g into plane waves can be preferable, since it does not involve the somewhat artificial introduction of exponentials.

A function G with domain R^n is called a *plane wave* function, if its level surfaces form a family of parallel planes, that is if G can be expressed in the form $G = G(s)$, where the scalar argument

$$s = x \cdot \xi = \sum_k \xi_k x_k \tag{2.70}$$

is a linear combination of the independent variables. The exponential functions above are plane wave functions with $G = e^{is}$. Assume that the differential operator $P(D,\tau)$ is homogeneous of degree m, and thus agrees with its principal part $p_m(D,\tau)$. Let $P(0,1) = 1$ and the degree m of P be even. We notice that for any function $G(s)$ of a scalar argument s we have in

$$u(x,t) = G(x \cdot \xi + \lambda t) \tag{2.71}$$

a solution of

$$P(D,\tau)u = P(\xi,\lambda)G^{(m)}(x \cdot \xi + \lambda t). \tag{2.72}$$

In particular u will be a solution of

$$P(D,\tau)u = 0 \tag{2.73a}$$

if ξ,λ satisfy the algebraic equation

$$P(\xi,\lambda) = 0. \tag{2.73b}$$

We can find a linear combination of expressions (2.71) corresponding to the various roots $\lambda = \lambda_k$ of (2.73b), that satisfies the standard initial conditions for $t = 0$

$$\begin{aligned} \tau^k u &= 0 \quad \text{for } k = 0, 1, \ldots m-2 \\ \tau^{m-1} u &= g \end{aligned} \tag{2.74a}$$

for a prescribed plane wave function

$$g = g(s) = g(x \cdot \xi). \tag{2.74b}$$

In the case where $g(s)$ is an entire analytic function of s we easily verify that a solution is given by the Cauchy integral

$$u(x,t) = \frac{1}{2\pi i} \int_\Gamma \frac{G(x \cdot \xi + \lambda t)}{P(\xi,\lambda)} d\lambda, \tag{2.75a}$$

where the path of integration Γ in the complex λ-plane encloses all roots λ_k of (2.73b). Here G is to be chosen so that

$$G^{(m-1)}(s)=g(s). \tag{2.75b}$$

In the special case $G(s)=e^{is}$ we regain the formula

$$u(x,t) = e^{ix\cdot\xi}Z(\xi,t)\,i^{m-1} \tag{2.75c}$$

with Z given by (2.31).

We now restrict ourselves to the case that P is *strictly hyperbolic*, that is, that all roots λ_k of the homogeneous equation are real and distinct for $\xi\neq 0$. In that case the calculus of residues permits us to evaluate the integral (2.75a) for $\xi\neq 0$:

$$u(x,t)=\sum_{k=1}^{m}\frac{G(x\cdot\xi+\lambda_k t)}{P_\lambda(\xi,\lambda_k)}. \tag{2.76}$$

Now formula (2.76) was derived under the assumption that G is analytic. By continuity arguments or direct verification it follows immediately that (2.76) represents a solution of (2.73a), (2.74a,b), when G is only of class $C^m(\mathbb{R})$ and satisfies (2.75b). Of special interest for us will be the case when g is of the form

$$g=|(x-y)\cdot\xi|=|s-y\cdot\xi| \tag{2.77a}$$

depending on parameter vectors y,ξ. The corresponding solution u of (2.73a), (2.74a,b) is then given by

$$U(x-y,t,\xi)=\sum_{k=1}^{m}\frac{((x-y)\cdot\xi+\lambda_k t)^m\operatorname{sgn}((x-y)\cdot\xi+\lambda_k t)}{m!\,P_\lambda(\xi,\lambda_k)} \tag{2.77b}$$

where we have used (2.76)*

$$G(s)=\frac{(s-y\cdot\xi)^m\operatorname{sgn}(s-y\cdot\xi)}{m!}. \tag{2.77c}$$

We shall show that for odd n and for s sufficiently large the general $g(x)\in C_0^s(\mathbb{R}^n)$ can be decomposed into functions of the form (2.77a) with $|\xi|=1$ by a formula of the type

$$g(x)=\int_{|\xi|=1}dS_\xi\int dy\,|(x-y)\cdot\xi|\,q(y) \tag{2.78a}$$

with a suitable continuous function $q(y)$ of compact support. It follows then that

$$u(x,t)=\int_{|\xi|=1}dS_\xi\int dy\,U(x-y,t,\xi)\,q(y) \tag{2.78b}$$

solves our standard initial-value problem (2.14a,b,c).†

*Strictly speaking here $G\notin C^m(\mathrm{R})$. But the jump discontinuity in $G^{(m)}(s)$ for $s=y\cdot\xi$ is harmless.

†The U given by (2.77b) has continuous derivatives of order $\leqslant m$ with respect to its arguments, except for jumps in the mth derivatives. We use here that the P_λ in the denominators are bounded away from 0 for $|\xi|=1$, because $P(\xi,\lambda)$ has no multiple roots.

To arrive at (2.78a) we first decompose the function

$$r=|x-y| \tag{2.79a}$$

into plane waves of type (2.77a). This is achieved by writing

$$x-y=r\eta, \quad \text{where } |\eta|=1. \tag{2.79b}$$

Then

$$\int_{|\xi|=1}|(x-y)\cdot\xi|\,dS_\xi=r\int_{|\xi|=1}|\eta\cdot\xi|\,dS_\xi=c_n r. \tag{2.79c}$$

Here c_n is a positive constant independent of η, since a simultaneous rotation of η,ξ does not change the integral of $\eta\cdot\xi$ over the unit sphere, so that we can always bring about that η is the unit vector in the x_n-direction. By (1.9) of Chapter 4,

$$\Delta_x r^j=j(j+n-2)r^{j-2}.$$

It follows *for odd n* that there is a constant d_n such that

$$d_n\Delta_x^{(n-1)/2}r=\frac{r^{2-n}}{(2-n)\omega_n}=K(x,y) \tag{2.80}$$

is a fundamental solution of the Laplace equation with pole y (see Chapter 4, (1.15a)). Equivalently

$$k(x,y)=d_n r=c_n^{-1}d_n\int_{|\xi|=1}|(x-y)\cdot\xi|\,dS_\xi \tag{2.81}$$

is a fundamental solution* with pole x for the operator $\Delta_y^{(n+1)/2}$:

$$g(x)=\int k(x,y)\Delta_y^{(n+1)/2}g(y)\,dy. \tag{2.82}$$

Substituting for k its expression (2.81) we are led to a decomposition (2.78a) for g with

$$q(y)=c_n^{-1}d_n\Delta_y^{(n+1)/2}g(y). \tag{2.83}$$

With this q, formula (2.78b) solves the standard initial-value problem.

PROBLEMS

1. Give the value for $n=3$ of the constants c_n,d_n in formulas (2.79c), (2.80). [Answer: $c_3=2\pi$, $d_3=-1/8\pi$.]

2. Show that (2.78b) can be rewritten in the form

$$u(x,t)=c_n^{-1}d_n\int\Delta_y^j g(y)\,dy\int_{|\xi|=1}\Delta_y^k U(x-y,t,\xi)\,dS_\xi, \tag{2.84}$$

where

$$k=\min\left(\frac{n+1}{2},\frac{m}{2}\right),\qquad j=\frac{n+1}{2}-k \tag{2.85}$$

$$\Delta_y^k U(x-y,t,\xi)=\sum_{k=1}\frac{((x-y)\cdot\xi+\lambda_k t)^{m-2k}\operatorname{sgn}((x-y)\cdot\xi+\lambda_k t)}{(m-2k)!P_\lambda(\xi,\lambda_k)}.$$

*Formula (2.82) follows rigorously from (1.31) of Chapter 4 by integration by parts, provided $g\in C_0^s(\mathbb{R}^n)$ with $s\geqslant n+1$.

3. For the wave equation (1.1) in $n = 3$ and $n = 5$ dimensions identify the solution obtained from (2.84) with that given by formula (1.14), (1.22). [Hint: Show that here for $n = 3$

$$\Delta_y U = 1/c \quad \text{for } |(x-y)\cdot\xi| < ct$$

$$\Delta_y U = 0 \quad \text{for } |(x-y)\cdot\xi| > ct$$

$$\int_{|\xi|=1} \Delta_y U(x-y,t,\xi)\, dS_\xi = \begin{cases} 4\pi t/|x-y| & \text{for } |x-y| > ct \\ 4\pi/c & \text{for } |x-y| < ct. \end{cases} \tag{2.86}$$

Apply Green's formula.]

4. The "Radon transform" $G(\xi,p)$ of a function $g(x)$ defined for $x \in \mathbb{R}^n$ is the integral of g over the hyperplane with unit normal ξ and distance p from the origin:

$$G(\xi,p) = \int_{x\cdot\xi=p} g(x)\, dS_x \tag{2.87}$$

for $\xi \in \mathbb{R}^n$, $|\xi| = 1$, $p \in \mathbb{R}$. The "Radon problem" consists in determining g from G. (The problem is of importance in the reconstruction of objects from X-ray pictures.)

(a) Show that for $g \in C_0^2(\mathbb{R}^n)$

$$2G(\xi, x\cdot\xi) = \int |(x-y)\cdot\xi|\,\Delta_y g(y)\, dy. \tag{2.88}$$

(b) Show that for odd n and $g \in C_0^{n+1}(\mathbb{R}^n)$ formulae (2.81), (2.82) yield the solution

$$g(x) = 2c_n^{-1} d_n \Delta_x^{(n-1)/2} \int_{|\xi|=1} G(\xi, x\cdot\xi)\, dS_\xi \tag{2.89}$$

of the Radon problem.

(c) Show that for any n and $g \in C_0^{n+1}(\mathbb{R}^n)$ a solution of the Radon problem is given by

$$g(x) = (2\pi)^{(1-2n)/2} \int_0^\infty q^{n-1}\, dq \int_{|\xi|=1} e^{iqx\cdot\xi}\hat{G}(\xi,q)\, dS_\xi \tag{2.90}$$

where $\hat{G}(\xi,q)$ for fixed ξ denotes the Fourier transform of $G(\xi,p)$ with respect to p. [Hint: Show from (2.15) that $\hat{g}(q\xi) = (2\pi)^{(1-n)/2}\hat{G}(\xi,q)$].

(d) For odd n verify formula (2.89) from (2.90) with

$$c_n^{-1} d_n = \tfrac{1}{4}(2\pi i)^{1-n}. \tag{2.91}$$

[Hint: Use

$$\int_{|\xi|=1} e^{ixq\cdot\xi} G(\xi,q)\, dS_\xi = \int_{|\xi|=1} e^{-iqx\cdot\xi} G(\xi,-q)\, dS_\xi.] \tag{2.92}$$

(e) For solving the Radon problem for $n = 2$ describe the unit normal ξ by its polar angle θ, writing (2.87) as

$$G(\theta,p) = \int_{-\infty}^{\infty} g(p\cos\theta + t\sin\theta,\, p\sin\theta - t\cos\theta)\, dt.$$

Show that

$$\int_0^{2\pi} G(\theta,p)\,d\theta = 4\pi \int_p^{\infty} \frac{r}{\sqrt{r^2-p^2}} I(r)\,dr \tag{2.93}$$

where $I(r)$ is the average of g on the circle of radius r about the origin. To find an expression for $I(r)$ and hence for the value of $g(0,0)=I(0)$ multiply (2.93) for $s>0$ by $p/\sqrt{p^2-s^2}$ and integrate from s to ∞.

3. Symmetric Hyperbolic Systems

(a) *The basic energy inequality*

In this section we shall be concerned with a linear first order system of P.D.E.s for a column vector $u=u(x,t)=u(x_1,\ldots,x_n,t)$ with N components $u_1,\ldots,u_N$. Such a system can be written symbolically in the form

$$Lu = A(x,t)\tau u + \sum_{k=1}^{n} A^k(x,t)D_k u + B(x,t)u = w(x,t). \tag{3.1a}$$

Here $A,A^1,\ldots,A^n,B$ are given $N\times N$ square matrices, w a given N-vector, and $\tau,D_1,\ldots,D_n$ again stand for the differential operators

$$\tau=\frac{\partial}{\partial t}, \qquad D_1=\frac{\partial}{\partial x_1}, \qquad \ldots, \qquad D_n=\frac{\partial}{\partial x_n}.$$

As initial data we prescribe the values of u on the hyperplane $t=0$ in xt-space. By a trivial substitution on u and w we can always bring about that the initial condition becomes

$$u(x,0)=0. \tag{3.1b}$$

Following K. O. Friedrichs the system (3.1a) is called *symmetric hyperbolic*, if all of the matrices $A,A^1,\ldots,A^n$ (but not necessarily B) are symmetric, and moreover A is positive definite for all arguments (x,t) in question. We shall see that symmetric hyperbolic systems (for sufficiently regular $A,A^1,\ldots,A^n,B,w$) are indeed hyperbolic in the sense that the initial-value problem (3.1a,b) can be solved.

Many hyperbolic equations or systems can be reduced to symmetric hyperbolic form. Consider, for example, a single scalar hyperbolic second-order equation

$$v_{tt} = \sum_{i,k=1}^{n} a_{ik}(x,t)v_{x_ix_k} + \sum_{i=1}^{n} b_i(x,t)v_{x_i} + c(x,t)v_t + d(x,t)v, \tag{3.2}$$

where the a_{ik} form a positive definite symmetric matrix. We introduce here the vector u with the $N=n+2$ components

$$u_1=v_{x_1}, \qquad \ldots, \qquad u_n=v_{x_n}, \qquad u_{n+1}=v_t, \qquad u_{n+2}=v. \tag{3.3}$$

The $n+2$ equations

$$\sum_{k=1}^{n} a_{ik}\tau u_k - \sum_{k=1}^{n} a_{ik} D_k u_{n+1} = 0 \quad \text{for } i=1,\ldots,n, \tag{3.4a}$$

$$\tau u_{n+1} - \sum_{i,k=1}^{n} a_{ik} D_k u_i - \sum_{i=1}^{n} b_i u_i - c u_{n+1} - d u_{n+2} = 0, \tag{3.4b}$$

$$\tau u_{n+2} - u_{n+1} = 0 \tag{3.4c}$$

are consequences of (3.2), (3.3). One easily verifies that they constitute a symmetric hyperbolic system. Similarly in Chapter 2 we were able for $n=1$ to write general hyperbolic systems (3.1a) in "canonical form," where A becomes the unit matrix and A^1 a diagonal matrix. This clearly implies a reduction of the system to symmetric hyperbolic form.

Multiplying (3.1a) with the transposed vector u^{T} we find that*

$$\tau(u^{\mathrm{T}}Au) + \sum_{k=1}^{n} D_k(u^{\mathrm{T}}A^k u) + u^{\mathrm{T}}Cu = 2u^{\mathrm{T}}w, \tag{3.5a}$$

where

$$C = 2B - \tau A - \sum_{k=1}^{n} D_k A^k. \tag{3.5b}$$

Integrating (3.5a) over a region R in xt-space and applying the divergence theorem yields

$$\int_{\partial R} u^{\mathrm{T}}\left(A\frac{dt}{dv} + \sum_k A^k \frac{dx_k}{dv}\right) u\, dS = \int_R (-u^{\mathrm{T}}Cu + 2u^{\mathrm{T}}w)\, dx\, dt, \tag{3.6}$$

where $dx_1/dv,\ldots,dx_n/dv, dt/dv$ denote direction cosines of the exterior normal and dS the element of "area" of the boundary ∂R, while $dx\,dt = dx_1\ldots dx_n\,dt$ is the element of volume of R.

We define the slab R_λ in xt-space for $\lambda>0$ as the set

$$R_\lambda = \{(x,t) | x\in\mathbb{R}^n,\ 0\leqslant t\leqslant\lambda\}. \tag{3.7}$$

Let for a certain T the function $u(x,t)$ be a solution of (3.1a,b) of class $C^1(R_T)$, which is of compact support in x for each t in $0\leqslant t\leqslant T$. Applying (3.6) to $R=R_\lambda$ with $0<\lambda<T$ yields the *energy identity*

$$E(\lambda) = \int_{t=\lambda} u^{\mathrm{T}}Au\, dx = \int_0^\lambda dt \int (-u^{\mathrm{T}}Cu + 2u^{\mathrm{T}}w)\, dx. \tag{3.8}$$

We assume that the matrices $A, A^1,\ldots,A^n, B$ together with their derivatives

*We make use here of the symmetry of $A, A^1,\ldots,A^n$ which implies

$$\tau(u^{\mathrm{T}}Au) = (\tau u^{\mathrm{T}})Au + u^{\mathrm{T}}(\tau A)u + u^{\mathrm{T}}A(\tau u) = 2u^{\mathrm{T}}A(\tau u) + u^{\mathrm{T}}(\tau A)u$$

and analogous identities for $D_k(u^{\mathrm{T}}A^k u)$. We tacitly assume that $A, A^1,\ldots,A^n$ are in C^1, and B and w in C^0.

of any desired order are continuous and bounded uniformly in R_T. Moreover the matrix A shall be *uniformly* positive definite in the sense that there exists a $\mu>0$ such that

$$v^{\mathrm{T}}A(x,t)v \geqslant \mu v^{\mathrm{T}}v \tag{3.9}$$

for all (x,t) in R_T and all vectors v. Since C is bounded there will exist a constant $K>0$ such that

$$|v^{\mathrm{T}}C(x,t)v| \leqslant Kv^{\mathrm{T}}A(x,t)v \tag{3.10a}$$

for $(x,t)\in R_T$ and all v. Moreover, since A is symmetric and positive definite the inequality

$$2u^{\mathrm{T}}w \leqslant \mu u^{\mathrm{T}}u + \frac{1}{\mu}w^{\mathrm{T}}w \leqslant u^{\mathrm{T}}Au + \frac{1}{\mu^2}w^{\mathrm{T}}Aw \tag{3.10b}$$

holds for all vectors u,w. Thus by (3.8)

$$E(\lambda) \leqslant (K+1)\int_0^\lambda E(t)\,dt + \mu^{-2}\int_{R_\lambda} w^{\mathrm{T}}Aw\,dx\,dt \quad \text{for } 0\leqslant\lambda\leqslant T. \tag{3.11}$$

Writing this as

$$\frac{d}{d\lambda}e^{-(K+1)\lambda}\int_0^\lambda E(t)\,dt \leqslant e^{-(K+1)\lambda}\mu^{-2}\int_{R_T} w^{\mathrm{T}}Aw\,dx\,dt$$

we conclude that

$$\begin{aligned}\int_0^T E(\lambda)\,d\lambda &\leqslant \frac{e^{(K+1)T}-1}{K+1}\mu^{-2}\int_{R_T} w^{\mathrm{T}}Aw\,dx\,dt\\ &\leqslant \mu^{-2}Te^{(K+1)T}\int_{R_T} w^{\mathrm{T}}Aw\,dx\,dt.\end{aligned} \tag{3.12a}$$

We define the inner product of two vectors u,v on R_T by

$$(u,v) = \int_{R_T} u^{\mathrm{T}}Av\,dx\,dt, \tag{3.12b}$$

and denote by $\|u\| = \sqrt{(u,u)}$ the corresponding norm. Setting

$$\Gamma^2 = \mu^{-2}Te^{(K+1)T}, \tag{3.12c}$$

the estimate (3.12a) takes the form

$$(u,u) \leqslant \Gamma^2(w,w), \tag{3.13}$$

whenever u is a solution of (3.1a,b) of compact support in x.

Denote by $\underline{C}^s$ the space of functions $u\in C^s(R_T)$, that vanish on $t=0$, and have compact support in x. Then by (3.13) the *energy inequality*

$$\|u\| \leqslant \Gamma\|Lu\| \tag{3.14}$$

holds for all $u\in\underline{C}^1$.

The estimate (3.14) can be made the basis for an existence proof for the solution of the initial-value problem (3.1a, b) which we just indicate. A first step in this direction is to establish the existence of a *weak* solution of that problem. For that purpose one introduces in analogy to $\underline{C}^1$ the space $\bar{C}^1$ of functions $v \in C^1(R_T)$ that are of compact support in x and vanish for $t = T$. For any $u \in \underline{C}^1$, $v \in \bar{C}^1$ we derive by integration by parts Green's identity (see (4.5) of Chapter 3)

$$(A^{-1}v, Lu) = (A^{-1}\tilde{L}v, u) \tag{3.15}$$

where $\tilde{L}$, the adjoint of L, is defined by

$$\begin{aligned}\tilde{L}v &= -\tau(Av) - \sum_{k=1}^{n} D_k(A^k v) + B^{\mathrm{T}}v \\ &= -Lv + (C^{\mathrm{T}} + B - B^{\mathrm{T}})v\end{aligned}$$

(see (3.5a,b)). A function $u \in C^1(R_T)$ is a solution of (3.1a,b) if and only if

$$(A^{-1}v, w) = (A^{-1}\tilde{L}v, u) + \int_{t=0} v^{\mathrm{T}}Au\,dx \tag{3.16}$$

for all $v \in \bar{C}^1$. For by Green's identity (not using (3.1b), but using that v has compact support in x and vanishes for $t = T$)

$$(A^{-1}\tilde{L}v, u) = (A^{-1}v, w) + \int_{t=0} v^{\mathrm{T}}Au\,dx. \tag{3.17}$$

Taking for v in (3.16) first an arbitrary function of compact support in R_T and vanishing for $t = 0$, we see that u satisfies (3.1a). Subsequently we find from (3.16) that

$$0 = \int_{t=0} v^{\mathrm{T}}Au\,dx$$

for all $v \in \bar{C}^1$, which implies (3.1b).

This suggests replacing (3.1a,b) by the requirement (3.16) in some suitable function space. For that purpose we observe that $\bar{C}^1$ is an inner product space (see p. 117), if we define as inner product of two vectors v and v' the expression

$$\langle v, v'\rangle = (\tilde{L}v, \tilde{L}v'). \tag{3.18a}$$

Indeed this expression is linear and symmetric in v, v'. One only has to verify that the square of the corresponding norm $|||\cdot|||$ satisfies

$$|||v|||^2 = \langle v, v\rangle = (\tilde{L}v, \tilde{L}v) > 0 \tag{3.18b}$$

for $v \neq 0$. This is obvious, since $-\tilde{L}$ is again a symmetric hyperbolic operator; replacing t by $T - t$ the class $\bar{C}^1$ goes over into $\underline{C}^1$. It follows in analogy to (3.14) that there exists a constant $\tilde{\Gamma}$ such that

$$(v, v) \leqslant \tilde{\Gamma}^2(\tilde{L}v, \tilde{L}v) = \tilde{\Gamma}^2\langle v, v\rangle, \tag{3.19}$$

which implies (3.18b). We complete $\bar{C}^1$ into a Hilbert space H, by taking

Cauchy sequences of functions in $\bar{C}^1$ with respect to the norm $|||v|||$. Because of the inequality (3.19) we have for v in $\bar{C}^1$ and $w \in C^0(R_T)$ that

$$|(v,w)| \leqslant \|v\|\,\|w\| \leqslant \tilde{\Gamma}\|w\|\,|||v|||.$$

Since by (3.19) Cauchy sequences with respect to the norm $|||v|||$ also are Cauchy sequences with respect to the norm $\|v\|$, it follows that $(A^{-1}v,w)$ defines a bounded linear functional on H. By the representation theorem (p. 118) we can then find an element U in H such that

$$(A^{-1}v,w) = \langle v,U\rangle = (\tilde{L}v,\tilde{L}U).$$

Obviously then $u = A\tilde{L}U$ satisfies (3.16) for all $v \in H$. This u, which belongs to the set of square integrable functions in R_T, can be considered a weak solution of the initial-value problem (3.1a,b). It remains, of course, to show that u can be identified with a *strict* solution in the ordinary sense, at least for sufficiently regular w. Here we shall not go into a proof of this fact, given in a classical paper by K. O. Friedrichs,* but shall instead give below an existence proof based on a completely different approach (also due to Friedrichs), namely the method of finite differences.

In connection with proving regularity of solutions it is important to have "a priori" estimates for the values of a solution u and of its derivatives *at each point* (x,t) of R_T. Such estimates can be obtained by first deriving estimates for the integrals of the squares of a sufficient number of derivatives. Those in turn can easily be obtained by using the differential equations satisfied by the derivatives. We find from (3.1a) and any $i = 1,\dots,n$ that

$$D_i w = D_i Lu = LD_i u + (D_i A)\tau u + \sum_k (D_i A^k) D_k u + (D_i B)u. \tag{3.20a}$$

Substituting still for τu its expression from (3.1a) we see that $D_i u$ satisfies an equation of the form

$$LD_i u = \sum_k a_{ik} D_k u + b_i u + D_i w + e_i w \tag{3.20b}$$

with certain square matrices a_{ik}, b_i, e_i, which by assumption are bounded in R_T. There exists then an M such that

$$\|LD_i u\| \leqslant M\left(\sum_k \|D_k u\| + \|u\| + \|w\|\right) + \|D_i w\|. \tag{3.20c}$$

Assume that $u \in \underline{C}^2$. Then $D_i u \in \underline{C}^1$, and we find from (3.14) that

$$\|D_i u\| \leqslant \Gamma M\left(\sum_k \|D_k u\| + \|w\|\right) + M\Gamma^2\|w\| + \Gamma\|D_i w\|.$$

*Symmetric hyperbolic linear differential equations, *Comm. Pure Appl. Math.* 7, (1954), 345–392.

Summing over i yields

$$\sum_i \|D_i u\| \leqslant \Gamma M n \sum_k \|D_k u\| + nM\Gamma(1+\Gamma)\|w\| + \Gamma \sum_i \|D_i w\|. \qquad (3.20d)$$

Now by (3.12c) the constant Γ can be made arbitrarily small by choosing T sufficiently small. Hence for sufficiently small T we have $\Gamma Mn < 1/2$, and conclude from (3.20d) that

$$\sum_i \|D_i u\| \leqslant 2nM\Gamma(1+\Gamma)\|w\| + 2\Gamma \sum_i \|D_i w\|. \qquad (3.21)$$

A similar estimate can then be obtained for $\|\tau u\|$ directly from (3.1a). Forming next second space derivatives of (3.1a) and proceeding in this manner, we find that for sufficiently small T we can estimate all $D^\alpha u$ for $|\alpha| \leqslant s$ in terms of the $\|D^\beta w\|$ for $|\beta| \leqslant s$ (using, of course, the existence and uniform boundedness of the derivatives of the coefficients of L). Moreover by (3.11), (3.12b) we have for any $u \in \underline{C}^1$ and any λ between 0 and T that

$$\int_{t=\lambda} u^{\mathrm{T}} A u\, dx = E(\lambda) \leqslant (K+1)\|u\|^2 + \mu^{-2}\|Lu\|^2.$$

Then for $u \in \underline{C}^s$ and $|\alpha| \leqslant s$ by (3.9)

$$\int_{t=\lambda} (D^\alpha u)^{\mathrm{T}} (D^\alpha u)\, dx \leqslant \mu^{-1} \int_{t=\lambda} (D^\alpha u)^{\mathrm{T}} A (D^\alpha u)\, dx$$
$$\leqslant \mu^{-1}(K+1)\|D^\alpha u\|^2 + \mu^{-3}\|LD^\alpha u\|^2.$$

This inequality permits us to estimate for any component u_j of u and any α with $|\alpha| \leqslant s$

$$\int_{t=\lambda} |D^\alpha u_j|^2\, dx$$

in terms of* the $\|D^\beta w\|$ with $|\beta| \leqslant s$.

The transition from estimates for integrals of squares to pointwise estimates is furnished by one of the *Sobolev inequalities*. This inequality can easily be derived from *Parseval's identity*, known from the theory of Fourier integrals, that connects the square integrals of a function g with that of its Fourier transform $\hat{g}$:

$$\int |g(x)|^2\, dx = \int |\hat{g}(\xi)|^2\, d\xi. \qquad (3.22)$$

Let

$$s = \left[\frac{n}{2}\right] + 1, \qquad (3.23)$$

where generally $[\nu]$ for a real number ν denotes the largest integer $\leqslant \nu$.

*We use here that $LD^\alpha u$ differs from $D^\alpha w$ by a differential operator of order $|\alpha|$.

Thus s is the smallest integer exceeding $n/2$. By (2.16) and the Cauchy–Schwarz inequality

$$|g^2(x)| = (2\pi)^{-n}\left|\int e^{ix\cdot\xi}(1+|\xi|)^{-s}(1+|\xi|)^s\hat{g}(\xi)\,d\xi\right|^2$$
$$\leqslant c_n\int(1+|\xi|)^{2s}|\hat{g}(\xi)|^2\,d\xi,$$

where

$$c_n = (2\pi)^{-n}\int(1+|\xi|)^{-2s}\,d\xi < \infty.$$

By (2.20)

$$(1+|\xi|)^{2s}|\hat{g}(\xi)|^2 \leqslant 2^{2s}n^s\left(\sum_{|\alpha|\leqslant s}|(i\xi)^\alpha\hat{g}(\xi)|\right)^2$$
$$\leqslant d_n\sum_{|\alpha|\leqslant s}|(i\xi)^\alpha\hat{g}(\xi)|^2 = d_n\sum_{|\alpha|\leqslant s}|\widehat{D^\alpha g}|^2,$$

with

$$d_n = 2^{2s}n^s\sum_{|\alpha|\leqslant s}1.$$

Using (3.22) with g replaced by $D^\alpha g$ it follows that

$$|g^2(x)| \leqslant c_n d_n\sum_{|\alpha|\leqslant s}\int|D^\alpha g(y)|^2\,dy. \tag{3.24}$$

Applied to the function $g = u(x,t)$ for a fixed t between 0 and T, inequality (3.24) permits us to estimate $|u(x,t)|$ in terms of the $\|D^\beta w\|$ with $|\beta| \leqslant s$, where s is given by (3.23).

Problems

1. (a) Write the system (3.4a,b,c) in matrix notation and show that it is symmetric hyperbolic.
   (b) Do the same for Maxwell's equations (Chapter 1, (2.6a)).

2. Show that a symmetric hyperbolic system (3.1a) with constant $A, A^1, \ldots, A^n$ and $B = 0$ satisfies the Gårding hyperbolicity condition.

3. (a) Prove Sobolev's inequality (3.24) for $n = 1$ from the identity

$$g^2(x) = \int_x^{x+1}\left[g(y) - \int_x^y g'(z)\,dz\right]^2 dy \tag{3.25a}$$

in the form

$$g^2(x) \leqslant 2\int_0^2 (g^2(y) + g'^2(y))\,dy \quad \text{for } 0 < x < 2. \tag{3.25b}$$

[Hint: Assume first that $0 < x \leqslant 1$. Use $(a-b)^2 \leqslant 2a^2 + 2b^2$ and Cauchy's inequality. Same for $1 \leqslant x < 2$ by symmetry.]

(b) Let Γ denote the cube

$$\Gamma = \{x \mid x \in \mathbb{R}^n; 0 \leqslant x_k \leqslant 2 \quad \text{for } k = 1, \ldots, n\}. \tag{3.25c}$$

Prove the Sobolev-type inequality for a bounded region

$$g^2(x) \leqslant 2^n \sum_{|\alpha| \leqslant n} \int_\Gamma (D^\alpha g(y))^2 \, dy \quad \text{for } x \in \Gamma. \tag{3.25d}$$

[Hint: Induction over n.]

4. Sobolev's inequality for a bounded region.

Definition. A "conical sector" Γ in $\mathbb{R}^n$ is the intersection of a ball with a cone from its center:

$$\Gamma = \{y \mid y = x + t\xi; 0 \leqslant t \leqslant h; \xi \in \sigma\} \tag{3.26a}$$

where σ is a relatively open subset of the unit sphere in $\mathbb{R}^n$. We call x the *vertex* and h the *radius* of Γ. The *solid angle* ω of Γ is the "area" (($n - 1$)-dimensional measure) of σ. An open set $\Omega \subset \mathbb{R}^n$ has the *cone property* if there exist positive numbers h,ω such that each x in Ω is vertex of a conical sector $\Gamma \subset \Omega$ of radius h and solid angle ω.

Show that for any $\Omega \subset \mathbb{R}^n$ with the cone property there exists a C (depending only on Ω) such that for any $g \in C^s(\Omega)$, $s = [n/2] + 1$, and for any $x \in \Omega$ we have

$$|g(x)| \leqslant C\|g\|_s. \tag{3.26b}$$

Here we define

$$\|g\|_s = \sqrt{\int_\Omega \sum_{|\alpha| \leqslant s} (D^\alpha g(y))^2 \, dy}. \tag{3.26c}$$

[Hint: For $\phi \in C^s(\mathbb{R})$ with $\phi(t) = 0$ for $t > h$

$$\phi(0) = \frac{(-1)^s}{(s-1)!} \int_0^h t^{s-1} \phi^{(s)}(t) \, dt. \tag{3.26d}$$

Thus by Cauchy, since $s - 1 = (2s - n - 1)/2 + (n - 1)/2$ and here $(2s - n - 1)/2 > -1$,

$$\phi^2(0) = 0\left(\int_0^h t^{n-1}(\phi^{(s)}(t))^2\right) dt.$$

Let $\zeta(t) \in C^\infty(\mathbb{R})$ with $\zeta(0) = 1$, $\zeta(t) = 0$ for $t > h$. Take in (3.26d)

$$\phi(t) = \zeta(t) g(x + t\xi).$$

Estimate $\phi^{(s)}$ in terms of derivatives of g. Integrate (3.26d) with respect to ξ over the set σ.]

5. Estimates for solutions in bounded regions.

Definition. A hyper-surface $S\colon t = \phi(x)$ is called "space-like" with respect to the operator L of (3.1a) if the matrix

$$A(x,t) - \sum_{k=1}^{n} \phi_{x_k}(x)A^k(x,t) \tag{3.27a}$$

is positive definite for all (x,t) on S.

(a) Write the wave equation $\Box u = 0$ as a symmetric hyperbolic system. Show that the definition of "space-like" agrees with the one given by (1.54).

(b) For fixed positive a,T and for $0 < \lambda < T$ consider the "truncated cone"

$$R_\lambda = \{(x,t) \mid |x| \leqslant a(T - t)/T; 0 \leqslant t \leqslant \lambda\} \tag{3.27b}$$

bounded by the planes $t = 0$, $t = \lambda$ and the conical surface

$$S_\lambda = \left\{(x,t) \mid t = T - \frac{T}{a}|x|; 0 \leqslant t \leqslant \lambda\right\}. \tag{3.27c}$$

We call R_λ space-like if S_λ is space-like, that is

$$A(x,t) + \frac{T}{a|x|}\sum_k x_k A^k(x,t)$$

is positive definite for all (x,t) on S_λ. Let $A, A^1, \ldots, A^n$ be symmetric, A positive definite, and R_λ be space-like. Let $u \in C^1(R)$ be a solution of

$$Lu = w \quad \text{for } (x,t) \in R_\lambda \tag{3.27d}$$

$$u = f(x) \quad \text{for } t = 0, \quad |x| \leqslant a. \tag{3.27e}$$

Set

$$E(\mu) = \int_{\sigma_\mu} u^T A u \, dx \tag{3.27f}$$

where σ_μ is the cross section

$$\sigma_\mu = \{(x,t) \mid (x,t) \in R_\lambda; t = \mu\} \tag{3.27g}$$

of R_λ. Show that for $0 < \mu < \lambda$, [see (3.5b)]

$$E(\mu) \leqslant E(0) + \int_{R_\mu} (-u^T C u + 2u^T w) \, dx \, dt. \tag{3.27h}$$

(c) Show that there exists a constant k (depending only on upper bounds for the matrices $A^{-1}, A, A^1, \ldots, A^n, B$ and their first derivatives in R_λ) such that

$$\int_{\sigma_\mu} u^T u \, dx \leqslant k\left[\iint_{R_\mu} w^T w \, dx \, dt + \int_{|x|<a} f^T f \, dx\right] \tag{3.27i}$$

for $0 \leqslant \mu \leqslant \lambda$. (This implies that u in R_λ is determined uniquely by the values of w in R_λ and of f on $|x| \leqslant a$.) [Hint: Estimate the forms $\xi^T A \xi$, $\xi^T C \xi$, using

$$\inf_{|\xi|=1} \xi^T A \xi = \left(\sup_{|\xi|=1} \xi^T A^{-1} \xi\right)^{-1}$$

to obtain from (3.27h) an estimate

$$\phi'(\mu) \leqslant \gamma(\phi(\mu) + \psi(\mu) + \phi'(0)) \tag{3.27j}$$

with a certain γ, where

$$\phi(\mu) = \iint_{R_\mu} u^T u \, dx \, dt; \qquad \psi(\mu) = \iint_{R_\mu} w^T w \, dx \, dt.$$

Since $\phi,\phi',\psi,\psi' \geqslant 0$, (3.27j) implies ("Gronwall's lemma") $\phi'(\mu) \leqslant \gamma e^{\gamma\mu}(\psi(\mu) + \phi'(0))$.]

(d) For $f = f(x)$, $w = w(x,t)$ and an integer $m \geqslant 0$ define

$$\|f\|_m = \sqrt{\sum_{|\alpha|\leqslant m} \int_{|x|\leqslant a} (D^\alpha f^T)(D^\alpha f)\, dx} \tag{3.27k}$$

$$\|w(\mu)\|_m = \sqrt{\sum_{|\alpha|\leqslant m} \int_{\sigma_\mu} (D^\alpha w^T)(D^\alpha w)\, dx}. \tag{3.27l}$$

Show that there exists a constant k_m depending on upper bounds for the matrices $A^{-1},A,A^1,\ldots,A^n,B$ and their derivatives of orders $\leqslant m+1$ such that for $0 < \mu < \lambda$

$$\|u(\mu)\|_m^2 \leqslant k_m\left(\int_0^\mu \|w(\gamma)\|_m^2 \, d\gamma + \|f\|_m^2\right). \tag{3.27m}$$

[Hint: Show that for $|\alpha| \leqslant m$ we have $LD^\alpha u = D^\alpha w + L_\alpha u$, where L_α is an operator of order $\leqslant m$. Apply Gronwall's lemma with

$$\phi(\mu) = \int_0^\mu \|u(\gamma)\|_m^2 \, d\gamma; \; \psi(\gamma) = \int_0^\mu \|w(\gamma)\|_m^2 \, d\gamma.$$

(e) Let s be defined as in (3.23), and let $m > 0$. Let $A,A^1,\ldots,A^n$, $B \in C^{m+s+1}(R_\lambda)$, $w \in C^{m+s}(R_\lambda)$. Show that there exists a constant K_m (depending on upper bounds for $A^{-1},A,A^1,\ldots,A^n,B$ and their derivatives of orders $\leqslant m+s+1$ in R_α) such that for a solution $u \in C^m(R)$ of (3.27d,e) the inequalities

$$|D^\alpha u(x,t)| \leqslant K_m\left[\sup_{\substack{|\beta|\leqslant m+s \\ R_\lambda}} |D^\beta w(x,t)| + \sup_{\substack{|\beta|\leqslant m+s \\ |x|<a}} |D^\beta f(x)|\right] \tag{3.27n}$$

hold for any $(x,t) \in R_\lambda$, $|\alpha| \leqslant m$. [Hint: Use (3.2.b).]

(b) *Existence of solutions by the method of finite differences**

We follow here the ideas and notations used on a trivial example in Chapter 1, p. 6. We cover $(n+1)$-dimensional xt-space by a lattice. We take three positive quantities h, k, T, fixed for the moment, and consider the set Σ of points (x,t) for which

$$x = (x_1,\ldots,x_n) = (\alpha_1 h,\ldots,\alpha_n h), \qquad t = mk, \; 0 \leqslant t \leqslant T.$$

* ([18], [25])

Here $\alpha_1,\ldots,\alpha_n,m$ shall be integers. It is convenient to combine the α_j into a multi-index $\tilde{\alpha}=(\alpha_1,\ldots,\alpha_n)$, where "$\tilde{\ }$" shall indicate that the components α_j of $\tilde{\alpha}$ range over *all* integers, in contrast to the common multi-indices α whose components shall continue to be nonnegative. Then Σ consists of the points

$$x=\tilde{\alpha}h, \qquad t=mk \quad \text{with } 0\leqslant m\leqslant T/k. \tag{3.28}$$

We define operators E_0, E_j corresponding to shifts to neighboring points:

$$E_j u(x_1,\ldots,x_n,t)=u(x_1,\ldots,x_j+h,\ldots,x_n,t) \quad \text{for } j=1,\ldots,n, \tag{3.29a}$$

$$E_0 u(x_1,\ldots,x_n,t)=u(x_1,\ldots,x_n,t+k). \tag{3.29b}$$

Obviously E_j has the inverse E_j^{-1}, where

$$E_j^{-1}u(x_1,\ldots,x_n,t)=u(x_1,\ldots,x_j-h,\ldots,x_n,t). \tag{3.29c}$$

More generally combining $(E_1,\ldots,E_n)$ into a symbolic vector E, we can write

$$E^{\tilde{\alpha}}u(x,t)=u(x+\tilde{\alpha}h,t)=u(x_1+\alpha_1 h,\ldots,x_n+\alpha_n h,t). \tag{3.29d}$$

We next define *divided difference operators* δ_0, δ_j by

$$\delta_j=\frac{E_j-1}{h} \quad \text{for } j=1,\ldots,n$$

$$\delta_0=\frac{E_0-1}{k} \tag{3.30a}$$

so that, for example,

$$\delta_0 u(x,t)=\frac{u(x,t+k)-u(x,t)}{k}. \tag{3.30b}$$

The operators $E_j, E_k, \delta_j, \delta_k$ commute. For functions with continuous second derivatives these difference quotients approximate the corresponding derivatives, and we have by Taylor's formula

$$\delta_j u(x,t)=D_j u(x,t)+\mathrm{O}(h), \qquad \delta_0 u(x,t)=\tau u(x,t)+\mathrm{O}(k). \tag{3.31}$$

It would appear natural to replace the differential equation (3.1a) for the vector u by the difference equation

$$A\delta_0 v+\sum_{j=1}^{n} A^j\delta_j v+Bv=w. \tag{3.32}$$

However in order to ensure *stability*, we have to select a more complicated difference scheme. We replace the space derivatives $D_j u$ by *central* difference quotients $(2h)^{-1}(E_j - E_j^{-1})v$, and, following Friedrichs, use instead of (3.32) the system of difference equations

$$\Lambda v = \frac{1}{k} A\left(E_0 - \frac{1}{2n} \sum_{j=1}^{n} \left(E_j + E_j^{-1}\right)\right) v + \frac{1}{2h} \sum_{j=1}^{n} A^j \left(E_j - E_j^{-1}\right) v + Bv = w. \tag{3.33a}$$

Here it is understood that the argument of all functions is the same point (x,t), only shifted as indicated explicitly. The operator Λ is defined by (3.33a). Equation (3.33a) is to hold for those (x,t) for which (x,t) and $(x,t+k)$ belong to Σ, restricting t to the interval $0 \leqslant t \leqslant T-k$. Since the matrix $A(x,t)$ is nondegenerate, we can solve (3.33a) for $E_0 v = v(x,t+k)$ in terms of $w(x,t)$ and the $v(y,t)$ with

$$\sum_j |y_j - x_j| \leqslant h.$$

Thus formula (3.33a) constitutes a recursion formula that permits us to find v at the time $t+k$ if v and w are known at the time t. If we add the initial condition

$$v(x,0) = 0, \tag{3.33b}$$

then there exists trivally for given w a unique solution $v(x,t)$ of (3.33a,b) in Σ. Obviously the value of v at any point (x,t) is determined already by the values of $w(y,s)$ at a finite number of points (y,s), namely those for which

$$\sum_{j=1}^{n} |y_j - x_j| < \frac{h}{k}(t-s) \qquad 0 \leqslant s < t. \tag{3.34}$$

The domain of dependence of $v(x,t)$ on w here has the shape of a pyramid with vertex $(x,t-k)$. If, as we shall assume, $w(y,s)$ has compact support in y for each s, then $v(x,t)$ has compact support in x.

We are relieved of the burden of proving *existence* for v. Instead we shall have to show that v for $h,k \to 0$ can be used to approximate a function $u(x,t)$ defined in R_T, which is a solution of (3.1a,b). For that we shall need the "discrete" analogues of the energy inequality (3.14), of similar inequalities for higher derivatives, and of Sobolev's inequality. We shall make use of the same symmetry and regularity assumptions on the matrices $A, A^1, \ldots, A^n, B$ as before on page 140.

We write (3.33a) as

$$AE_0 v = \sum_{j=1}^{n} \left(a^j E_j + b^j E_j^{-1}\right) v - kBv + kw, \tag{3.35a}$$

where the matrices a^j, b^j are defined by

$$a^j = \frac{1}{2n} A - \frac{k}{2h} A^j, \qquad b^j = \frac{1}{2n} A + \frac{k}{2h} A^j. \tag{3.35b}$$

The matrices a^j, b^j are symmetric. Since A is positive definite, the same holds for a^j, b^j provided the ratio k/h is sufficiently small. More precisely, using the fact that the A^j are bounded and A is uniformly positive definite there exists a positive λ (held fixed in what follows) so small that the matrices a^j, b^j are positive definite for all (x,t) in R_T, when the space and time steps are connected by

$$k = \lambda h. \tag{3.36}$$

Now for a positive definite symmetric matrix a and any vectors v, w the inequality

$$2v^{\mathrm{T}}aw \leqslant 2\sqrt{v^{\mathrm{T}}av}\,\sqrt{w^{\mathrm{T}}aw} \leqslant v^{\mathrm{T}}av + w^{\mathrm{T}}aw \tag{3.37}$$

holds.* (Compare (3.10b).) Multiplying (3.35a) by $2E_0v^{\mathrm{T}}$ from the left, and using (3.37) for $a = a^j, b^j$ we get

$$\begin{aligned}2(E_0v)^{\mathrm{T}}A(E_0v) \leqslant{} & (E_0v)^{\mathrm{T}}\left(\sum_{j=1}^{n}(a^j+b^j)\right)(E_0v)\\ & + \sum_{j=1}^{n}\left((E_jv)^{\mathrm{T}}a^j(E_jv) + (E_j^{-1}v)^{\mathrm{T}}b^j(E_j^{-1}v)\right)\\ & - 2k(E_0v)^{\mathrm{T}}Bv + 2k(E_0v)^{\mathrm{T}}w.\end{aligned} \tag{3.38}$$

By (3.35b)

$$\sum_{j=1}^{n}(a^j+b^j) = A. \tag{3.39}$$

Moreover

$$\begin{aligned}(E_jv)^{\mathrm{T}}a^j(E_jv) &= E_j(v^{\mathrm{T}}a^jv) - (E_jv)^{\mathrm{T}}(E_ja^j - a^j)(E_jv)\\ &= E_j(v^{\mathrm{T}}a^jv) - h(E_jv)^{\mathrm{T}}(\delta_ja^j)(E_jv).\end{aligned}$$

Now by assumption the derivatives, and hence also the difference quotients, of the a^j are bounded uniformly for $(x,t) \in R_T$. Using (3.9), we can (after applying E_j) find a constant K such that

$$(E_jv)^{\mathrm{T}}(\delta_ja^j)(E_jv) \leqslant K(E_jv)^{\mathrm{T}}(E_jA)(E_jv) = KE_j(v^{\mathrm{T}}Av).$$

Thus†

$$(E_jv)^{\mathrm{T}}a^j(E_jv) = E_j(v^{\mathrm{T}}a^jv) + O\big(hE_j(v^{\mathrm{T}}Av)\big).$$

*For a = unit matrix, relation (3.37) is just the Cauchy–Schwartz inequality + the elementary inequality $2xy \leqslant x^2 + y^2$ valid for real numbers x, y. The case of more general a goes over into the previous one by writing $a = c^{\mathrm{T}}c$ with a suitable matrix c. This amounts to writing the positive definite quadratic form $v^{\mathrm{T}}av$ as a sum of squares.

†We use the customary notation $F = O(G)$ to indicate that there exists a constant K such that $|F| \leqslant KG$ for all quantities F, G in question.

Similarly

$$(E_j^{-1}v)^{\mathrm{T}}b^j(E_j^{-1}v)=E_j^{-1}(v^{\mathrm{T}}b^jv)+\mathrm{O}(hE_j^{-1}(v^{\mathrm{T}}Av))$$

$$(E_0v)^{\mathrm{T}}A(E_0v)=E_0(v^{\mathrm{T}}Av)+\mathrm{O}(kE_0(v^{\mathrm{T}}Av))$$

$$2k(E_0v)^{\mathrm{T}}(Bv-w)=\mathrm{O}(kE_0(v^{\mathrm{T}}Av)+k(v^{\mathrm{T}}Av)+k(w^{\mathrm{T}}Aw)).$$

We then find from (3.38) that

$$E_0(v^{\mathrm{T}}Av)=\sum_{j=1}^{n}\left(E_j(v^{\mathrm{T}}a^jv)+E_j^{-1}(v^{\mathrm{T}}b^jv)\right)$$
$$+O\left[h\sum_{j=1}^{n}(E_j+E_j^{-1})(v^{\mathrm{T}}Av)\right.$$
$$\left.+k(v^{\mathrm{T}}Av)+kE_0(v^{\mathrm{T}}Av)+k(w^{\mathrm{T}}Aw)\right]. \tag{3.40}$$

We sum this inequality for a fixed $t=mk$ over all x of the form $\tilde{\alpha}h$. In analogy to the $E(\lambda)$ of (3.8) we introduce the *energy sum*

$$\eta(t)=h^n\sum_{\tilde{\alpha}}v^{\mathrm{T}}(\tilde{\alpha}h,t)A(\tilde{\alpha}h,t)v(\tilde{\alpha}h,t)=h^n\sum_{\tilde{\alpha}}E^{\tilde{\alpha}}(v^{\mathrm{T}}Av), \tag{3.41}$$

where $\tilde{\alpha}$ ranges over all vectors with integers as components. Define similarly

$$\zeta(t)=h^n\sum_{\tilde{\alpha}}E^{\tilde{\alpha}}(w^{\mathrm{T}}Aw). \tag{3.42}$$

Since, by assumption, $w(x,t)$ is of compact support in x, our sums only contain a finite number of nonvanishing terms. Obviously we arrive at the same sum in (3.41), if $v^{\mathrm{T}}Av$ is replaced by $E_j(v^{\mathrm{T}}Av)$ or $E_j^{-1}(v^{\mathrm{T}}Av)$, since such a shift does not affect* the set of points over which we sum $v^{\mathrm{T}}Av$. Thus we obtain from (3.41), (3.40) that

$$\eta(t+k)=h^n\sum_{\tilde{\alpha}}\sum_{j=1}^{n}(v^{\mathrm{T}}a^jv+v^{\mathrm{T}}b^jv)$$
$$+O(h\eta(t)+k\eta(t)+k\eta(t+k)+k\zeta(t))$$
$$\leqslant\eta(t)+K((h+k)\eta(t)+k\eta(t+k)+k\zeta(t)) \tag{3.43}$$

with a certain constant K. If $k=\lambda h$ is so small that $Kk<\frac{1}{2}$, we can solve (3.43) for $\eta(t+k)$, and obtain an inequality of the form

$$\eta(t+k)\leqslant e^{Ck}\eta(t)+k\gamma\zeta(t)$$

*This observation plays the same role as did the integration by parts in the derivation of (3.8).

with certain constants C,γ. Since $\eta(0)=0$, it follows for $t=mk\leqslant T$ that

$$\eta(t)\leqslant k\gamma\big(\zeta(t-k)+e^{Ck}\zeta(t-2k)+\cdots+e^{(m-1)kC}\zeta(0)\big)$$

$$\leqslant k\gamma e^{CT}\sum_{\nu=0}^{m}\zeta(\nu k). \tag{3.44}$$

In analogy to (3.12b) we define the norm $\|w\|$ of the vector $w(x,t)$ for the lattice Σ by

$$\|w\|^2=h^nk\sum_{(x,t)\in\Sigma}(w(x,t))^{\mathrm{T}}A(x,t)w(x,t)=k\sum_{0\leqslant m\leqslant T/k}\zeta(mk). \tag{3.45}$$

Summing (3.44) over all m with $0\leqslant m\leqslant T/k$ we obtain the energy estimate

$$\|v\|^2\leqslant\gamma Te^{CT}\|w\|^2=\gamma Te^{CT}\|\Lambda v\|^2 \tag{3.46}$$

in analogy to (3.14), (3.12c).

The next step consists in deriving similar energy estimates for the difference quotients of v. (Compare (3.20a, b, c, d).) These are obtained by applying the operator δ_r for $r=1,\ldots,n$ to (3.33a). We make use of the rule for *differencing a product* of two functions U, V

$$\delta_r(UV)=U(\delta_rV)+(\delta_rU)(E_rV) \tag{3.47a}$$

and of the identities

$$\frac{1}{k}\left(E_0-\frac{1}{2n}\sum_{j=1}^{n}\left(E_j+E_j^{-1}\right)\right)v=\left(\delta_0-\frac{1}{2n}\lambda^{-1}\sum_{j=1}^{n}\delta_j\left(1-E_j^{-1}\right)\right)v, \tag{3.47b}$$

$$\frac{1}{2h}\left(E_j-E_j^{-1}\right)=\frac{1}{2}\left(1+E_j^{-1}\right)\delta_j. \tag{3.47c}$$

We then find for δ_rv an equation of the form $\Lambda\delta_rv=w^r$ where w^r is a linear combination of $\delta_rw, E_rw, \delta_rv, E_rv, E_r\delta_sv, E_rE_j^{-1}\delta_sv$ with $j,s=1,\ldots,n$. [We express $E_r\delta_0v$ in terms of these quantities by (3.33a).] Applying the estimate (3.46) to δ_rv instead of v, and summing over r, we arrive at an inequality of the type

$$\sum_{r=1}^{n}\|\delta_rv\|^2=O\left(\|w\|^2+\sum_{r=1}^{n}\|\delta_rw\|^2\right), \tag{3.48}$$

provided T is sufficiently small. Repeating this procedure we arrive at estimates for the norms of the higher difference quotients:

$$\sum_{|\alpha|\leqslant s}\|\delta^\alpha v\|^2=O\left(\sum_{|\alpha|\leqslant s}\|\delta^\alpha w\|^2\right), \tag{3.49}$$

where we have combined the operators $\delta_1,\ldots,\delta_n$ into a vector δ, and write

$$\delta^\alpha=(\delta_1)^{\alpha_1}\ldots(\delta_n)^{\alpha_n}. \tag{3.50}$$

Using (3.49), (3.9), we find then for any $t = mk$ between 0 and T and any α with $|\alpha| \leqslant s$ that also

$$h^n \sum_{\tilde{\beta}} \left(\delta^\alpha v(\tilde{\beta}h,t)\right)^{\mathrm{T}}\left(\delta^\alpha v(\tilde{\beta}h,t)\right) \leqslant \mu^{-1} h^n \sum_{\tilde{\beta}} (\delta^\alpha v)^{\mathrm{T}} A (\delta^\alpha v)$$

$$= O\left(\|\Lambda \delta^\alpha v\|^2\right) = O\left(\sum_{|\gamma| \leqslant s} \|\delta^\gamma w\|^2\right). \tag{3.51}$$

From the l_2-estimates (3.51) we pass to pointwise estimates by developing a difference analogue to Sobolev's inequality (3.24). This we can derive without recourse to Fourier transforms. (Compare problem 3, p. 146.) We start with the case $n = 1$ of a function $g(x)$ of a scalar argument x. For a nonnegative integer r we have the identity

$$g(x) = g(x+rh) - \sum_{\nu=0}^{r-1} \left(g(x+(\nu+1)h) - g(x+\nu h)\right)$$

$$= g(x+rh) - h \sum_{\nu=0}^{r-1} \delta g(x+\nu h),$$

where, of course, $\delta g(x)$ stands for $(g(x+h) - g(x))/h$. Squaring we get the estimate

$$g^2(x) \leqslant 2g^2(x+rh) + 2rh^2 \sum_{\nu=0}^{r-1} (\delta g(x+\nu h))^2$$

by Cauchy–Schwartz. Summing over $r = 0, 1, \ldots, p-1$ yields

$$pg^2(x) \leqslant 2 \sum_{r=-\infty}^{\infty} g^2(x+rh) + p^2h^2 \sum_{r=-\infty}^{\infty} (\delta g(x+rh))^2.$$

We choose here for p the integer determined by

$$\frac{1}{h} \leqslant p < \frac{1}{h} + 1.$$

For h sufficiently small, say $h < \sqrt{2} - 1$, we have $p^2h^2 < 2$, and hence

$$g^2(x) \leqslant 2h \sum_{r=-\infty}^{\infty} (g(x+rh))^2 + (\delta g(x+rh))^2. \tag{3.52}$$

For x that are multiples of h we can replace x by 0 on the right. Next for a function $g(x_1, x_2)$ we find by repeated application of (3.52) that

$$g^2(x_1,x_2) \leqslant 2h \sum_{r_1=-\infty}^{\infty} \left[(g(r_1h,x_2))^2 + (\delta_1 g(r_1h,x_2))^2\right]$$

$$\leqslant 4h^2 \sum_{r_1,r_2=-\infty}^{\infty} \left[(g(r_1h,r_2h))^2 + (\delta_1 g)^2 + (\delta_2 g)^2 + (\delta_1\delta_2 g)^2\right].$$

Generally for any n and for x of the form $\tilde{\gamma}h$

$$g^2(x) \leqslant 2^n h^n \sum_{|\alpha| \leqslant n} \sum_{\tilde{\beta}} \left(\delta^\alpha g(\tilde{\beta}h)\right)^2. \tag{3.53}$$

It follows from (3.51) that for a solution of (3.33a, b) in Σ and any α

$$|\delta^\alpha v(x,t)|^2 = O\left(\sum_{|\beta| \leqslant |\alpha|+n} \|\delta^\beta w\|^2\right), \tag{3.54}$$

provided T (for given α) is sufficiently small.

We shall need estimates of the type (3.54) which do not depend on the particular values h which we shall let tend to 0. For that purpose the expression $\|\delta^\beta w\|$ defined by (3.45) as a sum over Σ involving difference quotients of w will have to be replaced by an integral over R_T involving derivatives of w. Let w have components $w_1, \ldots, w_n$. Since A is bounded we have

$$\begin{aligned} \|\delta^\beta w\|^2 &= h^n k \sum_{(x,t)\in\Sigma} \left(\delta^\beta w(x,t)\right)^{\mathrm{T}} A(x,t)\left(\delta^\beta w(x,t)\right) \\ &= O\left(h^n k \sum_{r=1}^{n} \sum_{(x,t)\in\Sigma} \left(\delta^\beta w_r(x,t)\right)^2\right). \end{aligned} \tag{3.55}$$

For the scalar w_r we have by the mean value theorem

$$\min_{|y-x| \leqslant h} D_j w_r(y,t) \leqslant \delta_j w_r(x,t) \leqslant \max_{|y-x| \leqslant h} D_j w_r(y,t).$$

It follows by induction for $|\beta| \leqslant s$ that

$$\min_{|y-x| \leqslant sh} D^\beta w_r(y,t) \leqslant \delta^\beta w_r(x,t) \leqslant \max_{|y-x| \leqslant sh} D^\beta w_r(y,t),$$

that is, that

$$\delta^\beta w_r(x,t) = D^\beta w_r(y,t)$$

for some y with $|y-x| \leqslant sh$. Thus

$$h^n k \sum_{(x,t)\in\Sigma} \left(\delta^\beta w_r(x,t)\right)^2$$

is essentially a Riemann sum for the integral

$$\int_{R_T} \left(D^\beta w_r(x,t)\right)^2 dx\,dt.$$

More precisely, for w_r of class C^s and of compact support in R_T the sum will converge to the integral for $h \to 0$. Consequently by (3.54)

$$\max_{(x,t)\in\Sigma} |\delta^\alpha v(x,t)|^2 = 0\left(\sum_{|\beta| \leqslant |\alpha|+n} \int_{R_T} |D^\beta w(x,t)|^2 dx\,dt\right) \tag{3.56}$$

for all sufficiently small h. We get similar estimates for the mixed space-time difference quotients $\delta_0^i\delta^\alpha$ by solving (3.33a) for $\delta_0 v$, using (3.47b, c). In what follows we shall assume that $w(x,t)$ is of class $C^{n+2}(R_T)$ and of compact support in x. Then v and all its difference quotients of orders $\leqslant 2$ are bounded on Σ uniformly independently of h. This implies that v and its difference quotients of orders $\leqslant 1$ are uniformly Lipschitz on Σ, with a Lipschitz constant that does not depend on h.

The rest is simple. We now *refine* our lattice Σ indefinitely, choosing $h=2^{-q}$, $k=\lambda 2^{-q}$ with $q=1,2,3,4,\ldots$, while λ is held fixed. We denote by Σ_q the lattice determined by these h,k and by $v^q(x,t)$ the solution of (3.33a, b) defined on Σ_q. The Σ_q form a monotone increasing sequence of denumerable sets in $\mathbb{R}^{n+1}$. Their union σ is again denumerable. The function v^q is defined on all sets $\Sigma_{q'}$ for $q'\leqslant q$. The same holds for the difference quotients $\delta_0^i\delta^\alpha v^q$ (where the operator $\delta_0^i\delta^\alpha$ is formed with respect to the lattice Σ_q). These difference quotients are bounded uniformly on $\Sigma_{q'}$ for $q'\leqslant q$ when $i+|\alpha|\leqslant 2$, and are uniformly Lipschitz, when $i+|\alpha|\leqslant 1$. From the boundedness it follows that there exists a subsequence S of natural numbers q such that

$$\lim_{\substack{q\in S\\ q\to\infty}} \delta_0^i\delta^\alpha v^q(x,t)=u^{i,\alpha}(x,t) \tag{3.57}$$

exists for $i+|\alpha|\leqslant 1$ and all $(x,t)\in\sigma$. Moreover the $u^{i,\alpha}$ are again uniformly bounded and Lipschitz on σ. Since the set σ is dense in R_T, we can immediately extend the $u^{i,\alpha}$ to all of the R_T as Lipschitz continuous functions with the same Lipschitz constant as had been found for the $\delta_0^i\delta^\alpha v^q$.

If we can prove that for $(x,t)\in R_T$ and $i+|\alpha|=1$

$$u^{i,\alpha}(x,t)=\tau^i D^\alpha u^{0,0}(x,t) \tag{3.58}$$

then the difference equation $\Lambda v^q=w$ will in the limit for $q\to\infty$ in S go over into the differential equation $Lu=w$ for $u=u^{0,0}$, and we have solved our initial-value problem (3.1a, b). Indeed rewriting (3.33a) with the help of (3.47b,c) and observing that

$$\delta_j\left(1-E_j^{-1}\right)v^q=E_j^{-1}2^{-q}\delta_j^2 v^q$$

$$\tfrac{1}{2}\left(1+E_j^{-1}\right)\delta_j v^q=\delta_j v^q-\tfrac{1}{2}E_j^{-1}2^{-q}\delta_j^2 v^q$$

and using the uniform boundedness of $\delta_j^2 v^q$ immediately yields the transition from (3.33a) to (3.1a).

We prove (3.58) for $i=1$, $\alpha=0$. The argument is the same for $i=0$, $|\alpha|=1$. Consider at first two fixed points (x,t) and $(x,t+c)$ of σ. There exists then a q' such that (x,t) and $(x,t+c)$ belong to Σ_q for all $q\geqslant q'$. Prescribe an $\varepsilon>0$. We can find a $q''>q'$ such that

$$|u(x,t)-v^q(x,t)|<\varepsilon, \qquad |u(x,t+c)-v^q(x,t+c)|<\varepsilon$$

for all $q>q''$ belonging to S. (Here u stands for $u^{0,0}$.) Thus

$$\left|\frac{u(x,t+c)-u(x,t)}{c}-\frac{v^q(x,t+c)-v^q(x,t)}{c}\right|<\frac{2\varepsilon}{c}$$

for $q>q''$, $q\in S$. Here c is a multiple of $k=\lambda 2^{-q}$, say $c=mk$. Then

$$\left|\frac{v^q(x,t+c)-v^q(x,t)}{c}-\delta_0 v^q(x,t)\right|=\left|\frac{1}{m}\sum_{\nu=0}^{m-1}\delta_0 v^q(x,t+\nu k)-\delta_0 v^q(x,t)\right|$$

$$=\left|\frac{k}{m}\sum_{\nu=0}^{m-1}\sum_{\mu=0}^{\nu-1}\delta_0^2 v^q(x,t+\mu k)\right|$$

$$\leqslant Mmk=Mc,$$

where M is an upper bound for the second difference quotients of v^q. Hence

$$\left|\frac{u(x,t+c)-u(x,t)}{c}-\delta_0 v^q(x,t)\right|\leqslant\frac{2\varepsilon}{c}+Mc.$$

Letting first q tend to ∞ in S and then ε tend to 0 we find that

$$\left|\frac{u(x,t+c)-u(x,t)}{c}-u^{1,0}(x,t)\right|\leqslant Mc$$

whenever (x,t) and $(x,t+c)$ belong to σ. By continuity of u and $u^{1,0}$ this inequality holds then for any $(x,t),(x,t+c)$ in R_T. Letting here c tend to 0 yields the desired relation

$$\tau u(x,t)=u^{1,0}(x,t).$$

This completes the existence proof for a solution $u(x,t)$ of (3.1a, b) for $0\leqslant t\leqslant T$, where T is sufficiently small, under the assumption that $w\in C^{n+2}(R_T)$ and that w has compact support with respect to x.*

Problems

1. Show that for the solution u of (3.1a, b) constructed here the domain of dependence of $u(x,t)$ on $w(x,t)$ is contained in the pyramid

$$\sum_{j=1}^{n}|y_j-x_j|\leqslant\frac{1}{\lambda}(t-s)\qquad 0\leqslant s\leqslant t. \tag{3.59}$$

2. Show that (3.1a, b) has a solution $u(x,t)$ of class $C^s(R_T)$ for sufficiently small T, if w is of class $C^{s+n+1}(R_T)$ and of compact support in x.

3. Let u be a solution of (3.1a, b) of class $C^{n+2}(R_T)$ and of compact support in x. Let $v(x,t)$ be the solution of (3.33a,b) defined on the lattice Σ. Show that $u(x,t)-v(x,t)=O(h)$ for (x,t) on Σ, provided T and k/h are sufficiently small. (This implies

*The fact that the domain of dependence of $u(x,t)$ on w is finite (see problem 5, p. 171 and problem 1 below) shows that the assumption of compact support for w is unessential.

that for w sufficiently regular we have $v^q \to u$ for $q \to \infty$ in any manner, not just for q in a "suitable" subsequence S. Thus, in principle, we have a "construction" for u, a way to find u numerically.) [Hint: Show that

$$\delta_j u - D_j u = h\int_0^1 \theta\, d\theta \int_0^1 dr\, D_j^2 u(x+\theta rh) \quad \text{for } j=1,\ldots,n \tag{3.60}$$

with a similar formula for $\delta_0 u - D_0 u$. This implies that $\Lambda u - \Lambda v = hW$, where W is of class $C^n(R_T)$ and has compact support in x. Apply (3.56).]

(c) *Existence of solutions by the method of approximation by analytic functions (Method of Schauder)*

Existence of a local solution for a linear analytic non-characteristic Cauchy problem is guaranteed by the Cauchy–Kowalevski theorem. One is tempted to solve a non-analytic Cauchy problem by approximating data and coefficients by analytic functions. Generally though, the solutions of the approximating problem "run away," they do not converge to a solution of the desired problem. Convergence however can be enforced, if suitable a priori estimates for the solution are available, as is the case for symmetric hyperbolic systems. The resulting existence proof requires less manipulation than the preceding proof by finite differences.

We are interested in solutions u of the system

$$Lu = A(x,t)\tau u + \sum_{k=1}^{n} A^k(x,t)D_k u + B(x,t)u = 0 \tag{3.61a}$$

with prescribed initial values P on a hyper-plane t = const. = μ in xt-space:

$$u(x,t) = P(x) \quad \text{for } t = \mu. \tag{3.61b}$$

We first assume that the matrices $A, A^1, \ldots, A^n, B$ (fixed for the moment) are *real analytic* in a closed cylinder

$$\Sigma = \{(x,t) \mid |x| \leqslant b; \quad 0 \leqslant t \leqslant T\} \tag{3.61c}$$

and that $A, A^1, \ldots, A^n$ are symmetric and A positive definite in Σ. We first need existence for this Cauchy problem in a domain about the initial manifold whose size does not depend on the initial data, as long as the data are polynomials:

Lemma. *There exists a positive δ (depending on A, A^k, B, b, T, but not on μ, P) such that* (3.62a,b) *has a real analytic solution defined for*

$$|x| \leqslant b; \mu - \delta \leqslant t \leqslant \mu + \delta \tag{3.61d}$$

for any μ with $0 \leqslant \mu \leqslant T$ and any polynomial vector $P(x)$.

PROOF. Using that A is positive definite, and hence non-degenerate, and introducing the new unknown $v = u - P$ we can bring (3.61a,b) into the "standard form" for Cauchy–Kowalevski

$$\tau v = -\sum_k A^{-1}A^k D_k v - A^{-1}Bv - A^{-1}LP \tag{3.61e}$$

$$v = 0 \quad \text{for } t = \mu. \tag{3.61f}$$

Then v exists in a domain (3.61d), where δ depends only on the class $C_{M,r}$ to which the coefficients in (3.61e) belong (see p. 77), that is on the "size" r of the complex neighborhood Ω of Σ to which $A^{-1}A^1, \ldots, A^{-1}A^n, A^{-1}, B$, $A^{-1}LP$ can be extended, and on an upper bound M of these quantities in Ω; (see p. 72). Now r and M suitable for $A^{-1}A^1, \ldots, A^{-1}A^n, A^{-1}B$ will apply to $A^{-1}LP$ as well, since P is a polynomial, provided we replace P by P/c with a sufficiently large constant c. But if U is the solution of (3.61a,b) with P replaced by P/c, then $u = cU$ is the solution of (3.61a,b) because of the linearity of L. Thus δ does not depend on the particular polynomial P. □

Take now for a certain a with $0 < a < b$ the truncated cone

$$R_\delta = \{(x,t) \mid |x| \leqslant a(T - t)/T; \qquad 0 \leqslant t \leqslant \delta\} \tag{3.62a}$$

contained in Σ. Assume that R_δ is space-like in the sense that its conical boundary S_δ is space-like (see (3.27c)). Then the estimates (3.27n) apply to the solution u of the Cauchy problem

$$Lu = 0 \tag{3.62b}$$

$$u(x,0) = P(x) \tag{3.62c}$$

in its known domain of existence:

$$|D^\alpha u(x,t)| \leqslant K_m \sup_{\substack{|\beta| \leqslant m+s \\ |x| \leqslant a}} |D^\beta P(x)| \tag{3.62d}$$

for $(x,t) \in R_\delta$, $|\alpha| \leqslant m$ and s given by (3.23).

Take now any $f(x) \in C^\infty$ for $|x| \leqslant a$. For every m we can find a sequence of polynomials $P^j(x)$ such that

$$\lim_{j \to \infty} D^\alpha P^j(x) = D^\alpha f(x)$$

uniformly for $|x| \leqslant a$ and $|\alpha| \leqslant m + s$; (see p. 213). If u^j denotes the solution of (3.62b,c) with P replaced by P^j, we find from (3.62d) that the u^j for $j \to \infty$ converge to a solution $u \in C^m(R_\delta)$ of

$$Lu = 0 \quad \text{for } (x,t) \in R_\delta \tag{3.62e}$$

$$u(x,0) = f(x). \tag{3.62f}$$

Since m is arbitrary and u is determined uniquely by its initial values (hence does not depend on the particular sequence P_j), it follows that $u \in C^\infty(R_\delta)$. In particular $g(x) = u(x,\delta) \in C^\infty$ for $|x| \leqslant a(T - \delta)/T$. We can apply the same argument to the Cauchy problem

$$Lu = 0 \quad \text{for } (x,t) \in R_{2\delta},\ t \geqslant \delta \tag{3.62g}$$

$$u(x,\delta) = g(x) \quad \text{for } |x| \leqslant a(T - \delta)/T. \tag{3.62h}$$

The solution of this problem yields an extension to the solution u of (3.62e,f) to $R_{2\delta}$. Proceeding in this manner we see that for $f \in C^\infty$ problem (3.62e,f) has a *global* solution in the cone

$$|x| \leqslant a(T - t)/T, \qquad 0 \leqslant t < T \tag{3.62i}$$

for L with real analytic coefficients in Σ.

Finally, let $A, A^1, \ldots, A^n, B$ just be of class C^{s+3} in the cylinder Σ, and let the conical surface

$$S = \{(x,t) \,|\, |x| = a(T - t)/T, 0 \leqslant t < T\} \tag{3.62j}$$

be space-like with respect to L. We can approximate the coefficient matrices by matrices with polynomial elements uniformly with their derivatives of orders $\leqslant s + 3$ in Σ. Let $L_1, L_2, \ldots$ denote the sequence of corresponding approximating operators. Let $0 < \lambda < T$. Let $f \in C^\infty$ for $|x| \leqslant a$, and u^j be the solution in R_λ of $L_j u^j = 0$ with initial values f. For sufficiently large j the truncated cone R_λ will be space-like with respect to L_j, and by (3.27n)

$$|D^\alpha u^j(x,t)| \leqslant K_2 \sup_{\substack{|\beta| \leqslant 2+s \\ |x| \leqslant a}} |D^\beta f(x)| \tag{3.62k}$$

for $(x,t) \in R_\lambda$, $|\alpha| \leqslant 2$. Thus the u^j and their first and second derivatives are bounded uniformly in R_λ. We can select a sub-sequence for which the u^j and their first derivatives converge uniformly in R_λ. The limit u then represents a solution $u \in C^1(R_\lambda)$ of $Lu = 0$ with initial values f. Since λ is arbitrary in the interval $(0, T)$, u exists in the whole cone (3.62i). If f is only known to be of class C^{2+s}, we can approximate f by functions in C^∞ and, using (3.62k), obtain the same result. We have proved:

Theorem. *Let the matrices $A, A^1, \ldots, A^n, B$ belong to C^{s+3} in the cylinder* (3.61c). *Let $A, A^1, \ldots, A^n$ be symmetric and A positive definite in the cone* (3.62i). *Let the conical surface* (3.62j) *be space-like. Let $f \in C^{s+2}$ for $|x| \leqslant a$. Then there exists a solution $u \in C^1$ of $Lu = 0$ in the cone with initial values f.*

6 Higher-Order Elliptic Equations with Constant Coefficients*

Here we can only indicate how some of the notions developed for the Laplace equation apply to more general elliptic equations. We shall restrict ourselves to a single linear homogeneous mth-order equation with constant real coefficients for a scalar function $u(x)=u(x_1,\dots,x_n)$. We write the equation in the familiar form

$$P(D)u=0, \tag{0.1}$$

where

$$P(D)=\sum_{|\alpha|=m} A_\alpha D^\alpha \tag{0.2}$$

is an mth-degree form in $D=(D_1,\dots,D_n)$ with constant real coefficients A_α. Equation (0.1) is *elliptic* (see p. 58), if there are no real characteristic surfaces, or, more precisely, if

$$P(\xi)\neq 0 \quad \text{for all real } \xi\neq 0. \tag{0.3}$$

We are only interested in the case when the dimension n is at least 2 and when the form P does not vanish identically. In that case the order m of the elliptic equation must be even. For taking any real vector ζ with $P(\zeta)\neq 0$ we can find a vector η in $\mathbb{R}^n$ which is independent of ζ. Then $P(\eta+t\zeta)$ is an mth-degree polynomial in t with leading coefficient $P(\zeta)\neq 0$. Such a polynomial for odd m has at least one real root t, corresponding to a real $\xi=\eta+t\zeta\neq 0$ with $P(\xi)=0$. Accordingly we shall set $m=2\mu$ with a positive integer μ.

*([1], [2], [12], [15], [23], [24], [30])

The expression $P(\xi)/|\xi|^m$ is homogeneous of degree 0; it does not vanish on the sphere $|\xi| = 1$ and hence does not change sign. We can assume that P has a positive minimum on $|\xi| = 1$. Thus ellipticity implies the existence of a $c > 0$ such that

$$P(\xi) \geqslant c|\xi|^m = c|\xi|^{2\mu} \quad \text{for all } \xi \in \mathbb{R}^n. \tag{0.4}$$

For convenience we shall write the partial differential equation (0.1) as

$$Lu = (-1)^\mu P(D)u = (-1)^\mu \sum_{|\alpha|=2\mu} A_\alpha D^\alpha u = 0. \tag{0.5}$$

1. The Fundamental Solution for Odd n

A fundamental solution for L with pole y is a function $K(x,y)$ satisfying

$$LK = \delta_y \tag{1.1}$$

in the distribution sense (see p. 91). For odd dimensions n it is easy to obtain such an K by the method of decomposition into plane waves described in Chapter 5, Section 5(d). We start with the observation that by (2.81) of Chapter 5 the function

$$k(x,y) = d_n r = d_n|x-y| \tag{1.2}$$

is a fundamental solution for the operator $\Delta^{(n+1)/2}$:

$$\int (\Delta_x^{(n+1)/2} v(x)) k(x,y)\, dx = v(y)$$

for every "test function" $v \in C_0^\infty(\mathbb{R}^n)$. If then $G(x,y)$ is a solution of

$$LG(x,y) = r = |x-y| \tag{1.3}$$

which has integrable derivatives of orders $\leqslant n+m$, we have in

$$K(x,y) = d_n \Delta_x^{(n+1)/2} G(x,y) \tag{1.4}$$

a solution of (1.1). For if v is a test function, and $\tilde{L}$ (here $= L$) is the operator adjoint to L, we have by Green's identity

$$\begin{aligned}
\int (\tilde{L}v) K\, dx &= d_n \int (\tilde{L}v) \Delta_x^{(n+1)/2} G\, dx \\
&= d_n \int (\Delta_x^{(n+1)/2} \tilde{L}v) G\, dx \\
&= d_n \int (\tilde{L}\Delta_x^{(n+1)/2} v) G\, dx = d_n \int (\Delta_x^{(n+1)/2} v)(LG)\, dx \\
&= \int (\Delta_x^{(n+1)/2} v) d_n r\, dx = v(y).
\end{aligned} \tag{1.4a}$$

Identity (2.79c) of Chapter 5 suggests the construction first of a solution $\Gamma(x,y,\xi)$ of

$$L\Gamma = |(x-y)\cdot\xi|. \tag{1.5}$$

For then

$$G(x,y)=c_n^{-1}\int_{|\xi|=1}\Gamma(x,y,\xi)\,dS_\xi \tag{1.6}$$

will be a solution of (1.3). Trivially

$$\Gamma=\frac{|(x-y)\cdot\xi|^{2\mu+1}}{(2\mu+1)!\,P(\xi)} \tag{1.7}$$

satisfies (1.5), leading to

$$G(x,y)=c_n^{-1}\int_{|\xi|=1}\frac{|(x-y)\cdot\xi|^{2\mu+1}}{(2\mu+1)!\,P(\xi)}\,dS_\xi. \tag{1.8}$$

Setting $x-y=r\eta$ with $|\eta|=1$, we have

$$G(x,y)=c_n^{-1}r^{2\mu+1}\int_{|\xi|=1}\frac{|\eta\cdot\xi|^{2\mu+1}}{(2\mu+1)!\,P(\xi)}\,dS_\xi. \tag{1.9}$$

Clearly the integral in (1.9) is of class $C^{2\mu+1}$ in the vector η. Actually it is C^∞, even real analytic, in η for $|\eta|=1$. This can be seen by applying a suitable orthogonal transformation to the variable of integration ξ. Choose a fixed unit vector ζ. Let η be any unit vector linearly independent of ζ. Then η and ζ span a 2-dimensional subspace π of $\mathbb{R}^n$. There exists a unique orthogonal transformation T (depending on η and ζ) of $\mathbb{R}^n$ which leaves all vectors orthogonal to π fixed, rotates π in itself, and takes ζ into η. Clearly C is determined by algebraic conditions, and must be analytic in η for real unit vectors $\eta\neq\pm\xi$. (For an explicit expression for T see problem 1 below.) We replace ξ in (1.9) by $T\xi$, which changes neither the domain of integration nor the element of surface dS_ξ. Since

$$\eta\cdot T\xi=T\zeta\cdot T\xi=\zeta\cdot\xi,$$

it follows that

$$G(x,y)=c_n^{-1}r^{2\mu+1}\int_{|\xi|=1}\frac{|\zeta\cdot\xi|^{2\mu+1}}{(2\mu+1)!\,P(T\xi)}\,dS_\xi.$$

Since T is real analytic in η for unit vectors $\eta\neq\pm\zeta$, the same follows for the integral. Since ζ was an arbitrary unit vector, we see that G is analytic in η for all real unit vectors $\eta=(x-y)/|x-y|$. Thus the only singularities of $G(x,y)$ for real arguments occur for $r=0$, that is, $x=y$.

By (1.8) $G(x,y)$ is homogeneous of degree $2\mu+1$ in $x-y$. A derivative of G of order k will be homogeneous of degree $2\mu+1-k$, hence integrable for $k\leqslant 2\mu+n=m+n$. In particular

$$K(x,y)=d_nc_n^{-1}\Delta_x^{(n+1)/2}\int_{|\xi|=1}\frac{|(x-y)\cdot\xi|^{2\mu+1}}{(2\mu+1)!\,P(\xi)}\,dS_\xi \tag{1.10}$$

is homogeneous of degree $2\mu-n=m-n$, and hence is of the form

$$K(x-y)=r^{m-n}\phi(\eta) \quad \text{for } x-y=r\eta, \tag{1.11}$$

where $\phi(\eta)$ depends analytically on the vector η for real η with $|\eta|$ near 1. We can draw up to μ of the Laplace operators in (1.10) under the integral sign, and obtain

$$K(x,y)=d_n c_n^{-1}\int_{|\xi|=1}\frac{|(x-y)\cdot\xi|^{m-n}}{(m-n)!\,P(\xi)}\,dS_\xi \tag{1.12a}$$

for $n<m$, and

$$K(x,y)=d_n c_n^{-1}\Delta_x^{(n+1-m)/2}\int_{|\xi|=1}\frac{|(x-y)\cdot\xi|}{P(\xi)}\,dS_\xi \tag{1.12b}$$

for $n>m$.

When the number n of dimensions is even, the analogue of formula (2.79c) of Chapter 5 for decomposition of functions into plane waves becomes more complicated. The result that a fundamental solution becomes singular like r^{m-n} need not hold. Terms of the order $r^{m-n}\log r$ may occur. The simplest example is the fundamental solution $(2\pi)^{-1}\log r$ for the Laplace equation in two dimensions.

PROBLEMS

1. Prove that the orthogonal transformation T described on p. 158 is given by

$$T\xi=\xi-\frac{a+b}{1+c}\zeta+\frac{(1+2c)b-a}{1+c}\eta, \tag{1.13a}$$

where

$$a=\xi\cdot\eta, \qquad b=\xi\cdot\zeta, \qquad c=\eta\cdot\zeta. \tag{1.13b}$$

(Observe that T actually stays analytic in η for $\eta=\zeta$, that is, $c=1$.)

2. Let the A_α in (0.2) stand for square matrices with n rows and columns, so that (0.1) is an mth-order system for a vector u with n components. A fundamental solution matrix F satisfies $LF=\delta_y I$ where I is the unit matrix. Prove that for odd $n>m$ formula (1.12b) represents a fundamental solution matrix if in the integrand we write P^{-1} instead of $1/P$.

3. Use problem 2 to find a fundamental solution for the equations of elastic equilibrium

$$0=\mu\Delta u_i+(\lambda+\mu)\frac{\partial}{\partial x_i}(\operatorname{div}u) \tag{1.14}$$

(see (2.7) of Chapter 1). [Hint: Involved is the computation of the integrals

$$q_{ik}(\eta)=\int_{|\xi|=1}|\eta\cdot\xi|\xi_i\xi_k\,dS_\xi \tag{1.15}$$

for $|\eta|=1$. Introduce

$$Q(\eta,\zeta)=\sum_{i,k} q_{ik}(\eta)\zeta_i\zeta_k.$$

This is a quadratic form in ζ. Show that for $|\zeta|=1$ Q only depends on $\eta\cdot\zeta$, which implies that

$$Q(\eta,\zeta)=a|\zeta|^2+b(\eta\cdot\zeta)^2$$

for $|\eta|=1$, with constants a,b. Evaluate Q for η=unit vector in the x_3-direction.].

4. Find a fundamental solution of (1.14) by finding a solution of the equations

$$\mu\,\Delta u_i+(\lambda+\mu)\frac{\partial}{\partial x_i}(\operatorname{div})\,u)=F_i(x) \tag{1.16}$$

for $F_i\in C_0^\infty$, using Poisson's formula, (1.28) of Chapter 4, and the equations $(\lambda+2\mu)\Delta(\operatorname{div} u)=\operatorname{div} F$, $\Delta r=2/r$. [Answer: The fundamental solution matrix has the elements

$$-\frac{1}{4\pi\mu}\left(r^{-1}\delta_{ik}-\frac{\lambda+\mu}{2(\lambda+2\mu)}\frac{\partial^2 r}{\partial x_i\partial x_k}\right)\quad].$$

5. (a) With K given by (1.4) for odd n, and with M defined as in Lagrange's identity (4.5) of Chapter 3, show that under suitable regularity assumptions

$$u(y)=\int_\Omega K(x,y)\tilde{L}u(x)\,dx+\int_{\partial\Omega} M(u,K,\zeta)\,dS_x$$

for $y\in\Omega$, in analogy to (1.17), Chapter 4. [Hint: Write $u(x)=\phi(x)u(x)+(1-\phi(x))u(x)$ where $\phi\in C_0^\infty(\Omega)$ and $\phi(y)=1$. Apply (1.4a) to ϕu and Lagrange's identity to $(1-\phi)u$.]

(b) Show that solutions u of an elliptic equation (0.5) with constant coefficients for odd n are real analytic in the interior of their domain of definition.

(c) Show the same for even n, using Hadamard's method of descent. [Hint: Add a term ξ_{n+1}^m to $P(\xi)$.]

6. (Local fundamental solution of linear elliptic equations with analytic coefficients for odd n). Let

$$L=\sum_{|\alpha|\leqslant m} A_\alpha(x)D^\alpha$$

with $A_\alpha(x)$ analytic at z and satisfying

$$\sum_{|\alpha|=m} A_\alpha(x)\xi^\alpha\neq 0\quad\text{for real } \xi\neq 0.$$

Assume that $n+1\leqslant m$. For real unit vectors ξ denote by $\Gamma_j(x,y,\xi)$ for $j=0,1$ the solution of

$$L\Gamma_j=(-1)^j(x-y)\cdot\xi$$

with zero Cauchy data on the hyperplane $(x-y)\cdot\xi=0$. Let

$$\Gamma(x,y,\xi)=\begin{cases}\Gamma_0(x,y,\xi) & \text{for } (x-y)\cdot\xi\geqslant 0\\ \Gamma_1(x,y,\xi) & \text{for } (x-y)\cdot\xi\leqslant 0.\end{cases}$$

Then

$$K(x,y) = \tfrac{1}{4}(2\pi i)^{1-n}\Delta_y^{(n+1)/2}\int_{|\xi|=1}\Gamma(x,y,\xi)\,dS_\xi$$

is a fundamental solution for L with pole y for x,y near z. [Hint: $\Gamma(x,y,\xi)$ is piecewise real analytic and of class C^m in x,y,ξ (see p. 77) for all real x,y near z and real ξ with $|\xi|$ near 1. For any $\phi(x)\in C_0^\infty(\mathbb{R}^n)$ with support in a small ball with center z we have

$$\int(\tilde{L}\phi(x))\Gamma(x,y,\xi)\,dx = \int\phi(x)|(x-y)\cdot\xi|\,dx.$$

Use (2.81), (2.91) of Chapter 5.]

2. The Dirichlet Problem

The Dirichlet problem for the equation (0.5) of order $m=2\mu$ consists in finding a solution u in the bounded region Ω for prescribed values of u and its first $\mu-1$ normal derivatives on $\partial\Omega$, or equivalently for *consistently* prescribed values of u and its derivatives of orders $\leqslant \mu-1$ on $\partial\Omega$. We can instead solve an equation of the form

$$Lu=w(x) \quad \text{for } x\in\Omega \tag{2.1}$$

with w prescribed in Ω, where u satisfies the homogeneous Dirichlet conditions

$$D^\alpha u=0 \quad \text{for } x\in\partial\Omega \text{ and } |\alpha|\leqslant\mu-1. \tag{2.2}$$

We shall show how at least a *weak* solution of this problem can be obtained by following the Hilbert-space approach used in Chapter 4, Section 5. The method used here for homogeneous L with constant coefficients can be generalized to more general linear elliptic operators with variable coefficients. The key elements are the Gårding inequality and the Lax–Milgram lemma. The passage from weak to strong solutions will not be attempted here.

We denote by $C_0^\infty(\Omega)$ the set of functions $u\in C^\infty(\mathbb{R}^n)$ whose support is contained in a compact subset of Ω. Using as norm the expression

$$\|u\|_\mu = \sqrt{\int_\Omega\sum_{|\alpha|\leqslant\mu}|D^\alpha u|^2\,dx} \tag{2.3}$$

we can complete $C_0^\infty(\Omega)$ into a Hilbert space $H_0^\mu(\Omega)$ which we shall consider as the proper set of functions satisfying the Dirichlet condition (2.2) in the generalized sense.*

* $H^\mu(\Omega)$ can be defined as the completion with respect to the norm (2.3) of the set of functions in $C^\infty(\Omega)$ with bounded norm $\|u\|_\mu$. We can identify $H^0(\Omega)$ with $H_0^0(\Omega)$ or with the set $L_2(\Omega)$ of u that are square integrable over Ω in the sense of Lebesgue. For "nice" Ω and "nice" u conditions (2.2) are equivalent to membership in $H_0^\mu(\Omega)$. [See problem 9 p. 198 and the remarks on pp. 200–201.]

We need to define an analogue $B(u, v)$ to the bilinear form (5.13) of Chapter 4. For that purpose we decompose each multi-index α with $|\alpha| = 2\mu$ in a trivial (nonunique) way into $\alpha = \beta + \gamma$, where $|\beta| = |\gamma| = \mu$. Writing

$$A_\alpha = B_{\beta,\gamma} \tag{2.4}$$

we have

$$Lu = (-1)^\mu P(D)u = (-1)^\mu \sum_{|\beta|=|\gamma|=\mu} B_{\beta,\gamma} D^{\beta+\gamma} u = w \tag{2.5}$$

with certain $B_{\beta,\gamma}$. Replacing, if necessary, $B_{\beta,\gamma}$ by $\frac{1}{2}(B_{\beta,\gamma} + B_{\beta,\gamma})$ (again denoted by $B_{\beta,\gamma}$), we can bring about that

$$B_{\beta,\gamma} = B_{\gamma,\beta}. \tag{2.5a}$$

For any solution $u \in C^{2\mu}(\Omega)$ of (2.5) and any $v \in C_0^\infty(\Omega)$ we find by repeated integration by parts that

$$B(v,u) = (v,w) \tag{2.6}$$

where B denotes the bilinear functional

$$B(v,u) = \int_\Omega \sum_{|\beta|=|\gamma|=\mu} B_{\beta,\gamma}(D^\beta v)(D^\gamma u)\, dx \tag{2.7}$$

and

$$(v,w) = \int_\Omega vw\, dx. \tag{2.7a}$$

It is clear that for $u \in C^{2\mu}(\Omega)$ the partial differential equation (2.5) can be replaced by the requirement that (2.6) holds for all "test functions" $v \in C_0^\infty(\Omega)$. Here by Cauchy's inequality

$$|B(v,u)| \leqslant K\|v\|_\mu \|u\|_\mu \tag{2.8}$$

with a suitable constant K. This inequality permits us to extend the domain of $B(v,u)$ to all v,u in $H_0^\mu(\Omega)$. We just take Cauchy sequences of v,u in $C_0^\infty(\Omega)$ with respect to the norm (2.3) and observe that the corresponding values of B converge by (2.8). Similarly, given $w \in C^0(\overline{\Omega})$, the expression

$$\phi(v) = (v,w) \tag{2.9}$$

represents a linear functional satisfying

$$|\phi(v)| \leqslant \|v\|_0 \|w\|_0 \leqslant \|v\|_\mu \|w\|_0. \tag{2.10}$$

This functional also can be extended to $H_0^\mu(\Omega)$ and is bounded because of (2.10). The extended B is then again bilinear and satisfies (2.6). Our modified version of the Dirichlet problem is then to find a $u \in H_0^\mu(\Omega)$ such that

$$\phi(v) = B(v,u) \tag{2.11}$$

for all $v \in H_0^\mu(\Omega)$. The existence of u follows from the representation theorem on p. 118, if we can prove the existence of a positive constant k such that

$$k\|u\|_\mu^2 \leqslant B(u,u) \tag{2.12}$$

for all $u \in H_0^\mu(\Omega)$. For by (2.5a) the form B is symmetric:

$$B(v,u) = B(u,v). \tag{2.13}$$

Moreover $B(u,u) > 0$ for $u \neq 0$ by (2.12). Thus $\sqrt{B(u,u)} = |||u|||$ can be used as a norm on $H_0^\mu(\Omega)$. Because of (2.12), (2.8)

$$k\|u\|_\mu^2 \leqslant |||u|||^2 \leqslant K\|u\|_\mu^2. \tag{2.14}$$

Thus the new norm is "equivalent" to the old one, in the sense that boundedness of functionals is the same in both norms. This applies in particular to the functional $\phi(v)$, and the existence of u in (2.11) follows.

It remains to establish (2.12). This is achieved most easily by Fourier transformation, making use of Parseval's identity ((3.22) of Chapter 5) between a function $g(x)$ and its transform $\hat{g}(\xi)$. More generally we have for two real-valued functions u, v in $C_0^\infty(\bar{\Omega})$

$$\begin{aligned} 4\int u(x)v(x)\,dx &= \int \left[(u+v)^2 - (u-v)^2 \right] dx \\ &= \int \left[|\hat{u}(\xi) + \hat{v}(\xi)|^2 - |\hat{u}(\xi) - \hat{v}(\xi)|^2 \right] d\xi \\ &= 2\int (\hat{u}\bar{\hat{v}} + \bar{\hat{u}}\hat{v})\,d\xi = 4\,\mathrm{Re}\int \hat{u}\bar{\hat{v}}\,d\xi. \end{aligned} \tag{2.15}$$

It follows for $u \in C_0^\infty(\bar{\Omega})$ from (2.7), ((2.18) of Chapter 5), (2.5), (0.4) that

$$\begin{aligned} B(u,u) &= \sum_{|\beta|=|\gamma|=\mu} B_{\beta,\gamma} \int (D^\beta u)(D^\gamma u)\,dx \\ &= \mathrm{Re} \sum_{|\beta|=|\gamma|=\mu} B_{\beta,\gamma} \int ((i\xi)^\beta \hat{u})(\overline{(i\xi)^\gamma \hat{u}})\,d\xi \\ &= \int \sum_{|\beta|=|\gamma|=\mu} B_{\beta,\gamma} \xi^{\beta+\gamma} |\hat{u}(\xi)|^2\,d\xi \\ &= \int \sum_{|\alpha|=2\mu} A_\alpha \xi^\alpha |\hat{u}(\xi)|^2\,d\xi \\ &= \int P(\xi)|\hat{u}(\xi)|^2\,d\xi \geqslant c\int |\xi|^{2\mu} |\hat{u}(\xi)|^2\,d\xi. \end{aligned}$$

Since

$$\sum_{|\beta|=\mu} |(i\xi)^\beta \hat{u}|^2 \leqslant \sum_{|\beta|=\mu} |\xi|^{2\mu} |\hat{u}|^2 = C|\xi|^{2\mu} |\hat{u}|^2 \tag{2.15a}$$

with a certain constant C, we find that

$$B(u,u) \geqslant \frac{c}{C} \int \sum_{|\beta|=\mu} |(i\xi)^\beta \hat{u}(\xi)|^2 d\xi$$

$$= \frac{c}{C} \int \sum_{|\beta|=\mu} (D^\beta u(x))^2 dx. \tag{2.16}$$

Assume that Ω is contained in the cube $|x_i| \leqslant a$ for $i=1,\ldots,n$. Applying repeatedly Poincaré's inequality ((5.19a) of Chapter 4) we have for $0 \leqslant \nu \leqslant \mu$

$$\sum_{|\gamma|=\nu} \int (D^\gamma u)^2 dx \leqslant \sum_{|\gamma|=\nu} (2a)^{2\mu-2\nu} \int (D_1^{\mu-\nu} D^\gamma u)^2 dx$$

$$\leqslant (2a)^{2\mu-2\nu} \sum_{|\beta|=\mu} \int (D^\beta u)^2 dx$$

and thus by (2.3), (2.16)

$$\|u\|_\mu^2 = \sum_{\nu=0}^{\mu} \sum_{|\gamma|=\nu} \int (D^\gamma u)^2 dx$$

$$\leqslant \sum_{\nu=0}^{\mu} (2a)^{2\mu-2\nu} \sum_{|\beta|=\mu} \int (D^\beta u)^2 dx \leqslant \frac{1}{k} B(u,u), \tag{2.17}$$

where

$$\frac{1}{k} = \frac{C}{c} \sum_{\nu=0}^{\mu} (2a)^{2\mu-2\nu}. \tag{2.18}$$

This establishes the inequality (2.12) for $u \in C_0^\infty(\bar{\Omega})$. It holds then clearly also in the completed space $H_0^\mu(\Omega)$, thus proving the existence of a weak solution u of the Dirichlet problem.

We give some indications for the analogous existence proof for more general linear elliptic equations $Lu=w$ of order 2μ, with coefficients depending on x. It is then natural to write the differential equation in the form

$$Lu = \sum_{\substack{|\beta| \leqslant \mu \\ |\nu| \leqslant \mu}} (-1)^{|\beta|} B_{\beta,\gamma}(x) D^{\beta+\gamma} u \tag{2.19}$$

and to introduce the bilinear form

$$B(v,u) = \int \sum_{\substack{|\beta| \leqslant \mu \\ |\gamma| \leqslant \mu}} \left(D^\beta B_{\beta,\gamma} v\right)(D^\gamma u)\, dx. \tag{2.20}$$

We assume that L is uniformly elliptic in the sense that there exists a constant $c > 0$ that

$$\sum_{|\beta|=|\gamma|=\mu} B_{\beta,\gamma}(x)\xi^{\beta+\gamma} \geqslant c|\xi|^{2\mu} \tag{2.20a}$$

for all $x \in \bar{\Omega}$, $\xi \in \mathbb{R}^n$, and that the $B_{\beta,\gamma}$ belong to $C^\mu(\bar{\Omega})$.

A weak solution of the Dirichlet problem can then again be defined as a $u \in H_0^\mu(\Omega)$ for which (2.11) holds for all $v \in H_0^\mu(\Omega)$. A difficulty about applying the representation theorem arises from the fact that now $B(v,u)$ cannot be expected to be symmetric in v and u; (symmetry for the special case considered earlier arises from the fact that there the differential operator L is formally selfadjoint). This difficulty disappears, if we apply the *Lax–Milgram lemma* which assures us that in a Hilbert space H_0^μ any bounded linear functional $\phi(v)$ can be written as $B(v,u)$ with the help of a suitable u in the same space, provided the given bilinear form B satisfies inequalities (2.8), (2.12) for some positive constants k, K. (For a proof see [12], [15].)

A further obstacle arises now from the circumstance that (2.12) just does not hold in general. If it did, the Dirichlet problem for the equation $\Delta u + u = 0$ would always be solvable, which is not the case. (See Problem 1 below.) What can be proved under suitable regularity assumptions is a weakened form of (2.12):

Gårding's inequality. There exists a positive constant k such that

$$k\|u\|_\mu^2 \leqslant B(u,u) + \int_\Omega u^2\,dx \tag{2.21}$$

holds for all $u \in H_0^\mu(\Omega)$.

The integral on the right-hand side generally cannot be omitted. As a consequence there is no existence for the solution of the Dirichlet problem without some further qualification. Instead one obtains with the help of Gårding's inequality a statement in the form of an *alternative*: Either there exists a nontrivial solution of $Lu = 0$ with Dirichlet data 0, or there exists a solution u of $Lu = w$ with Dirichlet data 0 for every sufficiently regular w. We only indicate how one arrives at this conclusion.

We are led to the alternative by introducing the bilinear form

$$B^*(v,u) = B(v,u) + (v,u). \tag{2.21a}$$

Then (2.11) becomes

$$B^*(v,u) = (v, w + u) \quad \text{for all } v \in H_0^\mu(\Omega). \tag{2.21b}$$

Here (2.8) and (2.12) are satisfied if B is replaced by B^* and K by $K + 1$. Consequently, there exists for every $F \in H_0^0(\Omega)$ an element $u = TF$ in $H_0^\mu(\Omega)$ such that

$$B^*(v,u) = (v,F) \quad \text{for all } v \in H_0^\mu(\Omega). \tag{2.21c}$$

By (2.21)

$$k\|TF\|_\mu^2 \leqslant B^*(TF,TF) = (TF,F) \leqslant \|TF\|_\mu \|F\|_0$$

or

$$\|TF\|_\mu \leqslant \frac{1}{k}\|F\|_0$$

so that T is a bounded operator, mapping $H_0^0(\Omega)$ into $H_0^\mu(\Omega)$. Our problem is to find a u in $H_0^\mu(\Omega)$ for which

$$u = Tu + Tw. \tag{2.21d}$$

By *Rellich's lemma* T defines a *compact* mapping of $H_0^0(\Omega)$ into itself,* provided Ω satisfies certain regularity conditions. By Riesz-Schauder the Fredholm alternative holds for the compact operator T on $H_0^0(\Omega)$: Equation (2.21d) either has a solution $u \in H_0^0(\Omega)$ for all $Tw \in H_0^0(\Omega)$ or the equation has for $Tw = 0$ a non-trivial solution $u \in H_0^0(\Omega)$.† For $w \in H_0^0(\Omega)$ a solution $u \in H_0^0(\Omega)$ of (2.21d) belongs to $H_0^\mu(\Omega)$. Thus the alternative holds for the problem of finding a $u \in H_0^\mu(\Omega)$ satisfying (2.21b), and hence also for our generalized Dirichlet problem.

PROBLEMS

1. Let $n=3$ and Ω be the ball $|x|<\pi$. Show that solution u of $\Delta u+u=w(x)$ with vanishing boundary values can only exist, if

$$\int_\Omega w(x)\frac{\sin|x|}{|x|}\,dx=0 \tag{2.22}$$

(compare problem 2d, p. 102).

2. Show that for odd n the weak solution u of (2.5) constructed, is a strict solution in the region Ω. [Hint: Use the fundamental solution, as on p. 122.]

3. Consider a homogeneous system of equations with constant real coefficients of the form

$$Lu=(-1)^\mu P(D)u=(-1)^\mu \sum_{|\alpha|=2\mu} A_\alpha D^\alpha u=w, \tag{2.23}$$

where u and w are vectors with N components and the A_α constant square matrices. Let L be *strongly elliptic* in the sense that there exists a positive c such that

$$\eta^{\mathrm{T}}P(\xi)\eta \geqslant c|\xi|^{2\mu}|\eta|^2 \tag{2.24}$$

for all real n-vectors ξ and N-vectors η. Write L in the form (2.5) and define

$$B(v,u)=\int \sum_{|\beta|=|\gamma|=\mu} (D^\beta v)^{\mathrm{T}} B_{\beta,\gamma}(D^\gamma u)\,dx.$$

* That is, for a sequence $u^k \in H_0^0(\Omega)$ with uniformly bounded norms $\|u^k\|_0$ and hence $\|Tu^k\|_\mu$ there exists a subsequence and an element $v \in H_0^0(\Omega)$ such that $\|v - Tu^k\|_0$ tends to zero for k in the subsequence. See [1], [12].

† [7].

Show that the "strong" Gårding inequality (2.12) holds for $u \in C_0^\infty(\bar{\Omega})$ with a suitable positive k. [Hint: Show that

$$B(u,u) = \text{Re} \int \hat{u}^T P(\xi) \bar{\hat{u}} \, d\xi \quad].$$

4. Show that the system (1.14) is strongly elliptic in the sense of problem 3 for $\mu > 0$, $\lambda + 2\mu > 0$.

5. For $u \in C_0^\infty(\Omega)$ define the norm

$$[u]_\nu = \sqrt{\sum_{|\alpha| = \nu} \int_\Omega (D^\alpha u)^2 \, dx} \tag{2.24a}$$

and extend it to $H_0^\mu(\Omega)$, so that

$$\|u\|_\mu^2 = \sum_{\nu=0} [u]_\nu^2 = [u]_\mu^2 + \|u\|_{\mu-1}^2. \tag{2.24b}$$

Show that for $\varepsilon > 0$ there exists a K (depending on ε,μ,Ω) such that

$$\|u\|_{\mu-1} \leqslant \varepsilon[u]_\mu + K\|u\|_0. \tag{2.24c}$$

[Hint: For $0 < |\alpha| = \nu < \mu$ set $\alpha = \beta + \gamma$ where $|\beta| = 1, |\gamma| = \nu - 1$. For $u \in C_0^\infty(\Omega)$

$$(D^\alpha u, D^\alpha u) = -(D^{\alpha+\beta} u, D^\gamma u) \leqslant [u]_{\nu+1}[u]_{\nu-1}.$$

Hence for $\varepsilon > 0$ there exists a $k = k(\varepsilon,\nu)$ such that

$$[u]_\nu \leqslant \varepsilon[u]_{\nu+1} + k[u]_{\nu-1}.$$

Apply to $[u]_{\nu-1}, [u]_{\nu-2}, \ldots,$ with different ε.]

6. Let $\chi(y) \in C^\infty(\mathbb{R}^n)$ and

$$0 \leqslant \chi(y) \quad \text{for all } y; \qquad \chi(y) = 0 \quad \text{for } |y| > 1 \tag{2.25a}$$

$$\int \chi(y) \, dy = 1. \tag{2.25b}$$

For $u \in C_0^\mu(\Omega)$ introduce the "mollified" functions

$$u_\varepsilon(x) = \int \chi(y) u(x + \varepsilon y) \, dy. \tag{2.25c}$$

Show that

(a) $u_\varepsilon(x)$ is defined and belongs to $C_0^\infty(\Omega)$ for all sufficiently small $\varepsilon > 0$.
(b) $\lim_{\varepsilon\to 0} D^\alpha u_\varepsilon(x) = D^\alpha u(x)$ uniformly for $x \in \Omega$ if $|\alpha| \leqslant \mu$.
(c) $\|u_\varepsilon(x)\|_\mu \leqslant \|u(x)\|_\mu$. [Use triangle inequality!]
(d) $u \in H_0^\mu(\Omega)$.

7. (Partition of unity). Let Ω be an open bounded set in R^n, and let $\Omega_1, \ldots, \Omega_N$ be open bounded sets whose union contains $\bar{\Omega}$. Show that there exist functions $\zeta_k(x) \in C_0^\infty(\Omega_k)$ for $k = 1, \ldots, N$ such that

$$\sum_{k=1}^N \zeta_k^2(x) = 1 \quad \text{for } x \in \bar{\Omega}. \tag{2.26}$$

[Hint: Let $B_{y,\rho}$ denote the ball of center y and radius ρ. For $\rho > 0$ construct a non-negative function $\chi_{y,\rho}(x) \in C^\infty(\mathbb{R}^n)$ for which

$$\chi_{y,\rho}(x) > 0 \quad \text{for } x \in B_{y,2\rho}; \qquad \chi_{y,\rho}(x) = 0 \quad \text{for } x \notin B_{y,3\rho}.$$

For each $y \in \overline{\Omega}$ select an integer $v = v(y)$ and a positive number $\rho = \rho(y)$ such that $B_{y,4\rho} \subset \Omega_v$. A finite number of the balls $B_{y,\rho(y)}$ covers Ω, say those with centers $y_1, \ldots, y_M$. For $j = 1, \ldots, M$ set

$$\xi_j(x) = \chi_{y,\rho}(x) \quad \text{where } y = y_j, \quad \rho = \rho(y_j),$$

and for $k = 1, \ldots, N$ set

$$\eta_k(x) = \sum_{v(y_j)=k} \xi_j(x).$$

Show that $\eta_k(x) \in C_0^\infty(\Omega_k)$, and that for $x \in \overline{\Omega}$ we have $\eta_k(x) \neq 0$ for some k. Take

$$\zeta_k(x) = \eta_k(x) \Big/ \sqrt{\sum_r \eta_r^2(x)}.]$$

8. (Proof of Gårding's inequality). Take $B(v,u)$ as in (2.20), c as in (2.20a), $[u]_\mu$ as in (2.24a), and C as in (2.15a).

(a) Show that for any $y \in \overline{\Omega}$

$$\int_\Omega \sum_{|\beta|=|\gamma|=\mu} B_{\beta,\gamma}(y)(D^\beta u(x))(D^\gamma u(x))\,dx \geqslant \frac{c}{C}[u]_\mu^2.$$

(b) Show that for given $B_{\beta,\gamma}(x)$

$$B(u,u) = \overline{B}(u,u) + O(\|u\|_\mu \|u\|_{\mu-1})$$

where

$$\overline{B}(u,u) = \int_\Omega \sum_{|\beta|=|\gamma|=\mu} B_{\beta,\gamma}(x)(D^\beta u(x))(D^\gamma u(x))\,dx.$$

(c) Given $\delta > 0$ cover $\overline{\Omega}$ by a finite number of open sets $\Omega_1, \ldots, \Omega_N$ so small that any two points of Ω_k have distance $< \delta$. Use the partition of unity of problem 7. Let y_k denote a point of $\Omega_k \cap \Omega$. Given ε show that for δ sufficiently small

$$\left| \overline{B}(u,u) - \sum_k \int_\Omega \sum_{|\beta|=|\gamma|=\mu} B_{\beta,\gamma}(y^k)\zeta_k^2(x)(D^\beta u(x))(D^\gamma u(x))\,dx \right| \leqslant \varepsilon[u]_\mu^2.$$

[$B_{\beta,\gamma}$ varies little in Ω_k for small δ. Notice that δ and hence also the ζ_k depend on ε.]

(d) Show that for fixed ε and ζ_k

$$\sum_k \int_\Omega \sum_{|\beta|=|\gamma|=\mu} B_{\beta,\gamma}(y^k)\zeta_k^2(x)(D^\beta u(x))(D^\gamma u(x))\,dx$$

$$= \sum_k \int_\Omega \sum_{|\beta|=|\gamma|=\mu} B_{\beta,\gamma}(y^k)(D^\beta \zeta_k u)(D^\gamma \zeta_k u)\,dx + O(\|u\|_\mu \|u\|_{\mu-1})$$

$$\geqslant \frac{c}{C} \sum_k [\zeta_k u]_\mu^2 + O(\|u\|_\mu \|u\|_{\mu-1})$$

$$= \frac{c}{C}[u]_\mu^2 + O(\|u\|_\mu \|u\|_{\mu-1})$$

$$\geqslant \left(\frac{c}{C} - \varepsilon\right)\|u\|_\mu^2 - O(\|u\|_0^2)$$

from which (2.21) follows. [Hint: Use (2.24c) and the inequality $ab \leqslant ca^2 + (1/4c)b^2$ valid for any a,b and $c > 0$.]

9. (Justification of the generalized Dirichlet condition for "nice" functions and domains).

Definition. The boundary $\partial\Omega$ of an open set $\Omega \subset \mathbb{R}^n$ is said to be of class C^s, if for every $x^1 \in \partial\Omega$ there exists a neighborhood ω and a $1-1$ mapping $y = f(x)$ of class C^s with non-vanishing Jacobian, mapping ω onto the ball $|y| < 1$, $\omega \cap \Omega$ onto the half-ball $|y| < 1$, $y_n > 0$, and x^1 onto 0. (Essentially the condition states that $\partial\Omega$ can be "straightened out" locally by a mapping in C^μ.)

Let Ω be an open bounded set in $\mathbb{R}^n$ with $\partial\Omega \in C^\mu$. Let $u \in C^\mu(\overline{\Omega})$.* Show that necessary and sufficient for u to belong to $H_0^\mu(\Omega)$ is that u and all its derivatives of orders $< \mu$ vanish on $\partial\Omega$. [Hint: We can find a finite number of points $x^1, \ldots, x^{N-1}$ on $\partial\Omega$ and neighborhoods $\omega_1, \ldots, \omega_{N-1}$ of those points that can be mapped by $y = f_k(x)$ onto balls as in the definition above, such that $\omega_1, \ldots, \omega_{N-1}$ cover $\partial\Omega$. We can find a further open set ω_N with $\overline{\omega}_N \subset \Omega$, such that $\omega_1, \ldots, \omega_N$ cover $\overline{\Omega}$. Define a corresponding partition of unity as in problem 7. Let $u_k(x) = \zeta_k^2(x)u(x)$, so that $u = u_1 + \cdots + u_N$. It is sufficient to prove the statement for each u_k. In the non-trivial case $k < N$ we transform u_k into a function $U_k(y)$ defined and of class C^μ in the closure of the half-space Σ: $y_n > 0$ and with support in $|y| < 1$. The relations $u_k \in H_0^\mu(\Omega)$ and $U_k \in H_0^\mu(\Sigma)$ are equivalent (use problem 6). Let δ denote differentiation with respect to y_n. For $\psi \in C^\infty(\mathbb{R}^n)$ and $1 \leqslant j \leqslant \mu$

$$\int_{\partial\Sigma} \psi(\delta^{j-1}U)\, dS = \int_{\Sigma} [\psi(\delta^j U) + (\delta\psi)(\delta^{j-1}U)]\, dy.$$

Here the righthand side vanishes for $U \in C_0^\mu(\Sigma)$, and hence also for $U \in H_0^\mu(\Sigma)$. Thus for $U \in H_0^\mu(\Sigma)$

$$\delta^j U = 0 \quad \text{for } y_n = 0 \quad \text{and } j = 0, \ldots, \mu - 1. \tag{2.27}$$

Conversely, let (2.27) hold. Let $\chi(s) \in C^\infty(\mathbb{R})$ with

$$\chi(s) = 1 \quad \text{for } s > 1; \qquad \chi(s) = 0 \quad \text{for } s < \tfrac{1}{2}.$$

Set $U_\varepsilon(y) = \chi(y_n/\varepsilon)U(y)$. Then $U_\varepsilon \in C_0^\mu(\Sigma)$ and $\lim_{\varepsilon \to 0} \|U_\varepsilon - U\|_\mu = 0$.]

3. More on the Hilbert Space H_0^μ and the Assumption of Boundary Values in the Dirichlet Problem

In discussing the Dirichlet problem we deal with objects u that sometimes are elements of a Hilbert space (Cauchy sequences), sometimes distributions (linear functionals), and sometimes ordinary functions with values at points. While in principle we deal with objects of quite different kinds they also can be just different aspects of the same u.

* This means that $u \in C^\mu(\Omega)$ and that each $D^\alpha u$ with $|\alpha| \leqslant \mu$ can be extended to a continuous function on $\overline{\Omega}$.

We start with the space $C_0^\infty(\Omega)$ of *test functions* with the inner product

$$(u,v) = \int_\Omega uv\,dx \tag{3.1a}$$

and the norm $\|u\|_0 = \sqrt{(u,u)}$. An element u of $H_0^0(\Omega)$ is a Cauchy sequence $u^k \in C_0^\infty(\Omega)$ with respect to the norm $\|\cdot\|_0$. Two Cauchy sequences $u^k,v^k \in C_0^\infty(\Omega)$ determine the same $u \in H_0^0(\Omega)$, iff $\|u^k - v^k\|_0 \to 0$ for $k \to \infty$.

For $u,v \in H_0^0(\Omega)$ we define

$$(u,v) = \lim_{k\to\infty} (u^k,v^k); \qquad \|u\|_0 = \lim_{k\to\infty} \sqrt{(u^k,u^k)} \tag{3.1b}$$

in terms of the Cauchy sequences u^k,v^k leading to u,v. Then

$$|(u,v)| \leqslant \|u\|_0\|v\|_0. \tag{3.1c}$$

We can identify any test function ψ with the element u of $H_0^0(\Omega)$ given by the Cauchy sequence $u^k = \psi$ for all k. Any $u \in H_0^0(\Omega)$ defines a linear functional

$$u[\psi] = (u,\psi) \tag{3.2}$$

on $C_0^\infty(\Omega)$.

PROBLEMS

1. Show that the functional $u[\psi]$ defined by $u \in H_0^0(\Omega)$ for test functions ψ is a *distribution* in the sense of p. 90.
2. Show that two elements u,v of $H_0^0(\Omega)$ are identical if the corresponding distributions are identical.
3. Show that the Dirac "function" is a distribution that is not generated by any $u \in H_0^0(\Omega)$.
4. A sequence $u^k \in H_0^0(\Omega)$ is said to converge "strongly" to an element $u \in H_0^0(\Omega)$, if $\|u^k - u\|_0 \to 0$ for $k \to \infty$. Show that strong convergence of the u^k implies "weak" convergence in the sense of distributions which requires that

$$\lim_{k\to\infty} (u^k,\psi) = (u,\psi)$$

uniformly for all ψ that have their support in the same compact subset of Ω and have a common upper bound for $|\psi|$.

It can happen that the distribution $u[\psi]$ determined by $u \in H_0^0(\Omega)$ agrees with a function $U \in C^0(\Omega)$ in the sense of p. 90, that is

$$u[\psi] = (u,\psi) = (U,\psi) \quad \text{for all } \psi \in C_0^\infty(\Omega). \tag{3.3a}$$

Then also

$$(u,\psi) = (U,\psi) \quad \text{for all } \psi \in C_0^0(\Omega) \tag{3.3b}$$

by problem 6, p. 196. We can find a sequence of test functions ζ_k such that

(a) $$0 \leqslant \zeta_k(x) \leqslant 1 \quad \text{for all } x \in \Omega. \tag{3.3c}$$

(b) for any compact subset σ of Ω

$$\zeta_k(x) = 1 \quad \text{for all } x \in \sigma \tag{3.3d}$$

provided k is sufficiently large.

Then by (3.3b)

$$\|\zeta_k U\|_0^2 = (\zeta_k U, \zeta_k U) \leqslant (\zeta_k U, U) = (\zeta_k U, u) \leqslant \|\zeta_k U\|_0 \|u\|_0.$$

Thus

$$\|\zeta_k U\|_0 < \|u\|_0. \tag{3.3e}$$

It follows that U is *square integrable* over Ω:

$$\begin{aligned}(U,U) &= \int_\Omega U^2(x)\,dx \\ &= \lim_{k\to\infty} \int_\Omega \zeta_k^2(x) U^2(x)\,dx \leqslant \|u\|_0^2.\end{aligned} \tag{3.3f}$$

We can identify U with u, since for any test functions u^k converging strongly to u by (3.3b,f)

$$\|u^k - U\|_0^2 = \|u^k - u\|_0^2 + (U,U) - (u,u) \leqslant |u^k - u|_0^2 \to 0$$

which also implies that

$$\|u\|^2 = \int_\Omega U^2(x)\,dx. \tag{3.3g}$$

We say $u \in C^0(\Omega)$ and do not distinguish between u and U. Conversely, every square integrable $U \in C^0$ defines a corresponding $u \in H_0^0$ as strong limit of the sequence of functions $\zeta_k U \in C_0^0$.

It follows that $H_0^0(\Omega)$ can be identified with $H^0(\Omega)$ the completion of the set of square integrable functions in $C^0(\Omega)$ with respect to the norm $\|\cdot\|_0$.

Functions in $H_0^0(\Omega)$ may possess *strong* derivatives* defined as follows:

Definition. For $u,v \in H_0^0(\Omega)$ we say $v = D^\alpha u$ in the strong sense, if there exist test functions u^k for which

$$\lim_{k\to\infty} \|u - u^k\|_0 = 0; \ \lim_{k\to\infty} \|v - D^\alpha u^k\|_0 = 0. \tag{3.4}$$

It follows immediately for $\psi \in C^\infty(\Omega)$ that

$$\begin{aligned} v[\psi] = (v,\psi) &= \lim_{k\to\infty} (D^\alpha u^k, \psi) = \lim_{k\to\infty} (-1)^{|\alpha|}(u^k, D^\alpha \psi) \\ &= (-1)^{|\alpha|}(u, D^\alpha\psi) = (-1)^{|\alpha|} u[D^\alpha \psi], \end{aligned}$$

so that $v = D^\alpha u$ in the weak (distribution) sense of p. 91.

* These have to be distinguished from the ordinary derivatives of functions defined at points, which will be called "strict" derivatives.

We defined $H_0^\mu(\Omega)$ as the completion of $C_0^\infty(\Omega)$ with respect to the norm $\|u\|_\mu$ given by (2.3). It follows immediately that a $u \in H_0^\mu(\Omega)$ possesses strong derivatives $D^\alpha u \in H_0^0(\Omega)$ for all $|\alpha| \leqslant \mu$.

Problems

1. Show that if a strong derivative $D^\alpha u$ of a $u \in H_0^0(\Omega)$ exists, it is determined uniquely.

2. Show that if $u \in H_0^0(\Omega)$ can be identified with a function $U \in C^{|\alpha|}(\Omega)$, and if u has the strong derivative $D^\alpha u$, then $D^\alpha u$ can be identified with the strict derivative $D^\alpha U$, which then is square integrable.

3. Let Ω be a bounded open set in $\mathbb{R}^n$ with $\partial\Omega \in C^{n+s+1}$. Let $u \in C^{n+s+1}(\Omega)$, and let all $D^\alpha u$ for $|\alpha| \leqslant n + s + 1$ be square integrable over Ω. Show that $u \in C^s(\overline{\Omega})$.* [Hint: We only have to show that $u \in C^s(\overline{\Omega})$ near any point x^1 of $\partial\Omega$. Apply the transformation $y = f(x)$, $x = F(y)$ from the definition of $\partial\Omega \in C^s$ in problem 9, p. 198. Let Γ be the cube

$$|y_i| < \frac{1}{2n} \quad \text{for } i = 1, \ldots, n - 1; \qquad \left|y_n - \frac{1}{n}\right| < \frac{1}{2n}$$

which is contained in the half-ball $|y| < 1$, $y_n > 0$. Then $U(y) = u(F(y)) \in C^{n+s+1}(\Gamma)$, and all y-derivatives of U of orders $\leqslant n + s + 1$ are square integrable over Γ. By Chapter 5, (3.25d) all y-derivatives of U of orders $\leqslant s + 1$ are bounded uniformly in Γ. Then the y-derivatives of U of orders $\leqslant s$ are continuous in $\overline{\Gamma}$.]

4. (a) Show that the space $H_0^0(\Omega)$ is *separable*, that is that there exists a countable set of elements that is dense in $H_0^0(\Omega)$. [Hint: Let test functions $\zeta_k(x)$ be defined as in (3.3c,d). Then the set of functions $\zeta_k(x)p(x)$, where p is a polynomial with rational coefficients, is countable and dense in $H_0^0(\Omega)$. (Use Weierstrass approximation theorem, problem 1, p. 213 in larger set).]

(b) $H_0^0(\Omega)$ contains a complete orthonormal sequence. This means that there is a sequence of elements $u^k \in H_0^0(\Omega)$ with

$$(u^j, u^k) = \delta_{jk}, \tag{3.5a}$$

such that for every $u \in H_0^\mu(\Omega)$

$$\lim_{N\to\infty} \left\| u - \sum_{k=1}^{N} (u,u^k)u^k \right\|_0 = 0. \tag{3.5b}$$

$$\sum_{k=1}^{\infty} (u,u^k)^2 = \|u\|_0^2. \tag{3.5c}$$

[Hint: Take a sequence $v^1, v^2, \ldots$ that is dense in $H_0^0(\Omega)$ and such that v^k is independent of $v^1, \ldots, v^{k-1}$. Define inductively u^k satisfying (3.5a) and such that u^k is dependent on $v^1, \ldots, v^k$.]

* The assumptions are unnecessarily strong. The statement is only intended to show that square integrability of sufficiently high derivatives assures regularity up to the boundary.

Show that

$$v^N = \sum_{k=1}^{N} (v^N,u^k)u^k$$

$$\|v^N - u\|_0^2 - \sum_{k=1}^{N} ((v^N,u^k) - (u,u^k))^2 = \left\| u - \sum_{k=1}^{N} (u,u^k)u^k \right\|_0^2$$

$$\left. = \|u\|_0^2 - \sum_{k=1}^{N} (u,u^k)^2 \right].$$

(c) (Weak compactness). Let $w^1, w^2, \ldots$ be a sequence in $H_0^0(\Omega)$ with $\|w^k\| \leqslant M$ for all k. Show that there exists a subsequence of the w^k converging weakly to an element $w \in H_0^0(\Omega)$; that is for w^j in the subsequence and any $\psi \in C_0^\infty(\Omega)$

$$\lim_{j\to\infty} (w^j,\psi) = (w,\psi).$$

[Hint: Let the u^k form a complete orthonormal set. Choose the subsequence of the w^j so that

$$\lim_{j\to\infty} (u^k,w^j) = c_k$$

exists for all k. Then $\sum_{k=1}^N c_k u^k$ for $N \to \infty$ has a strong limit w, since

$$\sum_{k=1}^{N} c_k^2 \leqslant \lim_{j\to\infty} \|w^j\|_0^2 \leqslant M^2.]$$

The Hilbert space approach to the Dirichlet problem provides a *generalized* solution $u \in H_0^\mu(\Omega)$. One is faced then with the difficult task of establishing regularity of u in the interior and on the boundary of Ω. We shall give some idea of a method that furnishes the desired regularity, and for simplicity will restrict ourselves to the Dirichlet problem for the Laplace equation.*

The problem is to find a solution u of $\Delta u = -w$ in an open bounded set Ω of $\mathbb{R}^n$, for which $u = 0$ on $\partial\Omega$. In Chapter 4 we constructed instead an element u of $H_0^1(\Omega)$ for which

$$B(u,v) = (w,v) \quad \text{for all } v \in H_0^1(\Omega). \tag{3.6a}$$

Here $B(u,v)$ and (w,v) are the bilinear forms defined by completion from the integral expressions

$$B(u,v) = \int_\Omega \sum_i u_{x_i} v_{x_i}\, dx; \qquad (u,v) = \int uv\, dx \tag{3.6b}$$

valid for test functions. Moreover we showed on p. 121 that u for $w \in C^1(\overline{\Omega})$ can be identified with a point function (again denoted by u) belonging to $C^2(\Omega)$. By definition u and its first derivatives are square integrable over Ω. If we could show (see problem 3b, p. 170) that all derivatives of u of orders $\leqslant$

* For more general linear elliptic equations with variable coefficients see L. Nirenberg, Remarks on strongly elliptic partial differential equations, *Comm. Pure Appl. Math.* **VIII** (1955), 649–675, and A. Friedman [12].

$n + 2$ are square integrable (assuming $\partial\Omega \in C^{n+2}$, $w \in C^{n+1}(\overline{\Omega})$) we could conclude that $u \in C^1(\overline{\Omega})$; it would follow then (see problem 9, p. 198) that u has boundary values 0 in the ordinary sense. Here we shall just show the square integrability for the second derivatives of u; the result for higher derivatives follows by similar arguments.

Let then $w \in C^1(\overline{\Omega})$, and let $u \in H_0^1(\Omega)$ be the solution of (3.6a). In the estimates below we consider $\|w\|_1$ and $\|u\|_1$ as known constants.

Denote by $\zeta(x)$ a fixed function of $C_0^\infty(\mathbb{R}^n)$. Using that $\Delta u = -w$ pointwise in any compact subset of Ω, we find for $v \in C_0^\infty(\Omega)$ by integration by parts

$$B(v,\zeta u) = \int_\Omega v(w\zeta - u\Delta\zeta - 2\sum_i (D_i\zeta)(D_iu))\,dx = 0(\|v\|_0). \tag{3.7a}$$

It is sufficient to show that the second derivatives of u are square integrable near any boundary point x^1 of Ω. Assume that $\partial\Omega \in C^2$. Then (see the definition in problem 9, p. 198) there exists a $1 - 1$ mapping $x = F(y)$ of class C^2 with non-vanishing Jacobian and inverse $y = f(x)$, which maps 0 onto x^1, the ball $|y| < 1$ onto a neighborhood ω of x^1 and the half-ball $|y| < 1$, $y_n > 0$ onto $\omega \cap \Omega$. We choose for $\zeta(x)$ a function in $C_0^\infty(\mathbb{R}^n)$ whose support is mapped into the ball $\{y \,|\, |y| < \frac{1}{2}\}$. Let Σ be the half-ball

$$\Sigma = \{y \,|\, y \in \mathbb{R}^n;\quad |y| < \tfrac{1}{2};\quad y_n > 0\}.$$

We have to show that the function

$$W(y) = \zeta(f(y))u(f(y)) \tag{3.7b}$$

has square integrable second derivatives. Here W is of class C^2 for $y_n > 0$ and vanishes for $|y| > \frac{1}{2}$. For functions $\psi(y)$ with support in Σ the $\|\cdot\|_2$-norms of $\psi(y)$ and of $\psi(f(x))$ as a function of x are comparable. Thus for $\psi(y) \in C_0^\infty(\Sigma)$

$$B(\psi(f),\zeta u) = O(\|\psi\|_0). \tag{3.7c}$$

Transforming B to y-coordinates, we have

$$B(\psi(f),\zeta u) = B(\psi, W) = \int_{y_n>0} \sum_{r,s} a_{rs}(y)(d_r\psi)(d_sW)\,dy. \tag{3.7d}$$

Here d_r stands for $\partial/\partial y_r$, D_i for $\partial/\partial x_i$, and

$$a_{rs} = \sum_i (D_iy_r)(D_iy_s)\frac{dx}{dy} \tag{3.7e}$$

(dx/dy = Jacobian). The $a_{rs}(y)$ are of class C^1 for $|y| < 1$.

We first estimate the $\|\cdot\|_1$-norms of the first "tangential" difference quotients of W, then those of the tangential derivatives. Let k denote one of the numbers $1,2,\ldots,n-1$, fixed for the present. We introduce the shift operator E^h whose action on a function of y defined for $y_n > 0$ consists in replacing y_k by $y_k + h$ in its arguments. For $h \neq 0$ we define the divided difference operator

$$\delta_h = \frac{1}{h}(E^h - 1). \tag{3.8a}$$

Both E_h and δ_h commute with the differentiation operators d_r. We observe that

$$\delta_h \psi = \int_0^1 E^{th} d_k \psi \, dt. \tag{3.8b}$$

For functions $\psi(y)$ of compact support in $y_n > 0$

$$\int E^h \psi \, dy = \int \psi \, dy; \qquad \int \delta_h \psi \, dy = 0; \qquad \|E^h \psi\|_\mu = \|\psi\|_\mu. \tag{3.8c}$$

It follows from (3.8b) that

$$\|\delta_h \psi\|_0 \leqslant \|d_k \psi\|_0 \leqslant \|\psi\|_1, \tag{3.8d}$$

where all integrals and norms refer to the half-space $y_n > 0$.

For $\psi \in C_0^1(\Sigma)$ we form $B(\psi, \delta_h W)$ and shift δ_h onto the first argument. We observe that

$$\begin{aligned} E^{-h}(a_{rs}(d_r\psi)(d_s\delta_h W)) = \delta_{-h}(a_{rs}(d_r\psi)(d_s w)) - a_{rs}(d_r\delta_{-h}\psi)(d_s W) \\ -(\delta_{-h}a_{rs})(E^{-h}d_r\psi)(d_s W). \end{aligned}$$

It follows, using (3.7c), (3.8d), that

$$\begin{aligned} B(\psi, \delta_h W) &= -B(\delta_{-h}\psi, W) + O(\|\psi\|_1 \|W\|_1) \\ &= O(\|\delta_{-h}\psi\|_0) + O(\|\psi\|_1) = O(\|\psi\|_1) \end{aligned}$$

for any $\psi \in C_0^1(\Sigma)$, since $\|W\|_1 = O(\|u\|_1)$. The same inequality holds then for any $\psi \in H_0^1(\Sigma)$. Since with W also $\delta_{-h} W$ belongs to $H_0^1(\Sigma)$, we conclude

$$B(\delta_h W, \delta_h W) = O(\|\delta_h W\|_1).$$

Using Poincaré's inequality (5.19a) of Chapter 4 it follows that $|\delta_h W|_1 = O(1)$, or that there exists a constant M such that

$$\|\delta_h W\|_1 \leqslant M \tag{3.8e}$$

independent of h for $h \neq 0$.

Relation (3.8e) implies that for any $i = 1, \ldots, n$

$$\|\delta_h d_i W\|_0 \leqslant M \quad \text{for } h \neq 0. \tag{3.9a}$$

This implies (see problem 4(c), p. 202) that there exists a sequence $h_1, h_2, \ldots$ tending to zero and an element V in $H_0^0(\Sigma)$ such that

$$\lim_{j \to \infty} (\psi, \delta_{h_j} d_i W) = (\psi, V)$$

for all $\psi \in C_0^\infty(\Sigma)$. But W has continuous second derivatives in Σ, so that

$$\lim_{j\to\infty} \delta_{h_j} d_i W = d_k d_i W$$

uniformly in the support of ψ. Hence

$$(\psi, d_k d_i W) = (\psi, V)$$

for all ψ in $C_0^\infty(\Sigma)$. This means that $d_k d_i W$ can be identified in Σ with a function $V \in H_0^0(\Sigma)$. Thus $d_k d_i W$ is square integrable over Σ for $u = 1,\ldots,n$ and $k = 1,\ldots,n-1$. The same holds then also for the missing second derivative $d_n^2 W$ which can be expressed in terms of the others from the differential equation $\Delta u = -w$. That equation transformed to y-coordinates becomes a linear second order equation; here the coefficient of $d_n^2 u$ is

$$\sum_i (D_i y_n)^2$$

which does not vanish, since the Jacobian of the mapping $y = f(x)$ is different from zero. Assuming that $\zeta(x) = 1$ near x^1, we can express $d_n^2 W$ in terms of quantities that are square integrable for $y_n > 0$ near the origin of y-space. This shows that all second derivatives of $W(y)$ are square integrable near $y = 0$, and hence that $u(x)$ has square integrable second derivatives near the arbitrary point x^1 on $\partial\Omega$.

7 Parabolic Equations

1. The Heat Equation*

(a) *The initial-value problem*

The equation of heat for a function $u = u(x_1, \ldots, x_n, t) = u(x,t)$ has the form

$$u_t = k\Delta u \tag{1.1}$$

with a positive constant *conductivity coefficient* k. For $n=3$ the equation is satisfied by the temperature in a heat-conducting medium. For $n=1$ it holds for the temperature distribution in a heat-conducting insulated wire. The same type of equation occurs in the description of diffusion processes. Applying a suitable linear substitution on x, t we transform (1.1) into

$$u_t = \Delta u \tag{1.2}$$

which will be used in the discussion to follow.

Equation (1.2) is parabolic. A characteristic hypersurface $\phi(x,t) = t - \psi(x) = 0$ has to satisfy the degenerate quadratic condition (see (2.24) of Chapter 3)

$$\sum_{k=1}^{n} \psi_{x_i}^2 = 0. \tag{1.3}$$

Thus the only characteristic surfaces are the planes $t = \text{const}$. Unlike the usual equations in mechanics (including the wave equation), equation (1.1) is not preserved when we replace t by $-t$. This indicates that the heat equation describes "irreversible" processes and makes a distinction between past and future (the "arrow of time"). More generally, (1.2) is preserved under linear substitutions $x' = ax$, $t' = a^2 t$, the same ones that

*([35], [29])

leave the expression $|x|^2/t$ invariant. Thus it is not surprising that the combination $|x|^2/t$ occurs frequently in connection with equation (1.2).

Important information is obtained by considering the exponential solutions

$$u = e^{i(\lambda t + x\cdot\xi)} \qquad (x\cdot\xi = x_1\xi_1 + \dots + x_n\xi_n)$$

with constant λ and $\xi = (\xi_1, \dots, \xi_n)$. Substitution in (1.2) yields the relation $i\lambda = -|\xi|^2$, and hence

$$u(x,t) = e^{ix\cdot\xi - |\xi|^2 t} \qquad (|\xi|^2 = \xi_1^2 + \dots + \xi_n^2). \tag{1.4}$$

For each fixed $t \geqslant 0$ equation (1.4) describes a plane wave function, constant on the planes $x\cdot\xi = \text{const.}$ with unit normal $\xi/|\xi|$, and repeating itself when x is replaced by $x + 2\pi\xi/|\xi|^2$. Thus the waves in (1.4) have *wave length*

$$L = \frac{2\pi}{|\xi|} \tag{1.5}$$

and *amplitude*

$$A = |u(x,t)| = e^{-|\xi|^2 t} = e^{-4\pi^2 t/L^2}. \tag{1.6}$$

The solutions (1.4) decay exponentially with time, except in the trivial case $\xi = 0$, $u = 1 = \text{const.}$

We have tacitly assumed in (1.4) that the vector ξ is real, so that u for fixed $t > 0$ is bounded uniformly for $x \in \mathbb{R}^n$. If we only consider our solutions in the half space $x_1 \geqslant 0$, it is natural to require boundedness just in that half space, which leads to the condition that $\xi_2, \dots, \xi_n$ should be real and $\operatorname{Im}\xi_1 \geqslant 0$. For example for a real positive λ

$$u = \exp(i\lambda t + i\xi_1 x_1) \tag{1.7a}$$

is such a solution when

$$\xi_1 = \sqrt{-i\lambda} = (-1+i)\sqrt{\frac{\lambda}{2}}\ . \tag{1.7b}$$

The corresponding "physical" real solution would be

$$v = \operatorname{Re} u = \cos\left(\lambda t - \sqrt{\frac{\lambda}{2}}\, x_1\right)\exp\left(-\sqrt{\frac{\lambda}{2}}\, x_1\right). \tag{1.7c}$$

For $n = 3$ we can interpret v as the temperature below ground in a flat earth represented by the half space $x_1 > 0$. Here v has boundary values $\cos\lambda t$ for $x_1 = 0$, which oscillate periodically with frequency λ and amplitude 1. The resulting temperature at depth x_1 still oscillates with frequency λ but with a *phase lag* $\sqrt{(1/2\lambda)}\, x_1$ and with an amplitude

$\exp(-\sqrt{(\lambda/2)}\ x_1)$ that decays exponentially with depth. Thus at the depth

$$x_1=\sqrt{\frac{2}{\lambda}}\ \log 2 \tag{1.7d}$$

the amplitude will have decreased to 1/2 its surface value. This "half-depth" is inversely proportional to $\sqrt{\lambda}$ or proportional to $\sqrt{P}$, where $P=2\pi/\lambda$ is the time period of v. Thus yearly surface variations of temperature can be expected to penetrate $\sqrt{365}$ times $=19$ times as deep as daily variations with the same amplitude.

The "pure" initial-value problem for the heat equation consists in finding a solution $u(x,t)$ of

$$u_t-\Delta u=0 \quad \text{for } x\in\mathbb{R}^n,\ t>0 \tag{1.8a}$$

$$u=f(x) \quad \text{for } x\in\mathbb{R}^n,\ t=0, \tag{1.8b}$$

where we require $u\in C^2$ for $x\in\mathbb{R}^n$, $t>0$, and $u\in C^0$ for $x\in\mathbb{R}^n$, $t\geqslant 0$. A *formal* solution is obtained immediately by Fourier transformation. Writing

$$f(x)=(2\pi)^{-n/2}\int e^{ix\cdot\xi}\hat{f}(\xi)\,d\xi, \tag{1.9a}$$

we would expect on the basis of (1.4) that

$$u(x,t)=(2\pi)^{-n/2}\int e^{ix\cdot\xi-|\xi|^2t}\hat{f}(\xi)\,d\xi \tag{1.9b}$$

satisfies (1.8a, b). Substituting

$$\hat{f}(\xi)=(2\pi)^{-n/2}\int e^{-iy\cdot\xi}f(y)\,dy$$

from Fourier's formula (2.15) of Chapter 5 and interchanging the integrations leads to

$$u(x,t)=\int K(x,y,t)f(y)\,dy, \tag{1.10a}$$

where

$$K(x,y,t)=(2\pi)^{-n}\int e^{i(x-y)\cdot\xi-|\xi|^2t}\,d\xi. \tag{1.10b}$$

The integral for K is easily evaluated by completing the squares in the exponent; introducing a new variable of integration η by

$$\xi=\frac{i(x-y)}{2t}+\frac{1}{\sqrt{t}}\eta,$$

we find that

$$K(x,y,t)=(2\pi)^{-n}\int e^{-|x-y|^2/4t}e^{-|\eta|^2}t^{-n/2}\,d\eta.$$

Using the well-known formula

$$\int e^{-|\eta|^2}\,d\eta = \left(\int_{-\infty}^{\infty} e^{-s^2}\,ds\right)^n = \pi^{n/2}, \tag{1.10c}$$

we arrive at

$$K(x,y,t) = (4\pi t)^{-n/2} e^{-|x-y|^2/4t}. \tag{1.10d}$$

We shall verify directly that the u from (1.10a, d) satisfies (1.8a, b), without trying to justify the steps in the formal derivation. In this we follow the model of the proof of Poisson's formula (Chapter 4, (3.9)).

Theorem. *Let $f(x)$ be continuous and bounded for $x \in \mathbb{R}^n$. Then*

$$\begin{aligned} u(x,t) &= \int K(x,y,t) f(y)\,dy \\ &= (4\pi t)^{-n/2} \int e^{-|x-y|^2/4t} f(y)\,dy \end{aligned} \tag{1.11}$$

belongs to C^∞ for $x \in \mathbb{R}^n$, $t > 0$, and satisfies $u_t = \Delta u$ for $t > 0$. Moreover u has the initial values f, in the sense that when we extend u by $u(x,0) = f(x)$ to $t = 0$, then u is continuous for $x \in \mathbb{R}^n$, $t \geqslant 0$.

The proof follows from basic properties of the kernel K:

(a) $K(x,y,t) \in C^\infty$ for $x \in \mathbb{R}^n, y \in \mathbb{R}^n, t > 0$. (1.12a)

(b) $\left(\dfrac{\partial}{\partial t} - \Delta_x\right) K(x,y,t) = 0$ for $t > 0$ (1.12b)

(c) $K(x,y,t) > 0$ for $t > 0$ (1.12c)

(d) $\int K(x,y,t)\,dy = 1$ for $x \in \mathbb{R}^n$, $t > 0$ (1.12d)

(e) For any $\delta > 0$ we have

$$\lim_{\substack{t \to 0 \\ t > 0}} \int_{|y-x| > \delta} K(x,y,t)\,dy = 0 \tag{1.12e}$$

uniformly for $x \in \mathbb{R}^n$.

Here (a), (c) are trivial from (1.10d), and so is (b) from (1.10b). Also, by (1.10d) substituting $y = x + (4t)^{1/2}\eta$

$$\int_{|y-x| > \delta} K(x,y,t)\,dy = \pi^{-n/2} \int_{|\eta| > \delta/\sqrt{4t}} e^{-|\eta|^2}\,d\eta. \tag{1.13}$$

When $\delta = 0$ this implies (d) by (1.10c), and implies (e) for $\delta > 0$.

Clearly these properties of K show that the u defined by (1.11) belongs to C^∞ and satisfies $u_t = \Delta u$ for $t > 0$. To prove that the extended u is continuous, at $t = 0$ we have to show that $u(x,t) \to f(\xi)$ for $x \to \xi$, $t \to 0$. For

$\varepsilon>0$ we can find a δ such that $|f(y)-f(\xi)|<\varepsilon$ for $|y-\xi|<2\delta$. Let $M=\sup|f(y)|$. Then for $|x-\xi|<\delta$

$$
\begin{aligned}
|u(x,t)-f(\xi)| &= \left|\int K(x,y,t)(f(y)-f(\xi))\,dy\right| \\
&\leqslant \int_{|y-x|<\delta} K(x,y,t)|f(y)-f(\xi)|\,dy + \int_{|y-x|>\delta} K(x,y,t)|f(y)-f(\xi)|\,dy \\
&\leqslant \int_{|y-\xi|<2\delta} K(x,y,t)|f(y)-f(\xi)|\,dy + 2M\int_{|y-x|>\delta} K(x,y,t)\,dy \\
&\leqslant \varepsilon\int K(x,y,t)\,dy + 2M\int_{|y-x|>\delta} K(x,y,t)\,dy < 2\varepsilon
\end{aligned}
$$

for t sufficiently small.

By the same type of argument one proves more generally that if $f(x)$ is measurable and satisfies an inequality

$$|f(x)| \leqslant Me^{a|x|^2} \tag{1.14}$$

for all x with fixed constants a, M then formula (1.11) defines a solution $u(x,t)$ of $u_t=\Delta u$ of class C^∞ for $x\in\mathbb{R}^n$ and $0<t<1/4a$. Here $u(x,t)\to f(\xi)$ for $x\to\xi$ and $t\to 0$ at every point ξ of continuity of f.

We point out some important features of the special solution (1.11), (not the only one), of the initial-value problem (1.8a, b). We observe that $u(x,t)$ for $t>0$ depends on the values of f at *all* points. Equivalently the values of f near one ξ a moment later affect the value of $u(x,t)$ at all x, though only imperceptibly at large distances. Thus effects here travel with *infinite* speed, indicating some limitation on the strict applicability of the heat equation to physical phenomena. We notice from (1.12c,d) that for bounded f

$$u(x,t) \leqslant \left(\int K(x,y,t)\,dy\right)\left(\sup_z f(z)\right) = \sup_z f(z),$$

and more generally that u satisfies the "maximum principle"

$$\inf_z f(z) \leqslant u(x,t) \leqslant \sup_z f(z) \quad \text{for } x\in\mathbb{R}^n,\ t>0. \tag{1.15}$$

Here for continuous bounded f the equals sign can hold only when f is constant.

The function u in (1.11) belongs to C^∞ for any $t>0$, even if the initial values f of u are only continuous, or even have jump discontinuities. More is true. For bounded continuous f the function $u(x,t)$ can be continued as an analytic function to all complex x,t with $\operatorname{Re} t>0$. We only have to replace $|x-y|^2$ in the exponent in formula (1.10d) for K by the *algebraic* expression $(x-y)\cdot(x-y)$. Then for complex $x=\xi+i\eta$, $t=\sigma+i\tau$, with

ξ,η,y,σ,τ real, $K(x,y,t)$ is analytic in x,t for $t\neq 0$, and moreover for $\sigma>0$

$$|K(x,y,t)|=(4\pi)^{-n/2}(\sigma^2+\tau^2)^{-n/4}\exp\left(-\operatorname{Re}\frac{(\xi-y+i\eta)\cdot(\xi-y+i\eta)}{4(\sigma+i\tau)}\right)$$

$$=(4\pi)^{-n/2}(\sigma^2+\tau^2)^{-n/4}\exp\left(\frac{|\eta|^2}{4\sigma}-\frac{|\sigma(\xi-y)+\tau\eta|^2}{4(\sigma^2+\tau^2)\sigma}\right)$$

$$=\left(1+\frac{\tau^2}{\sigma^2}\right)^{n/4}e^{|\eta|^2/4\sigma}K\left(\xi+\frac{\tau}{\sigma}\eta,y,\sigma+\frac{\tau^2}{\sigma}\right). \tag{1.16}$$

It follows then for bounded continuous f from (1.12d) that

$$|u(\xi+i\eta,\sigma+i\tau)|\leqslant\int|K(\xi+i\eta,y,\sigma+i\tau)|\,dy\sup_{z\in R^n}|f(z)|$$

$$=\left(1+\frac{\tau^2}{\sigma^2}\right)^{n/4}e^{|\eta|^2/4\sigma}\sup_{z\in R^n}|f(z)| \tag{1.17}$$

for $\sigma>0$. More precisely for bounded continuous f the function $u(x,t)$ and its first derivatives are represented by absolutely convergent integrals for complex x,t with $\operatorname{Re} t>0$. The analyticity of u follows (see p. 73).

Thus after an infinitesimal lapse of time a temperature distribution u (at least if represented by (1.11)) is perfectly *smooth*, though of course, by continuity, it will *approximate* the changes in the initial data. This smoothness of the future has as its counterpart that the past is likely to be rougher, as we shall see more precisely below. Values $u(x,0)=f(x)$ that are not analytic in x cannot have originated at all by conduction from a temperature distribution in the past.

Formula (1.11) represents only one out of infinitely many solutions of the initial-value problem (1.8a, b). The solution is not unique without further conditions on u, as is shown by examples of solutions $u\in C^\infty(\mathbb{R}^{n+1})$ of $u_t=\Delta u$, which vanish identically for $t<0$ but not for $t>0$.* Following Tychonoff we construct such u for the case $n=1$. They are obtained by formally solving the partial differential equation $u_t=u_{xx}$ for prescribed Cauchy data on the t-axis:

$$u=g(t),\qquad u_x=0\quad\text{for } x=0. \tag{1.18}$$

Writing u as a power series

$$u=\sum_{j=0}^{\infty}g_j(t)x^j$$

* Conditions on u that imply uniqueness for the initial-value problem can take the form of prescribing the behavior of $u(x,t)$ for large $|x|$, as shown below. We also mention a result of Widder (see p. 222) that there is at most one solution u which is nonnegative for $t\geqslant 0$ and all x. The assumption $u\geqslant 0$ is reasonable when u is the absolute temperature.

we find by substitution into $u_t = u_{xx}$ and comparison of coefficients of powers of x that

$$g_0 = g, \qquad g_1 = 0, \qquad g_j' = (j+2)(j+1)\, g_{j+2}.$$

This leads to the formal solutions

$$u(x,t) = \sum_{k=0}^{\infty} \frac{g^{(k)}(t)}{(2k)!} x^{2k}. \tag{1.19}$$

They will be actual solutions if the power series can be shown to converge sufficiently well. Choose now for some real $\alpha > 1$ the $g(t)$ defined by

$$g(t) = \begin{cases} \exp[-t^{-\alpha}] & \text{for } t > 0 \\ 0 & \text{for } t \leqslant 0. \end{cases} \tag{1.20}$$

(See problem 3, p. 73). There exists a $\theta = \theta(\alpha)$ with $0 < \theta$ such that for all $t > 0$

$$|g^{(k)}(t)| < \frac{k!}{(\theta t)^k} \exp\left[-\frac{1}{2} t^{-\alpha}\right]. \tag{1.23}$$

Since $k!/(2k)! < 1/k!$ we observe that for real $t > 0$ and any complex x

$$\begin{aligned} \sum_{k=0}^{\infty} \left| \frac{g^{(k)}(t)}{(2k)!} x^{2k} \right| &\leqslant \sum_{k=0}^{\infty} \frac{|x|^{2k}}{k!(\theta t)^k} \exp\left[-\frac{1}{2} t^{-\alpha}\right] \\ &= \exp\left[\frac{1}{t}\left(\frac{|x|^2}{\theta} - \frac{1}{2} t^{1-\alpha}\right)\right]. \end{aligned} \tag{1.24}$$

Thus, by comparison, the series (1.19) for u converges for real $t > 0$ and complex x, and, of course, trivially also for $t \leqslant 0$. Formula (1.24) shows that $\lim_{t \to 0} u(x,t) = 0$, uniformly in x for bounded complex x. The series (1.19) as a power series in x is majorised (see p. 67) by the power series for

$$U(x,t) = \begin{cases} \exp\left[\frac{1}{t}\left(\frac{x^2}{\theta} - \frac{1}{2} t^{1-\alpha}\right)\right] & \text{for } t > 0 \\ 0 & \text{for } t \leqslant 0. \end{cases}$$

Since $U(x,t)$ is bounded uniformly for bounded complex x and all real t, the series (1.19) converges uniformly in x, t for bounded x and real t, and the same holds for the series obtained by term by term x-differentiations. In particular the series

$$\sum_{k=2}^{\infty} \frac{g^{(k)}(t)}{(2k-2)!} x^{2k-2} = \sum_{k=0}^{\infty} \frac{g^{(k+1)}(t)}{(2k)!} x^{2k}$$

converges uniformly. Since this series is also obtained by formal differentiation of u with respect to t we find that $u_t = u_{xx}$. More generally the relation $(\partial/\partial t)^k u = (\partial/\partial x)^{2k} u$ holds, which implies that $u \in C^\infty(\mathbb{R}^{n+1})$. We observe that u is an entire analytic function of x for any real t, but is not analytic in t, since $u(0,t)$ vanishes for $t \leqslant 0$ but not for $t > 0$.

Problems

1. (a) Use formula (1.11) for $n = 1$ to prove *Weierstrass's approximation theorem*: A function $f(x)$ continuous on a closed interval $[a,b]$ can be approximated uniformly by polynomials. [Hint: Define $f(x) = f(b)$ for $x > b, f(x) = f(a)$ for $x < a$. Then $u(x,t) \to f(x)$ for $t \to 0$ uniformly for $a \leqslant x \leqslant b$, since u is continuous for $t \geqslant 0$. Approximate $K(x,y,t)$ by its truncated power series with respect to $x - y$.]
   (b) Show that $f \in C_0^m(\mathbb{R}^n)$ can be approximated by polynomials uniformly with its derivatives of order $\leqslant m$ in any bounded set containing the support of u.

2. Let $f(x)$ have uniformly bounded derivatives of orders $\leqslant s$. Show that the $u(x,t)$ given by (1.11) is of class C^s for $t \geqslant 0$ and all x. [Hint: Show that
$$D^\alpha u = \int K D^\alpha f \, dy.]$$

3. Let $f(x)$ be continuous in $\mathbb{R}^n$ and satisfy (1.14). Show that the u defined by (1.11) is analytic in x,t for all complex x and complex $t = \sigma + i\tau$ with
$$\frac{\sigma}{4(\sigma^2 + \tau^2)} > a. \tag{1.25}$$

4. Let $u_1(s,t), \ldots, u_n(s,t)$ be n solutions of $u_t = u_{ss}$. Prove that
$$u(x,t) = u(x_1, \ldots, x_n, t) = \prod_{k=1}^{m} u_k(x_k, t) \tag{1.26}$$
satisfies $u_t = \Delta u$.

5. Show that for $n = 1$ the solution of (1.8a, b) with $f(x) = 1$ for $x > 0$, $f(x) = 0$ for $x < 0$ is given by
$$u(x,t) = \frac{1}{2}\left[1 + \phi\left(\frac{x}{\sqrt{4t}}\right)\right], \tag{1.27a}$$
where $\phi(s)$ is the "error function"
$$\phi(s) = \frac{2}{\sqrt{\pi}} \int_0^s e^{-t^2} dt. \tag{1.27b}$$

6. Show that for $f(x)$ continuous and of compact support we have $\lim_{t\to\infty} u(x,t) = 0$ uniformly in x for the u given by (1.11).

7. For $n = 1$ let $f(x)$ be bounded, continuous, and positive for all real x.
   (a) Show that for the u given by (1.11)
$$|u(\xi + i\eta, t)| \leqslant e^{\eta^2/4t} u(\xi, t) \tag{1.28}$$
   for real ξ, η, t with $t > 0$. [Hint: (1.16).]

(b) Show that

$$|u_x(x,t)| \leqslant \frac{e^{1/2}}{\sqrt{2t}} \sup_{|y| \leqslant \sqrt{2t}} u(x+y,t) \tag{1.29}$$

(x,y,t real, $t>0$). [Hint: Use Cauchy's expression for $u_x(x,t)$ as an integral of u over the circle of radius $\sqrt{2t}$ and center x in the complex plane.] (This gives a means to estimate the maximum possible age t of an observed heat distribution u in terms of its maximum and its gradient, assuming that it has been positive and bounded for a time t.)

8. Find all solutions $u(x,t)$ of the one-dimensional heat equation $u_t = u_{xx}$ of the form

$$u = \frac{1}{\sqrt{t}} f\left(\frac{x}{2\sqrt{t}}\right).$$

[Hint: $f(z)$ has to satisfy a linear ordinary second-order equation, of which one solution $f(z) = e^{-z^2}$ is known, from $u = K(x,0,t)$. All others can then be found by quadratures.]

9. Show that the K defined by (1.10d) satisfies

$$K(x,y,t) < \pi^{-1/2}|x-y|^{-n}(n/2e)^{n/2} \tag{1.29a}$$

for $x,y \in \mathbb{R}^n$, $t > 0$.

10. Show that

$$\lim_{t\to 0} K(x,y,t) = \delta_y \tag{1.29b}$$

in the sense of distributions, p. 92.

11. (a) Let $f(x)$ be bounded and continuous for $x \in \mathbb{R}^n$ and satisfy

$$\int |f(y)|\, dy < \infty.$$

Show that there exists a solution $u(x,t)$ of (1.8a, b) for which

$$\lim_{t\to\infty} u(x,t) = 0.$$

(b) Let $n = 1$. Show that the same conclusion holds for $f \in C^2(\mathbb{R})$ that have period 2π and satisfy

$$\int_0^{2\pi} f(y)\, dy = 0.$$

[Hint: Use Fourier series for u.]

12. (a) Let $n = 1$ and μ be a positive constant. Let $u(x,t)$ be a positive solution of class C^2 of

$$u_t = \mu u_{xx} \quad \text{for } t > 0.$$

Show that $\theta = -2\mu u_x/u$ satisfies *Burgers' equation*

$$\theta_t + \theta\theta_x = \mu\theta_{xx} \quad \text{for } t > 0. \tag{1.29c}$$

(b) For $\phi \in C_0^2(\mathbb{R})$ find a solution of (1.29c) with initial values $\theta(x,0) = \phi(x)$, for which

$$\lim_{t \to \infty} \theta(x,t) = 0. \tag{1.29d}$$

(The "viscosity term" $\mu\theta_{xx}$ in (1.29c) prevents singularities that would occur for $\mu = 0$; compare p. 17.)

(b) *Maximum principle, uniqueness, and regularity*

Let ω denote an open bounded set of $\mathbb{R}^n$. For a fixed $T > 0$ we form the cylinder Ω in $\mathbb{R}^{n+1}$ with base ω and height T:

$$\Omega = \{(x,t) | x \in \omega, 0 < t < T\}. \tag{1.30a}$$

The boundary $\partial\Omega$ consists of two disjoint portions, a "lower" boundary $\partial'\Omega$, and an "upper" one $\partial''\Omega$ (see Figure 7.1):

$$\partial'\Omega = \{(x,t) | \text{either } x \in \partial\omega, 0 \leqslant t \leqslant T \text{ or } x \in \omega, t = 0\} \tag{1.30b}$$

$$\partial''\Omega = \{(x,t) | x \in \omega, t = T\}. \tag{1.30c}$$

As in the second-order elliptic case the maximum of a solution of the heat equation in Ω is taken on $\partial\Omega$; but a more subtle distinction between the forward and backwards t-directions makes itself felt:

Theorem. *Let u be continuous in $\bar{\Omega}$ and $u_t, u_{x_i x_k}$ exist and be continuous in Ω and satisfy $u_t - \Delta u \leqslant 0$. Then*

$$\max_{\bar{\Omega}} u = \max_{\partial'\Omega} u. \tag{1.31}$$

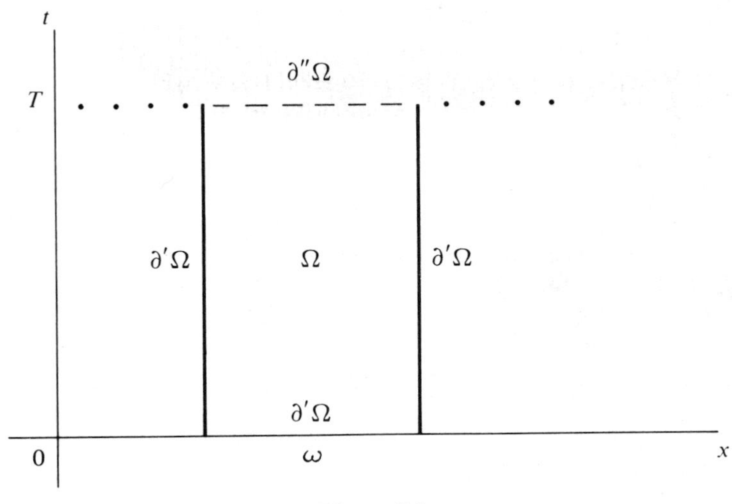

Figure 7.1

PROOF. Let at first $u_t - \Delta u < 0$ in Ω. Let Ω_ε for $0 < \varepsilon < T$ denote the set

$$\Omega_\varepsilon = \{(x,t) | x \in \omega,\ 0 < t < T - \varepsilon\}.$$

Since $u \in C^0(\bar{\Omega}_\varepsilon)$ there exists a point $(x,t) \in \bar{\Omega}_\varepsilon$ with

$$u(x,t) = \max_{\bar{\Omega}_\varepsilon} u.$$

If here $(x,t) \in \Omega_\varepsilon$ the necessary relations $u_t = 0$, $\Delta u \leqslant 0$ would contradict $u_t - \Delta u < 0$. If $(x,t) \in \partial'' \Omega_\varepsilon$ we would have

$$u_t \geqslant 0, \qquad \Delta u \leqslant 0$$

leading to the same contradiction. Thus $(x,t) \in \partial' \Omega_\varepsilon$, and

$$\max_{\bar{\Omega}_\varepsilon} u = \max_{\partial' \Omega_\varepsilon} u \leqslant \max_{\partial' \Omega} u.$$

Since every point of $\bar{\Omega}$ with $t < T$ belongs to some $\bar{\Omega}_\varepsilon$ and u is continuous in $\bar{\Omega}$, (1.31) follows. Let next $u_t - \Delta u \leqslant 0$ in Ω. Introduce

$$v(x,t) = u(x,t) - kt$$

with a constant positive k. Then $v_t - \Delta v = u_t - \Delta u - k < 0$ and

$$\max_{\bar{\Omega}} u = \max_{\bar{\Omega}} (v + kt) \leqslant \max_{\bar{\Omega}} v + kT = \max_{\partial' \Omega} v + kT \leqslant \max_{\partial' \Omega} u + kT.$$

For $k \to 0$ we obtain (1.31). □

The maximum principle immediately yields a uniqueness theorem

Theorem. *Let u be continuous in $\bar{\Omega}$ and $u_t, u_{x_i x_k}$ exist and be continuous in Ω. Then u is determined uniquely in $\bar{\Omega}$ by the value of $u_t - \Delta u$ in Ω and of u on $\partial' \Omega$.*

For the proof it is sufficient to consider the case where $u_t - \Delta u = 0$ in Ω and $u = 0$ on $\partial' \Omega$. Applying (1.31) to u and $-u$ we find that

$$\max_{\bar{\Omega}} u = \max_{\bar{\Omega}} (-u) = 0, \tag{1.32}$$

and hence that $u = 0$ in $\bar{\Omega}$.

We can extend the maximum principle and the uniqueness theorem to the case where Ω is the "slab"

$$\Omega = \{(x,t) | x \in \mathbb{R}^n, 0 < t < T\}, \tag{1.33}$$

if we assume that u satisfies a certain growth condition at infinity.

Theorem. *Let u be continuous for $x \in \mathbb{R}^n$, $0 \leqslant t \leqslant T$, and let $u_t, u_{x_i x_k}$ exist and be continuous for $x \in \mathbb{R}^n$, $0 < t < T$, and satisfy*

$$u_t - \Delta u \leqslant 0 \quad \text{for } 0 < t < T,\ x \in \mathbb{R}^n \tag{1.34a}$$

$$u(x,t) \leqslant Me^{a|x|^2} \quad \text{for } 0<t<T,\ x\in\mathbb{R}^n \tag{1.34b}$$

$$u(x,0)=f(x) \quad \text{for } x\in\mathbb{R}^n.$$

Then

$$u(x,t) \leqslant \sup_z f(z) \quad \text{for } 0\leqslant t\leqslant T,\ x\in\mathbb{R}^n. \tag{1.35}$$

It is clear that this theorem implies that the solution of the initial-value problem

$$u_t-\Delta u=0 \quad \text{for } 0<t<T \tag{1.36a}$$

$$u(x,0)=f(x) \tag{1.36b}$$

is unique provided we restrict ourselves to solutions satisfying

$$|u(x,t)| \leqslant Me^{a|x|^2} \quad \text{for } 0<t<T. \tag{1.36c}$$

This shows that for bounded continuous f formula (1.11) represents the only *bounded* solution u of (1.8a, b). Obviously the Tychonoff solution (1.19), (1.20), for which $u(x,0) = 0$, cannot satisfy an inequality of the type (1.36c). By (1.24) it does satisfy such an inequality with the constant a replaced by $1/\theta t$.

Proof of the theorem. It is sufficient to show (1.35) under the assumption that

$$4aT<1 \tag{1.37a}$$

For we can always divide the interval $0\leqslant t\leqslant T$ into equal parts, each of length $\tau<1/4a$, and conclude successively for $k=0,1,\ldots,T/\tau$ that

$$u(x,t)\leqslant \sup_y u(y,k\tau)\leqslant \sup_y u(y,0)$$

for $k\tau\leqslant t\leqslant(k+1)\tau$. Assume then (1.37a). We can find an $\varepsilon>0$ such that

$$4a(T+\varepsilon)<1. \tag{1.37b}$$

Given a fixed y we consider for constants $\mu>0$ the functions

$$\begin{aligned} v_\mu(x,t)&=u(x,t)-\mu(4\pi(T+\varepsilon-t))^{-n/2}\exp\left[|x-y|^2/4(T+\varepsilon-t)\right]\\ &=u(x,t)-\mu K(ix,iy,T+\varepsilon-t) \end{aligned} \tag{1.38}$$

defined for $0\leqslant t\leqslant T$. Since $K(x,y,t)$ as defined by (1.10d), with $|x-y|^2$ replaced by $(x-y)\cdot(x-y)$, satisfies $K_t=\Delta K$ for any complex x,y,t with $t\neq 0$, we find that

$$\frac{\partial}{\partial t}v_\mu-\Delta v_\mu=u_t-\Delta u\leqslant 0. \tag{1.39}$$

Consider the "circular" cylinder

$$\Omega=\{(x,t)\,|\,|x-y|<\rho,\ 0<t<T\} \tag{1.40}$$

of radius ρ. Then by (1.31)

$$v_\mu(y,t)\leqslant \max_{\partial'\Omega} v_\mu. \tag{1.41}$$

Here on the plane part of $\partial'\Omega$, since $\mu K > 0$,

$$v_\mu(x,0) \leqslant u(x,0) \leqslant \sup_z f(z). \tag{1.42a}$$

On the curved part $|x-y| = \rho$, $0 \leqslant t \leqslant T$ of $\partial'\Omega$ by (1.38), (1.34b), (1.37b)

$$\begin{aligned} v_\mu(x,t) &\leqslant Me^{a|x|^2} - \mu(4\pi(T+\varepsilon-t))^{-n/2}\exp\left[\rho^2/4(T+\varepsilon-t)\right] \\ &\leqslant Me^{a(|y|+\rho)^2} - \mu(4\pi(T+\varepsilon))^{-n/2}e^{\rho^2/4(T+\varepsilon)} \\ &\leqslant \sup_z f(z) \end{aligned}$$

for all sufficiently large ρ. Thus

$$\max_{\partial'\Omega} v_\mu \leqslant \sup f(z).$$

It follows from (1.41), (1.38) that

$$v_\mu(y,t) = u(y,t) - \mu(4\pi(T+\varepsilon-t))^{-n/2} \leqslant \sup f(z)$$

For $\mu \to 0$ we obtain (1.35). □

In order to derive regularity properties of a solution of the heat equation in a bounded region we make use of Green's identity, as was done for harmonic functions on p. 76. Let again Ω denote the cylindrical region (1.30a), where ω is a bounded open set in $\mathbb{R}^n$ with sufficiently regular boundary. Let $u, u_t, u_{x_i x_k}$ exist and be continuous in $\bar{\Omega}$ and satisfy $u_t - \Delta u = 0$. For an arbitrary function $v(x,t) \in C^2(\bar{\Omega})$ we find by integration by parts that

$$\begin{aligned} 0 &= \int_\Omega v(u_t - \Delta u)\,dx \\ &= -\int_\Omega u(v_t + \Delta v)\,dx + \int_{\substack{x\in\omega \\ t=T}} vu\,dx - \int_{\substack{x\in\omega \\ t=0}} vu\,dx \\ &\quad - \int_0^T dt \int_{x\in\partial\omega}\left(v\frac{du}{dn} - u\frac{dv}{dn}\right)dS_x. \end{aligned} \tag{1.43}$$

For a certain $\xi \in \omega$ and $\varepsilon > 0$ we choose

$$v(x,t) = K(x,\xi,T+\varepsilon-t), \tag{1.44}$$

so that $v_t + \Delta v = 0$. Then for $\varepsilon \to 0$

$$\int_{\substack{x\in\omega \\ t=T}} vu\,dx = \int_{x\in\omega} K(x,\xi,\varepsilon)u(x,T)\,dx \to u(\xi,T), \tag{1.44a}$$

since by the theorem of p. 209

$$w(\xi,\varepsilon) = \int_\omega K(\xi,x,\varepsilon)u(x,T)\,dx = \int_\omega K(x,\xi,\varepsilon)u(x,T)\,dx$$

is a solution of $w_\varepsilon - \Delta_\xi w = 0$ with initial values*

$$w(\xi,0) = u(x,T).$$

Since also $K(x,\xi,T+\varepsilon-t)$ is uniformly continuous in ε,x,t for $\varepsilon \geqslant 0$, $x \in \partial\omega$, $0 \leqslant t \leqslant T$ and for $x \in \omega$, $t=0$, we find from (1.43) that

$$u(\xi,T) = \int_\omega K(x,\xi,T)u(x,0)\,dx$$

$$+\int_0^T dt \int_{x\in\partial\omega}\left(K(x,\xi,T-t)\frac{du(x,t)}{dn} - u(x,t)\frac{dK(x,\xi,T-t)}{dn}\right)dS_x. \quad (1.45)$$

Here $K(x,\xi,s) \in C^\infty$ in x,ξ,s for $x,\xi \in \mathbb{R}^n$, $s \in \mathbb{R}$, $x \neq \xi$, if we extend K by $K(x,\xi,s) = 0$ for $s \leqslant 0$. Thus we can replace $\int_0^T dt$ by $\int_0^\infty dt$ in the last integral. It then becomes evident that $u(\xi,T) \in C^\infty$ in ξ,T for $\xi \in \omega$, $T > 0$.

We can also use (1.45) to extend $u(\xi,T)$ to complex ξ-arguments $\xi = \eta + i\zeta$ (with η,ζ real), keeping T real. The first integral in (1.45) trivially is an entire analytic function of ξ. Moreover for $0 \leqslant t < T$, $x \neq \eta$

$$K(x,\xi,T-t) = (4\pi(T-t))^{-n/2}\exp[-(x-\xi)\cdot(x-\xi)/4(T-t)]$$

is analytic in ξ and (see (1.16)) bounded in absolute value by

$$(4\pi(T-t))^{-n/2}\exp\left[\frac{|\zeta|^2-|x-\eta|^2}{4(T-t)}\right].$$

Thus $K(x,\xi,T-t)$ is bounded uniformly for complex $\xi = \eta + i\zeta$ as long as $|x-\eta|^2 - |\zeta|^2$ is bounded below by a positive constant. The same holds for dK/dn. In the second integral we first extend the t-integration from 0 to $T - \varepsilon$ and then let $\varepsilon \to 0$. Since (see p. 71) sequences of analytic functions which converge uniformly in a complex region have analytic limits, it follows that $u(\xi,t)$ is analytic in ξ, as long as $|x - \eta|^2 - |\zeta|^2 > 0$ for all $x \in \partial\omega$. This is certainly the case for complex ξ near a real point of ω.

It follows that $u(\xi,T)$ is real analytic for $\xi \in \omega$. More precisely $u(\xi,T)$ is analytic for those complex ξ for which $|\mathrm{Im}\,\xi|$ is less than the distance of $\mathrm{Re}\,\xi$ from $\partial\omega$. Hence a solution $u(x,t)$ of $u_t - \Delta u = 0$ is real analytic in x in any open set of $\mathbb{R}^{n+1}$ where u_t and the $u_{x_ix_k}$ are continuous. Moreover $u(x,t)$ will be an entire function of x if defined for all real x and for t restricted to an open interval.

As observed earlier analyticity of $u(x,t)$ with respect to t cannot be expected. This fits in with the idea that the future of a heat distribution does not depend exclusively on the past, but also on outside influences that cannot be predicted.

*The fact that x is integrated only over the region ω instead over all of $\mathbb{R}^n$ does not change the proof given on p. 210, as long as ξ is a fixed point of ω.

PROBLEM

Let u be a solution of the one-dimensional heat equation $u_t = u_{xx}$ in an open subset Ω of the xt-plane. Show that at a point of Ω there exist constants A, M such that

$$\left| \frac{\partial^k u}{\partial t^k} \right| \leqslant AM^k (2k)! \tag{1.46}$$

for all nonnegative integers k. This implies that $u(x,t)$ as a function of t is of Gevrey class 2, see p. 73. [Hint: Use that u is analytic in x.]

(c) *A mixed problem*

For $n=1$ let $u(x,t)$ be a solution of $u_t - u_{xx} = 0$ in a half strip

$$0 < x < L \qquad 0 < t. \tag{1.47}$$

We seek the u satisfying the boundary conditions

$$u(0,t) = u(L,t) = 0 \quad \text{for } t > 0 \tag{1.48a}$$

and initial condition

$$u(x,0) = f(x) \quad \text{for } 0 < x < L. \tag{1.48b}$$

Here u might represent the temperature in an insulated rod with the ends held at a constant temperature. This problem could be solved by Fourier expansion with respect to x, as was done for the wave equation in Chapter 2. A solution in closed form is obtained by *reflection*. We continue $f(x)$ to all x so that $f(x)$ and $f(L - x)$ are odd functions of x:

$$f(x) = -f(-x), \qquad f(x) = -f(2L - x). \tag{1.49}$$

We then solve the pure initial-value problem (1.8a, b) with the extended* function f by formula (1.11). The resulting $u(x,t)$ satisfies $u_t - u_{xx} = 0$ and the initial condition (1.48b). It also satisfies (1.48a) since $u(x,t) + u(-x,t)$ and $u(x,t) + u(2L - x, t)$ are again bounded solutions of the heat equation with initial values 0, and hence vanish identically by the uniqueness theorem.

Let $\phi(x)$ for $x \in \mathbb{R}$ be defined by

$$\phi(x) = \begin{cases} f(x) & \text{for } 0 < x < L \\ 0 & \text{for } x < 0 \text{ or } x > L. \end{cases}$$

Then the extended f satisfying (1.49) is given by

$$f(x) = \sum_{n=-\infty}^{\infty} (\phi(2nL + x) - \phi(2nL - x)). \tag{1.50}$$

*The extended f generally will have harmless jump discontinuities at the points $x = mL$, unless f satisfies the consistency conditions $f(0) = f(L) = 0$.

The corresponding solution (1.11) is

$$u(x,t)=\int_{-\infty}^{\infty}K(x,y,t)f(y)\,dy$$
$$=\int_{-\infty}^{\infty}K(x,y,t)\sum_{n=-\infty}^{\infty}(\phi(2nL+y)-\phi(2nL-y))\,dy.$$

Substituting $2nL+y=\xi$, respectively $2nL-y=\xi$, we obtain

$$u(x,t)=\int_{\infty}^{\infty}\phi(\xi)\sum_{n=-\infty}^{\infty}(K(x,\xi-2nL,t)-K(x,2nL-\xi,t))\,d\xi$$
$$=\int_0^L G(x,\xi,t)f(\xi)\,d\xi, \tag{1.51}$$

where by (1.10d)

$$G(x,\xi,t)=\sum_{n=-\infty}^{\infty}(K(x,\xi-2nL,t)-K(x,2nL-\xi,t)$$
$$=\frac{1}{\sqrt{4\pi t}}\sum_{n=-\infty}^{\infty}\left(e^{-(x-\xi+2nL)^2/4t}-e^{-(x+\xi-2nL)^2/4t}\right).$$

G can be expressed in terms of the classical theta-function

$$\vartheta_3(z,\tau)=\frac{1}{\sqrt{-i\tau}}\sum_{n=-\infty}^{\infty}\exp\left[-i\pi(z+n)^2/\tau\right]. \tag{1.52}$$

We find

$$G(x,\xi,t)=\frac{1}{2L}\left[\vartheta_3\left(\frac{x-\xi}{2L},\frac{i\pi t}{L^2}\right)-\vartheta_3\left(\frac{x+\xi}{2L},\frac{i\pi t}{L^2}\right)\right]. \tag{1.53}$$

Problems

1. Find the solution of the one-dimensional heat equation $u_t=u_{xx}$ with boundary conditions (1.48a,b) by expanding u into a Fourier sine series with respect to x. Show that interchanging integrations we are led to the representation

$$u(x,t)=\int_0^L G(x,\xi,t)f(\xi)\,d\xi$$

with

$$G(x,\xi,t)=-\frac{1}{\sqrt{4\pi t}}\vartheta_3\left(\frac{i(x+\xi)L}{2\pi t},\frac{iL^2}{\pi t}\right)e^{-(x+\xi)^2/4t}$$
$$+\frac{1}{\sqrt{4\pi t}}\vartheta_3\left(\frac{i(x-\xi)L}{2\pi t},\frac{iL^2}{\pi t}\right)e^{-(x-\xi)^2/4t}. \tag{1.54}$$

Show that the identity between the expressions (1.53), (1.54) is equivalent to the functional equation satisfied by the ϑ_3-function:

$$\vartheta_3\left(\frac{z}{\tau},-\frac{1}{\tau}\right)=\sqrt{\frac{\tau}{i}}\,e^{i\pi z^2/\tau}\vartheta_3(z,\tau). \tag{1.55}$$

2. Find a solution $u(x,t)$ of the one-dimensional heat equation $u_t - u_{xx} = 0$ in the quadrant $x > 0$, $t > 0$ satisfying the conditions

$$u(x,0) = 0, \qquad u(0,t) = h(t). \tag{1.56a}$$

[Hint: Apply formula (1.43) to the quadrant using

$$v = K(x, \xi, T + \varepsilon - t) - K(x, -\xi, T + \varepsilon - t). \tag{1.56b}$$

For $\varepsilon \to 0$, using (1.44a) we arrive at

$$u(\xi, T) = \int_0^T \frac{\xi}{\sqrt{4\pi}\,(T-t)^{3/2}} e^{-\xi^2/4(T-t)} h(t)\, dt.\Bigg] \tag{1.56c}$$

(d) *Non-negative solutions*

We saw on p. 211 that initial data alone do not determine the solution of the heat equation uniquely. Some additional information on u is needed. This can take the form of prescribed boundary data, or, in the absence of a boundary, of a restriction on the growth of $|u|$ for large $|x|$. It is remarkable that even a *unilateral* condition, a lower bound for u, can ensure uniqueness. The basic result here, due to D. V. Widder, concerns non-negative solutions.*

Theorem. *Let $u(x,t)$ be defined and continuous for $x \in \mathbb{R}$, $0 \leqslant t < T$, and let u_t, u_x, u_{xx} exist and be continuous for $x \in \mathbb{R}$, $0 < t < T$. Assume that*

$$u_t - u_{xx} = 0 \quad \text{for } x \in \mathbb{R}, \quad 0 < t < T \tag{1.57a}$$

$$u(x,0) = f(x) \tag{1.57b}$$

$$u(x,t) \geqslant 0 \quad \text{for } x \in \mathbb{R}, \quad 0 < t < T. \tag{1.57c}$$

Then $u(x,t)$ is determined uniquely for $x \in \mathbb{R}$, $0 < t < T$, is real analytic, and is represented by

$$u(x,t) = \int K(x,y,t) f(y)\, dy. \tag{1.57d}$$

Proof. We first derive an inequality between the two sides of (1.57d). For $a > 1$ define the cut-off function $\zeta^a(x)$ by

$$\zeta^a(x) = 1 \quad \text{for } |x| \leqslant a - 1; \qquad \zeta^a(x) = 0 \quad \text{for } |x| \geqslant a;$$

$$\zeta^a(x) = a - |x| \quad \text{for } a - 1 < |x| < a.$$

Consider the expression

$$v^a(x,t) = \int K(x,y,t) \zeta^a(y) f(y)\, dy. \tag{1.58a}$$

Since f and ζ^a are continuous, and $\zeta^a f$ has compact support, we know that

* e.g. temperatures measured on the Kelvin scale.

$v^a(x,t)$ is continuous for $x \in \mathbb{R}$, $0 \leqslant t < T$, and satisfies

$$v_t^a - v_{xx}^a = 0 \quad \text{for } x \in \mathbb{R}, \quad 0 < t < T \tag{1.58b}$$

$$v^a(x,0) = \zeta^a(x) f(x) \tag{1.58c}$$

(see p. 209). Let M be the maximum of $f(x)$ for $|x| \leqslant a$. Since ζ^a and f are non-negative and $\zeta^a \leqslant 1$, we have (see (1.29a)) for $|x| > a$

$$\begin{aligned} 0 \leqslant v^a(x,t) &\leqslant M \int_{-a}^{a} K(x,y,t)\,dy \\ &\leqslant \frac{2Ma}{\sqrt{2\pi e}} \frac{1}{|x| - a}. \end{aligned}$$

Let $\varepsilon > 0$ and let

$$\rho > a + \frac{2Ma}{\varepsilon\sqrt{2\pi e}}.$$

Then

$$\begin{aligned} &v^a(x,t) < \varepsilon \leqslant \varepsilon + u(x,t) \quad \text{for } |x| = \rho, \quad 0 < t < T \\ &v^a(x,0) \leqslant f(x) \leqslant \varepsilon + u(x,0) \quad \text{for } |x| \leqslant \rho. \end{aligned}$$

By the maximum principle

$$v^a(x,t) \leqslant \varepsilon + u(x,t)$$

for $|x| < \rho$, $0 \leqslant t < T$. Letting $\rho \to \infty$ we find the same inequality for all $x \in \mathbb{R}$, $0 \leqslant t < T$. It follows for $\varepsilon \to 0$ that

$$v^a(x,t) \leqslant u(x,t) \quad \text{for } x \in \mathbb{R}, \quad 0 \leqslant t < T. \tag{1.58d}$$

Since ζ^a and hence v^a are non-decreasing bounded functions of a we find from (1.58d) that

$$v(x,t) = \lim_{a \to \infty} v^a(x,t) = \int K(x,y,t) f(y)\,dy \tag{1.58e}$$

exists for $x \in \mathbb{R}$, $0 < t < T$ and that

$$0 \leqslant v(x,t) \leqslant u(x,t). \tag{1.58f}$$

Regularity of $v(x,t)$ can be deduced easily by considering $v^a(x,t)$ for complex arguments $x = \xi + i\eta$, $t = \sigma + i\tau$ where

$$0 < \sigma; \qquad \sigma + \frac{\tau^2}{\sigma} < T. \tag{1.59a}$$

It is clear that $v^a(x, t)$ is analytic in x, t in the set (1.59a) (see p. 73). Moreover, from (1.16), (1.58d) we have the estimate

$$|v^a(x,t)| \leqslant \left(1 + \frac{\tau^2}{\sigma^2}\right)^{1/4} e^{\eta^2/4\sigma} v^a\left(\xi + \frac{\tau}{\sigma}\eta, \sigma + \frac{\tau^2}{\sigma}\right)$$

$$\leqslant \left(1 + \frac{\tau^2}{\sigma^2}\right)^{1/4} e^{\eta^2/4\sigma} u\left(\xi + \frac{\tau}{\sigma}\eta, \sigma + \frac{\tau^2}{\sigma}\right).$$

Let Σ be a compact subset of the (x,t) in $\mathbb{C}^2$ satisfying (1.59a). The $v^a(x,t)$ and their first derivatives (see p. 73) are bounded in Σ independently of a. There exists then a sequence of values a tending to infinity for which the $v^a(x,t)$ converge to a function $v(x,t)$ analytic in Σ, and also corresponding derivatives converge. For real (x,t) with $0 < t < T$ the function $v(x,t)$ necessarily agrees with the one given by (1.58e). It follows that $v(x,t)$ for real (x,t) with $0 < t < T$ is real analytic, and moreover satisfies $v_t - x_{xx} = 0$, because of (1.58b). Finally the inequalities

$$v^a(x,t) \leqslant v(x,t) \leqslant u(x,t) \tag{1.59b}$$

combined with (1.57b), (1.58c) imply that $v(x,t)$ is continuous for $0 \leqslant t < T$ (even at $t = 0$!) and that $v(x,0) = f(x)$.

The function $w = u - v$ is continuous for $x \in \mathbb{R}$, $0 \leqslant t < T$, and $w_t, w_x w_{xx}$ exist and are continuous for $x \in \mathbb{R}$, $0 < t < T$. Moreover,

$$w_t - w_{xx} = 0 \quad \text{for } x \in \mathbb{R}, \quad 0 < t < T \tag{1.60a}$$

$$w(x,0) = 0 \tag{1.60b}$$

$$w(x,t) \geqslant 0 \quad \text{for } x \in \mathbb{R}, \quad 0 \leqslant t < T. \tag{1.60c}$$

It only remains to prove that w vanishes identically; we have reduced the theorem to the case where $f = 0$. We introduce

$$W(x,t) = \int_0^t w(x,s)\, ds. \tag{1.61a}$$

We shall show that W has the same properties as w, and, moreover, is convex in x, from which we conclude easily that $W = 0$. To begin with we know that $W(x,t)$ and $W_t = w$ are non-negative and continuous for $x \in \mathbb{R}$ and $0 \leqslant t < T$, and that $W(x,0) = 0$. The existence of W_x, W_{xx} is not obvious from (1.61a), since $w_x(x,s)$ need not exist for $s = 0$. We introduce the difference operator δ_h by

$$\delta_h w(x,t) = \frac{1}{h}(w(x + h,t) - w(x,t)) \tag{1.61b}$$

for $h \neq 0$. Then (see (3.8b), p. 204) for $h \neq 0$, $H \neq 0$

$$\delta_h w(x,t) = \int_0^1 w_x(x + zh, t)\,dz \tag{1.61c}$$

$$\begin{aligned}\delta_H w(x,t) - \delta_h w(x,t) &= \int_0^1 (w_x(x + zH, t) - w_x(x + zh, t))\,dz \\ &= \int_0^1 z\,dz \int_h^H w_{xx}(x + zp, t)\,dp \\ &= \int_0^1 z\,dz \int_h^H w_t(x + zp, t)\,dp.\end{aligned}$$

By integration with respect to t it follows for any $\varepsilon > 0$ that

$$\begin{aligned}\delta_H W(x,t) - \delta_h W(x,t) = \delta_H W(x,\varepsilon) - \delta_h W(x,\varepsilon) \\ + \int_0^1 z\,dz \int_h^H (w(x + zp, t) - w(x + zp, \varepsilon))\,dp.\end{aligned}$$

For $\varepsilon \to 0$ we find that

$$\delta_H W(x,t) - \delta_h W(x,t) = \int_0^1 z\,dz \int_h^H w(x + zp, t)\,dp \tag{1.61d}$$

since w and W are continuous and vanish for $t = 0$. Rewriting this equation as

$$W(x + H, t) = W(x,t) + H\delta_h W(x,t) + \int_0^1 z\,dz \int_h^H Hw(x + zp, t)\,dp \tag{1.61e}$$

we recognize that $W(x + H, t)$ is twice continuously differentiable with respect to H, and hence also with respect to x, and that

$$W_{xx}(x + H, t) = W_{HH} = \int_0^1 (2zw(x + zH, t) + Hz^2 w_x(x + zH, t))\,dz.$$

For $H \to 0$ we find that

$$W_{xx}(x,t) = w(x,t) = W_t(x,t).$$

From the fact that $W_{xx} = w \geqslant 0$ (or from (1.61e) for $h = -H$) we conclude that

$$2W(x,t) \leqslant W(x + H, t) + W(x - H, t).$$

Integrating this inequality with respect to H from 0 to x for a positive x we find

$$2xW(x,t) \leqslant \int_0^x W(x + H, t)\,dH + \int_0^x W(x - H, t)\,dH = \int_0^{2x} W(y,t)\,dy. \tag{1.62a}$$

Let $0 \leqslant s < t < T$. We have in $W(x,s+t)$ a non-negative solution of the heat equation with initial values $W(x,s)$. Applying (1.58e,f) to this solution with t replaced by $t-s$, we obtain

$$W(0,t) \geqslant \int K(0,y,t-s)W(y,s)\,dy \geqslant \int_0^{2x} K(0,y,t-s)W(y,s)\,dy$$

$$\geqslant \frac{\exp(-x^2/(t-s))}{\sqrt{4\pi(t-s)}} \int_0^{2x} W(y,s)\,dy.$$

Combined with (1.62a) this yields the inequality

$$W(x,s) \leqslant \sqrt{\frac{\pi(t-s)}{x^2}}\, e^{x^2/(t-s)} W(0,t). \tag{1.62b}$$

Replacing $W(x,t)$ by $W(-x,t)$ we see that this inequality is valid for negative x as well.

Take now any ε with $0 < \varepsilon < T/2$. Consider $W(x,s)$ for $x \in \mathbb{R}$, $0 \leqslant s \leqslant T - 2\varepsilon$. Here W is bounded for $|x| < \sqrt{\pi T}$, while by (1.62b) with $t = T - \varepsilon$ for $|x| > \sqrt{\pi T}$

$$W(x,s) \leqslant e^{x^2/\varepsilon} W(0, T-\varepsilon).$$

Hence W satisfies the assumptions of the uniqueness theorem on p. 216. It follows that $W(x,s) = 0$ for $x \in \mathbb{R}$, $0 \leqslant s \leqslant T - 2\varepsilon$, and hence also for $x \in \mathbb{R}$, $0 \leqslant s < T$, since ε is arbitrary. Thus $u(x,t) = v(x,t)$. □

Problems

1. Define for $x,y,t \in \mathbb{R}$, $t \neq 0$

$$K(x,y,t) = (4\pi|t|)^{-1/2} \exp(-(x-y)^2/4t). \tag{1.63a}$$

Show that

$$K(x,0,s+t) = \int K(x,y,t)K(y,0,s)\,dy \tag{1.63b}$$

holds

(a) when $s > 0$, $t > 0$
(b) when $0 < t < -s$.

2. Let $u(x,t)$ satisfy the assumptions of the theorem on p. 222. Show that

$$u(x,t) \leqslant \sqrt{1+c}\, e^{x^2/4ct} u(0,(1+c)t) \tag{1.64a}$$

for $0 < c$, $0 < t < T/(1+c)$.

3. Show that the theorem on p. 222 stays valid, if assumption (1.57c) is replaced by

$$u(x,t) \geqslant -c \quad \text{for } x \in \mathbb{R}, \quad 0 < t < T$$

with any constant c, or by

$$u(x,t) \geqslant -Me^{Ax^2} \quad \text{for } x \in \mathbb{R}, \quad 0 < t < T$$

with any constants M,A. [Hint: Consider $u^* = u(x,t) + mK(x,0,s + t)$ with suitable m,s.]

2. The Initial-Value Problem for General Second-Order Linear Parabolic Equations*

(a) *The method of finite differences and the maximum principle*

Restricting ourselves to one space variable x, we consider a linear equation of the form

$$Lu = u_t - a(x,t)u_{xx} - 2b(x,t)u_x - c(x,t)u = d(x,t). \tag{2.1}$$

We deal with solutions defined in a closed slab

$$\Omega = \{(x,t) | x \in \mathbb{R}, 0 \leqslant t \leqslant T\} \tag{2.2}$$

and satisfying the initial condition

$$u(x,0) = f(x). \tag{2.3}$$

To simplify statements we make the assumption that the coefficients a,b,c belong to $C^\infty(\Omega)$ and that they and each of their partial derivatives are bounded uniformly in Ω. (Actually only a small finite number of derivatives will be required in each theorem). Most important is the additional assumption (the "arrow of time"), that the coefficient $a(x,t)$ is positive and bounded away from 0 in Ω:

$$0 < \inf_{(x,t) \in \Omega} a(x,t). \tag{2.4}$$

All results derived here will be based on the method of finite differences. The model is the discussion of symmetric hyperbolic systems in Chapter 5, Section 3. However, the present situation is greatly simplified by the existence of a *maximum principle*, which allows us to work with the maximum norm rather than with L_2-norms.

Given two positive constants h,k we again consider the *lattice* Σ consisting of the points (x,t) with

$$x = nh, \qquad t = mk \qquad 0 \leqslant t \leqslant T \tag{2.4a}$$

with integers n,m. We replace (2.1) by the difference equation

$$\Lambda v = \frac{v(x,t+k) - v(x,t)}{k} - a(x,t)\frac{v(x+h,t) - 2v(x,t) + v(x-h,t)}{h^2}$$

$$-2b(x,t)\frac{v(x+h,t) - v(x-h,t)}{2h} - c(x,t)v(x,t) = d(x,t) \tag{2.5a}$$

* ([11], [18], [25])

for a function $v(x,t)$ defined in the lattice Σ, and satisfying the initial condition

$$v(x,0)=f(x) \quad \text{for } x=nh. \tag{2.5b}$$

Solving for $v(x,t+k)$ we write (2.5a) in the form

$$\begin{aligned} v(x,t+k)=&(\lambda a+h\lambda b)v(x+h,t)+(1-2\lambda a+h^2\lambda c)v(x,t)\\ &+(\lambda a-h\lambda b)v(x-h,t)+h^2\lambda d(x,t), \end{aligned} \tag{2.6}$$

where we have set

$$\lambda=k/h^2 \tag{2.7}$$

and a,b,c are taken at the point (x,t). Introduce the norm

$$\|g\|=\sup_x|g(x)| \tag{2.8}$$

for a bounded function g (here defined for $x=nh$ only). Introducing the shift operators E,η defined by

$$Ev(x,t)=v(x+h,t), \qquad \eta v(x,t)=v(x,t+k) \tag{2.9}$$

a solution of (2.6) clearly satisfies

$$|\eta v|\leqslant\big[|\lambda a+h\lambda b|+|1-2\lambda a+h^2\lambda c|+|\lambda a-h\lambda b|\big]\|v\|+k\|d\|. \tag{2.10}$$

Assume now that $\lambda=k/h^2$ is so small that the *stability condition*

$$2\lambda \sup_{(x,t)\in\Sigma} a(x,t)<1 \tag{2.11}$$

is satisfied. Then since (2.4) holds and b,c are bounded,

$$|\lambda a+h\lambda b|+|1-2\lambda a+h^2\lambda c|+|\lambda a-h\lambda b|=1+h^2\lambda c=1+kc$$

for all sufficiently small h. Setting

$$C=\max\Big(0,\ \sup_{\Omega} c(x,t)\Big), \tag{2.12}$$

we find from (2.10) that

$$\|\eta v\|\leqslant(1+kC)\|v\|+k\|d\|, \tag{2.13}$$

where $\|v\|$ and $\|d\|$ depend on $t=mk$. Introduce the norm

$$|||d|||=\sup_t\|d\|=\sup_{(x,t)\in\Sigma}|d(x,t)|. \tag{2.14}$$

Then iterating the inequality

$$\|\eta v\|\leqslant(1+kC)\|v\|+k|||d|||,$$

and using the initial condition (2.5b), we arrive for $t=mk$ at the estimate

$$\begin{aligned} \|v\|&\leqslant(1+kC)^m\|f\|+\frac{1}{C}\big((1+kC)^m-1\big)|||d|||\\ &\leqslant e^{mkC}\|f\|+\frac{1}{C}(e^{mkC}-1)|||d|||\\ &\leqslant e^{Ct}\|f\|+te^{Ct}|||d|||. \end{aligned}$$

This proves:

Lemma I. *If* $\lambda = k/h^2$ *satisfies the stability condition* (2.11) *and if* h *is sufficiently small, then the solution* v *of* (2.5a,b) *satisfies the "maximum principle"*

$$|||v||| \leqslant e^{CT}\|f\| + Te^{CT}|||d|||. \tag{2.15}$$

[This implies in the case $d=0$, $c \leqslant 0$ that the supremum of $|v|$ in Σ is the same as the supremum on the initial line.]

Let now $u(x,t)$ be a solution of (2.1), (2.3), for which u, u_t, u_x, u_{xx} are uniformly bounded and uniformly continuous for $(x,t) \in \Omega$. Choose a fixed λ for which (2.11) holds. Consider the monotone increasing sequence of lattices Σ_ν corresponding to the choices $h = 2^{-\nu}$, $k = \lambda 2^{-2\nu}$ for $\nu = 1, 2, \ldots$. Denote by v^ν the solution of (2.5a,b) corresponding to Σ_ν. If U is the union of all Σ_ν and (x,t) a point of U, then $v^\nu(x,t)$ is well defined for all sufficiently large ν. Under these assumptions we have

Lemma II. *For each* $(x,t) \in U$

$$\lim_{\nu \to \infty} v^\nu(x,t) = u(x,t). \tag{2.16}$$

Proof. By Taylor's theorem and the uniform continuity assumptions

$$\frac{u(x,t+k) - u(x,t)}{k} - u_t(x,t) \tag{2.17a}$$

$$\frac{u(x,+h,t) - u(x-h,t)}{k} - u_x(x,t) \tag{2.17b}$$

$$\frac{u(x+h,t) - 2u(x,t) + u(x-h,t)}{h^2} - u_{xx}(x,t) \tag{2.17c}$$

tend to 0 for $h \to 0$, $k \to 0$ uniformly for $(x,t) \in \Omega$. This shows that for a given $\varepsilon > 0$

$$|\Lambda u - d| < \varepsilon$$

for all $(x,t) \in \Sigma_\nu$ provided ν is sufficiently large. (Here the difference operator Λ is the one corresponding to the lattice Σ_ν.) Since v^ν satisfies (2.5a) it follows that

$$|\Lambda(u - v^\nu)| < \varepsilon.$$

Using that $u - v^\nu = 0$ for $t = 0$, and applying (2.15) to $u - v^\nu$, we find that

$$|u(x,t) - v^\nu(x,t)| \leqslant \varepsilon T e^{CT} \tag{2.18}$$

for all $(x,t) \in \Sigma_\nu$ provided ν is sufficiently large. This proves Lemma II. □

Since each $v^\nu(x,t)$ satisfies (2.15) for $(x,t)\in\Sigma_\nu$ and ν sufficiently large, we see from (2.16) that also

$$|u(x,t)|\leqslant e^{CT}\Big(\sup_{\mathbb{R}}|f|+T\sup_{\Omega}|d|\Big) \tag{2.19}$$

for all $(x,t)\in U$. The same inequality holds then for all $(x,t)\in\Omega$ because of the assumed continuity of u.

We notice that the stability condition (2.11) requires the "mesh ratio" k/h to tend to zero, as we refine the mesh and let h and k tend to 0. This requirement is to be expected from the Courant–Friedrichs–Lewy test (p. 7), if we want the solution of the difference scheme to approximate that of the differential equation. Indeed bounded h/k would correspond to a uniformly bounded domain of dependence for $v^\nu(x,t)$ on f, and hence also for $u(x,t)$, whereas the example of the heat equation shows that the domain of dependence of $u(x,t)$ on the initial values is the whole x-axis. That the stability test involves a numerical bound on ak/h^2 is plausible since this combination is dimensionless; [$a(x,t)$ is of dimension x^2/t by virtue of the differential equation (2.1)].

We can dispense with the uniform continuity assumptions in (2.19):

Theorem. *The inequality* (2.19) *holds for a solution u of* (2.1), (2.3), *if u,u_t,u_x,u_{xx} are continuous and u,u_x are uniformly bounded in Ω.*

PROOF. We introduce a "cutoff function" $\phi(x)\in C^\infty(\mathbb{R})$ for which $0\leqslant\phi\leqslant 1$ for all x, while $\phi=0$ for $|x|>2$ and $\phi=1$ for $|x|<1$. Then for $r>0$ the function

$$u^r(x,t)=\phi(x/r)u(x,t)$$

agrees with u for $|x|<r$ and vanishes for $|x|>2r$. u^r,u_t^r,u_x^r,u_{xx}^r are uniformly continuous and uniformly bounded in Ω. Moreover u^r satisfies

$$Lu^r=\phi d-2ar^{-1}\phi' u_x-(ar^{-2}\phi''+2br^{-1}\phi')u=d^*,$$

$$|u^r(x,0)|\leqslant|u(x,0)|=|f(x)|.$$

Thus by (2.19) applied to u^r

$$|u^r(x,t)|\leqslant e^{CT}\Big(\sup_{\mathbb{R}}|f|+T\sup_{\Omega}|d^*|\Big). \tag{2.20}$$

Since u,u_x are bounded uniformly we have

$$\lim_{r\to\infty}\sup_{\Omega}|d^*|=\sup_{\Omega}|d|,$$

and (2.19) follows for u from (2.20). □

The theorem trivially has the consequence that the solution of the initial-value problem (2.1), (2.3) is *unique*, if we restrict ourselves to solutions u for which u and u_x are bounded uniformly.

(b) *Existence of solutions of the initial-value problem*

Here we closely follow the pattern of the existence proof given in Chapter 5 for symmetric hyperbolic systems by the method of finite differences. We only need to sketch the arguments. We assume that the prescribed functions $f(x)$ and $d(x,t)$ together with their derivatives of orders $\leqslant 4$ are continuous and uniformly bounded for $x \in \mathbb{R}$ (respectively $(x,t) \in \Omega$). Without ambiguity we shall now use the notation

$$\|f\| = \sup_{\mathbb{R}} |f(x)|, \qquad |||d||| = \sup_{\Omega} |d(x,t)| \tag{2.21}$$

in the case of functions defined for all x (respectively all $(x,t) \in \Omega$). The suprema of these functions in a lattice do not exceed these values.

Let again v denote a solution of (2.5a,b), where it is assumed that (2.11) holds and that h is sufficiently small. We then have the estimate (2.15) for $v(x,t)$ in Σ. We can obtain analogous bounds for the difference quotients of v. Let $w(x,t)$ be defined in Σ by

$$w = \delta v = \frac{1}{h}(E-1)v = \frac{v(x+h,t) - v(x,t)}{h}. \tag{2.22}$$

Using the product rule

$$\delta(ab) = (Ea)\,\delta b + (\delta a)b, \tag{2.23}$$

we find from (2.6) that w satisfies the equation

$$\begin{aligned} w(x,t+k) &= E(\lambda a + h\lambda b)w(x+h,t) + E(1 - 2\lambda a + h^2\lambda c)w(x,t) \\ &\quad + E(\lambda a - h\lambda b)w(x-h,t) + \delta(\lambda a + h\lambda b)v(x+h,t) \\ &\quad + \delta(1 - 2\lambda a + h^2\lambda c)v(x,t) + \delta(\lambda a - h\lambda b)v(x-h,t) \\ &\quad + k\,\delta d(x,t) \\ &= E(\lambda a + h\lambda b)w(x+h,t) \\ &\quad + E\big(1 - 2\lambda a + h^2\lambda c + hE^{-1}\delta(\lambda a + h\lambda b)\big)w(x,t) \\ &\quad + E\big(\lambda a - h\lambda b - hE^{-1}\delta(\lambda a - h\lambda b)\big)w(x-h,t) \\ &\quad + k((\delta c)v + \delta d). \end{aligned}$$

Since δa, δb, δc, are bounded uniformly by $|||a_x|||$, $|||b_x|||$, $|||c_x|||$ respectively, we find for λ satisfying (2.11) and h sufficiently small that

$$\begin{aligned} |w(x,t+k)| &\leqslant (1 + kEc + 2k\,\delta b)\|w\| + k(|\delta c||v| + |\delta d|) \\ &\leqslant (1 + kC + 2k|||b_x|||)\|w\| + k(|||c_x|||\,|||v||| + |||d_x|||). \end{aligned}$$

Since $w(x,0) = \delta f$, it follows as before that

$$|||\delta v||| = |||w||| \leqslant e^{(C + 2|||b_x|||)T}(\|f_x\| + T(|||c_x|||\,|||v||| + |||d_x|||)), \tag{2.24}$$

where for $|||v|||$ we have the estimate (2.15) in terms of $\|f\|$ and $|||d|||$.

Similar estimates clearly can be obtained for the higher difference quotients $\delta^2 v$, $\delta^3 v$, $\delta^4 v$, and then also by (2.5a) for

$$\tau v = \frac{1}{k}(\eta - 1)v = \frac{v(x,t+k) - v(x,t)}{k}, \tag{2.25}$$

and for $\delta\tau v$, $\delta^2\tau v$, and $\tau^2 v$. All these are bounded uniformly.

As before we define the increasing sequence Σ_ν of lattices and the corresponding solutions v^ν defined in Σ_ν. Then v^ν, δv^ν, $\delta^2 v^\nu$, $\delta^3 v^\nu$, τv^ν, $\delta\tau^2 v^\nu$ are defined and bounded uniformly in Σ_ν, and hence also in Σ_μ for $\mu \leqslant \nu$. For a suitable subsequence of the integers ν

$$\lim_{\nu\to\infty} v^\nu(x,t) = u(x,t), \qquad \lim_{\nu\to\infty} \delta v^\nu(x,t) = u'(x,t),$$

$$\lim_{\nu\to\infty} \delta^2 v^\nu(x,t) = u''(x,t), \qquad \lim_{\nu\to\infty} \tau v^\nu(x,t) = \dot{u}(x,t)$$

exist for all (x,t) in the union U of the Σ_ν. Let (x,t), (y,t) be two points of U, and hence of Σ_ν, for all sufficiently large ν. Then for $0 < y - x = hn = n2^{-\nu}$

$$\left| \frac{v^\nu(y,t) - v^\nu(x,t)}{y-x} - \delta v^\nu(x,t) \right|$$

$$= \left| \frac{v^\nu(x+nh,t) - v^\nu(x,t)}{nh} - \frac{v^\nu(x+h,t) - v^\nu(x,t)}{h} \right|$$

$$= \left| \left(\frac{E^n - 1}{nh} - \frac{E-1}{h} \right) v^\nu \right|$$

$$= \left| \frac{(E-1)^2}{nh} (E^{n-2} + 2E^{n-3} + 3E^{n-4} + \cdots + n-1) v^\nu \right|$$

$$= \left| \frac{h}{n}(E^{n-2} + 2E^{n-3} + \cdots + n-1)\delta^2 v^\nu \right| \leqslant \frac{(n-1)h}{2} |||\delta^2 v^\nu|||$$

$$\leqslant \frac{|y-x|}{2} |||\delta^2 v^\nu||| \leqslant K|y-x|$$

with a constant K independent of ν. In the limit for $\nu\to\infty$ in the subsequence we obtain the inequality

$$\left| \frac{u(y,t) - u(x,t)}{y-x} - u'(x,t) \right| \leqslant C|y-x| \tag{2.26}$$

for all (x,t), (y,t) in U. Since v^ν, δv^ν, $\delta^2 v^\nu$, τv^ν are uniformly Lipschitz in Σ_ν, the limits $u, u', u'', \dot{u}$ are uniformly Lipschitz in U, and hence can be extended as continuous functions to all of Ω. By continuity then (2.26) holds for all (x,t), (y,t) in Ω. For $y \to x$ we find that $u_x(x,t)$ exists and that

$$u_x(x,t) = u'(x,t).$$

Similarly one verifies that $u_{xx}=u''$, $u_t=\dot{u}$, and that the differential equation (2.1) is satisfied. This proves the theorem:

Theorem. *If f,d and their derivatives of orders $\leqslant 4$ are continuous and bounded uniformly, the initial-value problem* (2.1), (2.3) *has a solution $u(x,t)$, for which u,u_t,u_x,u_{xx} are uniformly bounded and uniformly continuous in Ω.*

In conclusion we observe that there are many results we derived for the heat equation which one would expect to be valid for more general equations of the form (2.1). For example the solution of (2.1), (2.3) should be unique under the sole assumption that u,u_t,u_x,u_{xx} are continuous and that u is bounded uniformly. We would also expect the solution of the homogeneous equation $Lu=0$ to exist and to be in C^∞ for $0<t\leqslant T$, if f is only assumed to be continuous and bounded (always requiring a,b,c to be in C^∞ and bounded, and a to be positive and bounded away from zero). However, such questions are beyond the scope of this volume.

PROBLEMS

1. Let $u(x,t)$ be a solution of class C^2 of

$$u_t=a(x,t)u_{xx}+2b(x,t)u_x+c(x,t)u$$

in the rectangle

$$\Omega=\{(x,t)|0\leqslant x\leqslant L,\ 0\leqslant t\leqslant T\}.$$

Let $\partial'\Omega$ denote the "lower boundary" of Ω consisting of the three segments

$$\begin{aligned} &x=0, && 0\leqslant t\leqslant T\\ &0\leqslant x\leqslant L, && t=0\\ &x=L, && 0\leqslant t\leqslant T. \end{aligned}$$

(a) Prove that in case $c<0$ in Ω

$$|u(x,t)|\leqslant \sup_{\partial'\Omega}|u| \quad \text{for } (x,t)\in\Omega.$$

[Hint: Show $\max_\Omega u$ cannot be assumed on $\Omega-\partial'\Omega$ unless $\max_\Omega u\leqslant 0$.]

(b) Show that more generally

$$|u(x,t)|\leqslant e^{CT}\max_{\partial'\Omega}|u|, \tag{2.27}$$

where

$$C=\max\left(0,\ \max_{\Omega}|c|\right).$$

[Hint: Substitute $u=e^{\gamma t}v$, where $\gamma>C$, and apply part (a) to v.]

2. Let $u(x,t)$ denote the solution (1.11) of

$$Lu=u_t-u_{xx}=0 \quad \text{for } 0<t\leqslant T \tag{2.28a}$$

$$u(x,0)=f(x), \tag{2.28b}$$

where f is continuous. Let Σ denote the lattice of points (x,t) with x,t of the form $x=nh$, $t=mk$ and v the solution of

$$\Lambda v = \frac{v(x,t+k)-v(x,t)}{k} - \frac{v(x+h,t)-2v(x,t)+v(x-h,t)}{h^2} = 0 \quad (2.29)$$

with $v(x,0)=f(x)$.

(a) Show that for $\lambda = k/h^2 = 1/2$

$$v(nh,mk) = 2^{-m} \sum_{j=0}^{m} \binom{m}{j} f((n-m+2j)h) \quad (2.30)$$

and hence

$$\sup_{\Sigma} |v| \leqslant \sup_{\mathbb{R}} |f| .$$

(b) Show that for a fixed $f \in C_0^\infty(\mathbb{R})$, $0<\lambda\leqslant 1/2$ and $(x,t)\in\Sigma$

$$|u(x,t)-v(x,t)| = O(h^2).$$

Show that for the special value $\lambda = 1/6$ the better estimate

$$|u(x,t)-v(x,t)| = O(h^4)$$

holds. [Hint: Expand Λu by Taylor's theorem, using $u_t = u_{xx}$.]

(c) Let $0<\lambda\leqslant 1/2$ and Σ_ν be the lattice corresponding to $h=2^{-\nu}$, $k=\lambda 2^{-2\nu}$, and let v^ν be the corresponding solution of (2.29) with initial values f. Let U be the union of the Σ_ν for $\nu = 1,2,3,\ldots$. Show that for $(x,t)\in U$

$$\lim_{\nu\to\infty} v^\nu(x,t) = u(x,t) \quad (2.31)$$

provided f has bounded continuous derivatives of orders $\leqslant 4$. [Hint: Use the theorem on p. 187 and known properties of u.]

(d) Show that (2.31) holds assuming only that $0<\lambda\leqslant 1/2$ and that $f(x)$ is continuous and has compact support. [Hint: Approximate f uniformly by functions with bounded derivatives of orders $\leqslant 4$; use the maximum principle for the v^ν.]

8 H. Lewey's Example of a Linear Equation without Solutions

It is natural to believe that linear partial differential equations always have solutions, in fact so many that it is possible to impose additional conditions. [There are trivial *nonlinear* equations without solutions, e.g. the equation $\exp(u_x) = 0$.] Solutions may of course, cease to exist at a "singular" point, where the characteristic form vanishes, as is the case for the equation $xu_x + yu_y = 1$ which has no solution in a neighborhood of the origin, or the equation $xu_x + yu_y + u = 0$ which has only the trivial solution. (See Problem 2, p. 16). It was surprising therefore when H. Lewy (Annals of Mathematics 66 (1957), 155–158, see also [16]) constructed a linear equation without singular points that has no solution *anywhere.* His equation has the form $Lu = F$ where the linear first order differential operator L has complex-valued linear functions as coefficients, and F is a suitably chosen function of class C^∞. (For *analytic* F there would always be solutions by Cauchy–Kowalevski.) The function F in this example is not given explicitly; its existence is proved by a non-constructive argument. The single equation $Lu = F$ with complex coefficients for a complex-valued u is equivalent to a system of two equations with real coefficients for two real-valued functions.

Theorem. *Let L denote the differential operator acting on functions $u(x,y,z)$ defined by*

$$Lu = -u_x - iu_y + 2i(x + iy)u_z. \tag{1.1a}$$

There exists a function $F(x,y,z) \in C^\infty(\mathbb{R}^3)$ such that the equation

$$Lu = F(x,y,z) \tag{1.1b}$$

has no solution whose domain is an open set Ω in $\mathbb{R}^3$, with $u \in C^1(\Omega)$, and u_x, u_y, u_z Hölder continuous in Ω.

In the proof we first construct special F, for which every solution of (1.1b) must become singular at certain special points. By superposition we construct then an F such that every solution must become singular in a dense set of points. The proof is broken up into a number of lemmas.

Lemma I. *Let $\psi(z) \in C^\infty(\mathbb{R})$, where ψ is real-valued. Let for a certain $\delta > 0$ and $\zeta \in \mathbb{R}$*

$$\Omega = \{(x,y,z) \mid (x,y,z) \in \mathbb{R}^3; \quad x^2 + y^2 < \delta; \quad |z - \zeta| < \delta\}.$$

A solution $u \in C^1(\Omega)$ of

$$Lu = \psi'(z) \tag{1.2a}$$

can exist only if $\psi(z)$ is real analytic at ζ.

PROOF. Set

$$v(r,\theta,z) = e^{i\theta}\sqrt{r}\,u(\sqrt{r}\cos\theta, \sqrt{r}\sin\theta, z). \tag{1.2b}$$

Then $v \in C^1$ for $0 < r < \delta$, $\theta \in \mathbb{R}$, $|z - \zeta| < \delta$. Moreover, v has period 2π in θ. One easily verifies that

$$Lu = -2v_r - \frac{i}{r}v_\theta + 2iv_z = \psi'(z).$$

The function

$$V(z,r) = \int_0^{2\pi} v(r,\theta,z)\,d\theta$$

is defined and in C^1 for $0 < r < \delta$, $|z - \zeta| < \delta$, and moreover satisfies

$$V_z + iV_r = \int_0^{2\pi}\left(v_z - \frac{1}{2r}v_\theta + iv_r\right)d\theta = -\pi i\psi'(z).$$

Now the continuity of $u(x,y,z)$ implies that $v(r,\theta,z)$ is continuous for $0 \leqslant r < \delta$, $\theta \in \mathbb{R}$, $|z - \zeta| < \delta$, and that $v(0,\theta,z) = 0$. Then $V(z,r)$ is continuous for $0 \leqslant r < \delta$, $|z - \zeta| < \delta$, and vanishes for $r = 0$. It follows that the function $W = V(z,r) + i\pi\psi(z)$ is C^1 for $0 < r < \delta$, $|z - \zeta| < \delta$, satisfies $W_z + iW_r = 0$ and is continuous for $0 \leqslant r < \delta$, $|z - \zeta| < \delta$. Thus W is an analytic function of $z + ir$ for $r > 0$, $|z - \zeta| < \delta$, which is still continuous for $r = 0$ and has vanishing real part there. Equivalently the real and imaginary parts of $W(z,r)$ are conjugate harmonics. By *reflection*, that is by $W(z,-r) = -\overline{W(z,r)}$, we can extend W as analytic function of $z + ir$ to $|r| < \delta$, $|z - \zeta| < \delta$. (See Problem 5, p. 110). It follows that $\pi\psi(z)$, the imaginary part of $W(z,0)$, is real analytic for $|z - \zeta| < \delta$. □

Lemma II. *Let $\psi(z) \in C^\infty(\mathbb{R})$, where ψ is real valued. Let there exist a solution $u(x,y,z)$ of class C^1 of the equation*

$$Lu = \psi'(z - 2\eta x + 2\xi y) \tag{1.3a}$$

in a neighborhood of the point (ξ,η,ζ). Then, $\psi(z)$ is real analytic at $z = \zeta$.

PROOF. Set

$$U(X,Y,Z) = u(X + \xi, Y + \eta, Z + 2\eta X - 2\xi Y). \tag{1.3b}$$

Then $U(X,Y,Z)$ is of class C^1 in a neighborhood of $(0,0,\zeta)$ and satisfies

$$-U_X - iU_Y + 2i(X + iY)U_Z = \psi'(Z)$$

as is verified easily. Apply Lemma I. □

In what follows let $\psi(z)$ denote a fixed real-valued periodic function in $C^\infty(\mathbb{R})$ which is not real analytic (see problem 4, p. 69) at any real z. For a function $F(x,y,z) \in C^\infty(\mathbb{R}^3)$ and a multi-index $\alpha = (a,b,c)$ we write $D^\alpha F$ for $(\partial/\partial x)^a(\partial/\partial y)^b(\partial/\partial z)^c F$, and $|\alpha|$ for $a + b + c$. Let $Q_j = (\xi_j, \eta_j, \zeta_j)$ for $j = 1, 2, \ldots$, denote a sequence of points which is dense in $\mathbb{R}^3$, fixed in what follows. We set

$$c_j = 2^{-j}\exp(-\rho_j) \quad \text{where } \rho_j = |\xi_j| + |\eta_j|. \tag{1.4a}$$

Finally we introduce the bounded infinite sequences $\varepsilon = (\varepsilon_1, \varepsilon_2, \ldots)$ of real numbers ε_j. They form a vector space, with addition and multiplication by real scalars defined in an obvious manner, which becomes a Banach space B when referred to the norm

$$\|\varepsilon\| = \sup_j |\varepsilon_j|. \tag{1.4b}$$

Lemma III. *For any $\varepsilon \in B$ the series*

$$F_\varepsilon(x,y,z) = \sum_{j=1}^{\infty} \varepsilon_j c_j \psi'(z - 2\eta_j x + 2\xi_j y) \tag{1.4c}$$

and all its formal derivatives with respect to x,y,z converge uniformly, defining a function $F_\varepsilon \in C^\infty(\mathbb{R}^3)$.

PROOF. Since ψ is periodic,

$$M_k = \sup_z |\psi^{(k)}(z)| \tag{1.4d}$$

is finite for any k. Then

$$\begin{aligned} |D^\alpha \varepsilon_j c_j \psi'(z - 2\eta_j x + 2\xi_j y)| &\leqslant \|\varepsilon\| c_j M_{|\alpha|+1} \rho_j^{|\alpha|} \\ &\leqslant 2^{-j}\|\varepsilon\| M_{|\alpha|+1}\rho_j^{|\alpha|}\exp(-\rho_j) \leqslant 2^{-j}\|\varepsilon\| M_{|\alpha|+1}\left(\frac{|\alpha|}{e}\right)^{|\alpha|}. \end{aligned} \tag{1.4e}$$

This implies uniform convergence of the series for $D^\alpha F_\varepsilon$. □

Definition. For positive integers j,n let $\Omega_{j,n}$ denote the ball in $\mathbb{R}^3$ with center Q_j and radius $n^{-1/2}$, consisting of the points $P = (x,y,z)$ with

$$|P - Q_j|^2 = (x - \xi_j)^2 + (y - \eta_j)^2 + (z - \zeta_j)^2 < \frac{1}{n}. \tag{1.5a}$$

We denote by $E_{j,n}$ the subset of B consisting of those ε for which there exists a solution $u(P) = u(x,y,z)$ of class $C^1(\Omega_{j,n})$ of the equation

$$Lu = F_\varepsilon(x,y,z) \tag{1.5b}$$

for which

$$u(Q_j) = 0 \tag{1.5c}$$

$$|D^\alpha u(P)| \leqslant n \quad \text{for } |\alpha| \leqslant 1, \quad P \in \Omega_{j,n} \tag{1.5d}$$

$$|D^\alpha u(P) - D^\alpha u(Q)| \leqslant n|P - Q|^{1/n} \quad \text{for } |\alpha| = 1, \quad P \in \Omega_{j,n}, \quad Q \in \Omega_{j,n}. \tag{1.5e}$$

(Condition (1.5d) represents bounds for u and its first derivatives in $\Omega_{j,n}$, while (1.5e) prescribes a uniform Hölder condition on the first derivatives.)

Lemma IV. *The sets $E_{j,n}$ are closed subsets of B that are nowhere dense (i.e., have no interior points).*

PROOF. Let $\varepsilon^1, \varepsilon^2, \ldots$ be in $E_{j,n}$, let ε be in B, and

$$\lim_{k \to \infty} \|\varepsilon - \varepsilon^k\| = 0.$$

By (1.4e) with $\alpha = 0$

$$|F_\varepsilon - F_{\varepsilon^k}| \leqslant M_1 \|\varepsilon - \varepsilon^k\|.$$

Thus the F_{ε^k} converge to F_ε. Denote by u_k the solution u of $Lu = F_{\varepsilon^k}$ with the properties (1.5c,d,e). Since the u_k and their first derivatives are equi-bounded and equi-continuous in $\Omega_{j,n}$ there exists a subsequence of the u_k which converges uniformly to a function u together with its first derivatives. Then u must again satisfy (1.5c,d,e). Since also the Lu_k in the subsequence converge to Lu, the function u is a solution of (1.5b), and $u \in E_{j,n}$. This shows that $E_{j,n}$ is closed.

Let δ denote the bounded sequence all of whose elements are zero, except the j-th one which shall have the value $1/c_j$. Then

$$F_\delta = \psi'(z - 2\eta_j x + 2\xi_j y).$$

Let ε be an interior point of $E_{j,n}$. We can find a positive number θ so small that also

$$\varepsilon' = \varepsilon + \theta\delta \in E_{j,n}.$$

Let u, u' be the solutions of $Lu = F_\varepsilon$ respectively $Lu' = F_{\varepsilon'}$ with the properties guaranteed by the definition of $E_{j,n}$. Set $u'' = (u' - u)/\theta$. Then u'' is a solution of class C^1 of $Lu'' = F_\delta$ in a neighborhood of the point Q_j. This contradicts Lemma II, since ψ is not real analytic at ζ_j. □

Proof of the Theorem. Assume the theorem does not hold. There would exist for every $\varepsilon \in B$ an open set $\Omega \subset \mathbb{R}^3$ and a solution u of $Lu = F_\varepsilon$ in Ω with Hölder continuous first derivatives. Now Ω contains a point Q_j, since the sequence $Q_1, Q_2, \ldots$ is dense. Thus $\Omega_{j,n} \subset \Omega$ for all sufficiently large n. For n sufficiently large u will also satisfy (1.5d,e). It will satisfy as well (1.5c) if we replace u by $u - u(Q_j)$. But this means that $\varepsilon \in E_{j,n}$. Hence B is the union of all the $E_{j,n}$ with positive j, n. This contradicts Lemma IV, since the complete metric space B cannot be union of a countable set of closed nowhere dense subsets (Baire category argument! see [7]).

Problems

1. (a) Write equation (1.1b) as a system of two equations for the real and imaginary parts of u.
   (b) Show that the characteristic form of the system is semi-definite but not definite for all x,y,z.
   (c) Show that there are no real characteristic surfaces $\phi(x,y,z) = \text{const.}$ with grad $\phi \neq 0$.

Bibliography

[1] Agmon, Shmuel, Lectures on elliptic boundary value problems, Van Nostrand, 1965.

[2] Bers, Lipman, John, F., Schechter, M., Partial Differential Equations, Interscience Publishers, 1964.

[3] Bremermann, Hans, Distributions, Complex Variables and Fourier Transforms, Addison Wesley Publishers, 1965.

[4] Carroll, R. C., Abstract Methods in Partial Differential Equations, Harper and Row, 1969.

[5] Courant, R., and Friedrichs, K. O., Supersonic Flow and Shock Waves, 1948, (Repr.), Springer-Verlag.

[6] Courant, R., and Hilbert, D., Methods of Mathematical Physics, Interscience Publishers, Vol. I, 1953; Vol. II, 1962.

[7] Dunford, N. and Schwartz, J. T., Linear Operators, Part II, Interscience Publishers, 1963.

[8] Dym, H. and McKean, H. P., Fourier Series and Integrals, Academic Press, 1972.

[9] Epstein, B., Partial Differential Equations, McGraw Hill, 1962.

[10] Fichera, Gaetano, Linear Elliptic Differential Systems and Eigenvalue Problems, Lecture Notes in Mathematics, 8, Springer-Verlag, 1965.

[11] Friedman, Avner, Partial Differential Equations of Parabolic Type, Prentice Hall, 1964.

[12] Friedman, Avner, Partial Differential Equations, Holt, Rinehart and Winston, 1969.

[13] Garabedian, P. R., Partial Differential Equations, John Wiley & Sons, Inc., 1964.

[14] Helgason, S., The Radon Transform, Birkhäuser Boston, 1980.

[15] Hellwig, G., Partial Differential Equations, New York, Blaisdell, 1964.

[16] Hörmander, Lars, Linear Partial Differential Operators, Springer-Verlag, 1963.

[17] Hörmander, L., An Introduction to Complex Analysis in Several Variables, van Nostrand, 1966.

[18] Isaacson, E., and Keller, H. B., Analysis of Numerical Methods, John Wiley & Sons, Inc., 1966.

[19] John, Fritz, Plane Waves and Spherical Means Applied to Partial Differential Equations, Interscience Publishers, 1955; Springer Verlag, 1981.

[20] John, Fritz, Lectures on Advanced Numerical Methods, Gordon and Breach, 1967.

[21] Lax, P. D., Lectures on Hyperbolic Partial Differential Equations, Stanford University, 1963.

[22] Lax, P. D., Hyperbolic Systems of Conservationa Laws and the Mathematical Theory of Shock Waves, Regional Conference Series in Applied Mathematics 11, 1973.

[23] Lions, J. L., Équations différentielles opérationelles et problèmes aux limites, Springer, 1961.

[24] Lions, J. L., and Magenes, E., Non-homogeneous boundary value problems, Springer-Verlag, 1972.

[25] Meis, Th. and Marcowitz, U., Numerical Solution of Partial Differential Equations, Springer Verlag, 1981.

[26] Mikhlin, S. G., Linear Equations of Mathematical Physics, Holt, Rinehart and Winston, 1967.

[27] Petrovsky, I. G., Lectures on Partial Differential Equations, Interscience Publishers, 1954.

[28] Protter, M. H. and Weinberger, H. F., Maximum Principles in Differential Equations, Prentice Hall, 1967.

[29] Richtmyer, R. D. and Morton, K. W., Difference Methods for Initial Value Problems, 2nd Ed., Interscience Publishers, 1967.

[30] Schechter, M., Modern Methods in Partial Differential Equations. An Introduction, McGraw Hill, 1977.

[31] Smirnov, V. I., A Course of Higher Mathematics, Vol. IV, Translation, Addison Wesley, 1964.

[32] Stakgold, I., Boundary Value Problems of Mathematical Physics, MacMillan, 1967.

[33] Tikhonov, A. N. and Arsenin, V. Y., Solution of Ill-Posed Problems, Winston/Wiley, 1977.

[34] Treves, F., Basic Linear Partial Differential Equations, Academic Press, 1975.

[35] Widder, D. V., The Heat Equation, Academic Press, 1975.

Glossary

$\mathbb{R}$ = set of real numbers

$\mathbb{R}^n$ = real n-dimensional space (usually endowed with euclidean metric)

Z^n = set of $x = (x_1, \ldots, x_n)$ with x_k = integer $\geqslant 0$
$\mathbb{C}^n$ = set of $x = (x_1, \ldots, x_n)$ with x_k = complex number

For a set $\Omega \subset \mathbb{R}^n$ define:

$\bar{\Omega}$ = closure of Ω

$\partial\Omega$ = boundary of Ω

Ω^{int} = interior of Ω

$\omega \subset \Omega$ is *compact in* Ω if ω is a closed and bounded subset of Ω^{int}.

For a function f with domain $\Omega \subset \mathbb{R}^n$ define: supp$f$ = *support* of f = closure of subset of Ω where $f \neq 0$. f has *compact support* in Ω if $f = 0$ outside a closed and bounded subset of Ω^{int} (that is, if f vanishes in a neighborhood of $\partial\Omega$ and outside some ball.)

For $\Omega \subset \mathbb{R}^n$ define:

$C^0(\Omega)$ = set of functions f continuous in Ω

$C^s(\Omega)$ = set of f with continuous derivatives of orders $\leqslant s$ in Ω

$C^0(\bar{\Omega}) \cap C^s(\Omega^{\text{int}})$ = set of f continuous in the closure of Ω and with continuous derivatives or orders $\leqslant s$ in the interior.

$C_0^s(\Omega)$ = set of f with continuous derivatives of orders $\leqslant s$ in Ω and of compact support in Ω.

$\mathcal{D}$ = set of test functions = $C_0^\infty(\Omega)$

$C_{M,r}$ see p. 65

C^ω see p. 64

C^a see p. 70

For a matrix $a = (a_{ik})$

a^T = transpose of a = matrix $b = (b_{ik}) = (a_{ki})$ obtained by interchanging rows and columns of a.

$a>0$, if $a^T=a$ and the quadratic form $\sum_{i,k} a_{ik}\xi_i\xi_k$ is positive definite

O-notation: $u=O(v)$ if there exists a real constant K such that $|u|\leqslant Kv$ for all pairs u,v under consideration.

multiple-index notation: see pp. 54–55.

$\Delta=$ Laplacian $\Sigma\partial^2/\partial x_i^2$

$\Box=$ d'Alambertian $=\partial^2/\partial t^2-\Delta$

$\ll$ see p. 67

Index

Applied Mathematical Sciences

cont. from page ii

39. Piccini/Stampacchia/Vidossich: **Ordinary Differential Equations in $\mathbf{R}^n$.**
40. Naylor/Sell: **Linear Operator Theory in Engineering and Science.**
41. Sparrow: **The Lorenz Equations: Bifurcations, Chaos, and Strange Attractors.**
42. Guckenheimer/Holmes: **Nonlinear Oscillations, Dynamical Systems and Bifurcations of Vector Fields.**
43. Ockendon/Tayler: **Inviscid Fluid Flows.**
44. Pazy: **Semigroups of Linear Operators and Applications to Partial Differential Equations.**
45. Glashoff/Gustafson: **Linear Optimization and Approximation: An Introduction to the Theoretical Analysis and Numerical Treatment of Semi-Infinite Programs.**
46. Wilcox: **Scattering Theory for Diffraction Gratings.**
47. Hale et al.: **An Introduction to Infinite Dimensional Dynamical Systems—Geometric Theory.**
48. Murray: **Asymptotic Analysis.**
49. Ladyzhenskaya: **The Boundary-Value Problems of Mathematical Physics.**
50. Wilcox: **Sound Propagation in Stratified Fluids.**
51. Golubitsky/Schaeffer: **Bifurcation and Groups in Bifurcation Theory, Vol. I.**
52. Chipot: **Variational Inequalities and Flow in Porous Media.**
53. Majda: **Compressible Fluid Flow and Systems of Conservation Laws in Several Space Variables.**
54. Wasow: **Linear Turning Point Theory.**
55. Yosida: **Operational Calculus: A Theory of Hyperfunctions.**
56. Chang/Howes: **Nonlinear Singular Perturbation Phenomena: Theory and Applications.**
57. Reinhardt: **Analysis of Approximation Methods for Differential and Integral Equations.**
58. Dwoyer/Hussaini/Voigt (eds.): **Theoretical Approaches to Turbulence.**
59. Sanders/Verhulst: **Averaging Methods in Nonlinear Dynamical Systems.**
60. Ghil/Childress: **Topics in Geophysical Fluid Dynamics: Atmospheric Dynamics, Dynamo Theory and Climate Dynamics.**
61. Sattinger/Weaver: **Lie Groups and Algebras with Applications to Physics, Geometry, and Mechanics.**